LATOUR AND THE HUMANITIES

LATOUR and the HUMANITIES

EDITED BY
RITA FELSKI AND
STEPHEN MUECKE

Johns Hopkins University Press

Baltimore

© 2020 Johns Hopkins University Press
All rights reserved. Published 2020
Printed in the United States of America on acid-free paper
9 8 7 6 5 4 3 2 1

Johns Hopkins University Press
2715 North Charles Street
Baltimore, Maryland 21218-4363
www.press.jhu.edu

Library of Congress Cataloging-in-Publication Data

Names: Felski, Rita, 1956– editor. | Muecke, Stephen, 1951– editor.
Title: Latour and the humanities / edited by Rita Felski and Stephen Muecke.
Description: Baltimore : Johns Hopkins University Press, 2020. | Includes bibliographical
    references and index.
Identifiers: LCCN 2019057452 | ISBN 9781421439211 (hardcover) |
    ISBN 9781421438900 (paperback) | ISBN 9781421438917 (ebook)
Subjects: LCSH: Humanities—Philosophy. | Actor-network theory. |
    Latour, Bruno—Influence.
Classification: LCC AZ103 .L38 2020 | DDC 001.301—dc23
LC record available at https://lccn.loc.gov/2019057452

A catalog record for this book is available from the British Library.

Chapters 1–3, 6–8, 10, 13–14, 16–17, and the afterword were originally published
in a 2016 issue of *New Literary History*.

*Special discounts are available for bulk purchases of this book. For more information,
please contact Special Sales at specialsales@press.jhu.edu.*

Johns Hopkins University Press uses environmentally friendly book materials, including
recycled text paper that is composed of at least 30 percent post-consumer waste, whenever
possible.

Contents

PART TWO LATOUR AND THE DISCIPLINES

LATOUR AND THE HUMANITIES

Introduction

RITA FELSKI

THIS BOOK EXPLORES the relevance of Bruno Latour's work for the humanities from two angles. How might this work reinvigorate or reorient literary studies, art history, film studies, political theory, religious studies, and other disciplines? What new orientations or styles of thought does it offer? And how does it speak to "humanities discourse"—the various lamentations, perorations, jeremiads, diagnoses, and defenses of the humanities that have appeared in recent years, and that now constitute a genre in their own right? Most of these commentaries offer permutations of the same few themes: the humanities make us human; the humanities make us critical; the humanities are a vital counterweight to the soulless calculations and methods of the natural and social sciences.[1]

This collection of essays, based on a special issue of *New Literary History,* with six new contributions and an expanded introduction, offers a rather different take.[2] It brings together influential scholars from various disciplines who have been inspired by the work of Bruno Latour and actor-network-theory, or ANT. Born from the ferment of science and technology studies in the 1970s, ANT is an increasingly influential framework in the social sciences, with Latour being its most widely cited representative. Until recently, it has barely registered in humanistic fields, but the most lively approaches in these fields now include animal studies, thing theory, ecocriticism, and Anthropocene studies—all themes that resonate with ANT's stress on the codependence of human and nonhuman actors. At the level of method, meanwhile, Latour's work offers an alternative to the frameworks that have dominated the humanities in recent decades. It rejects the idea of a yawning gap between words and the world (the much-touted linguistic turn); insists on tracing empirical ties between phenomena rather than relying on theoretical shortcuts; and draws on a language of composition rather than critique, of making and doing rather than subverting and disrupting

What does this mean for the humanities? Rather than calling for separation, the essays in this volume argue for closer ties between the humanities and the natural and social sciences. Rather than stressing detachment, they speak of connection and entanglement. And rather than making heartwarming but tenuous claims about the humanities making us more human (more ethical, more empathic, more critical), they look closely at what the humanities *do* and their practical as well as intellectual ties to varying interests, institutions, and constituencies inside and outside the university. The result is a more down to earth but also more invigorating vision of how the humanities might look in the twenty-first century.

While our contributors share an interest in forging stronger links between Latour and the humanities, they do not shy away from disagreeing with some of his ideas as well as responding to what they take to be misleading accounts of ANT. We hope, then, to stimulate a more thorough and informed discussion of the intellectual and institutional issues at stake in brokering Latour to new audiences. Any translation—any transferal of ideas, things, practices into new arenas—involves losses as well as gains; there is a price to be paid, ANT insists, for every trade.

One potential source of confusion is Latour's advocacy of a flat ontology. ANT seeks to bypass long-standing hierarchies: culture/nature, subject/object, human/thing, and thought/matter. It is "flat" in striving to avoid *a priori* assumptions about what is foundational and what is peripheral, what really counts and what can be discounted. Such an approach—with its turn to a language of networks and relations rather than systems and foundations—has led to charges of relativism or conservatism. According to Bruce Robbins, for example, Latour "makes all causes seem equal, thereby erasing the differentials of power that determine in any given case who or what is in fact the major cause."[3] This criticism, however, misses its mark: that actors are treated symmetrically by ANT does not mean their effects are held to be equal or that differences of scale are denied. Quite the contrary. Networks, after all, differ massively in their size, persistence, and reach; the World Bank is a very different kind of collective to a local PTA. As Graham Harman—one of the most lucid explicators of Latour—puts it, "if all actors are equally real, not all are equally strong."[4] A driving concern of ANT is to describe, through painstaking empirical inquiry, what is often presupposed or taken for granted: *how and why* some associations of actors come to be far more stable, durable, or powerful than others.

Given Latour's increasing prominence as a staunch defender of science and an impassioned critic of climate change denialism—not just in his recent books but in world-wide public appearances and political speeches—it is becoming ever more implausible to portray him as an irrational postmodernist or a quietist who is only interested in traffic bumps and Berlin keys. Yet there are legitimate questions to be raised about how his work relates to existing scholarship on gender, race, and sexuality. Donna Haraway, for example, has reproached Latour for his lack of engagement with feminist science studies; there are also critiques of his lack of attention to the legacy of colonialism and his Eurocentrism.[5] The salient issue is whether such omissions are signs of a fundamental incommensurability between Latour's ideas and forms of humanistic scholarship. The growing interest in these ideas suggests that fruitful forms of collaboration are indeed possible. Maria Puig de la Bellacasa, for example, draws on and reworks Latour in making a feminist argument for "matters of care," while Kane Race has recently made a case for forging closer ties between ANT and queer theory. Such arguments indicate that Latour's work, rather than promoting value-free description or an anything-goes relativism, is highly amenable to articulating attachments and political commitments. And Oumelbanine Zhiri brings Latour's thought into dialog with postcolonial studies in order to question stark antinomies of East and West, tracing the significant role of Arab scholars—often reduced to passive victims or shadowy informants—in the making of early modern Orientalism.[6]

That his *Critical Inquiry* essay is his most well-known piece of writing among humanists has no doubt helped to cement Latour's reputation as Monsieur Anti-Critique. Placed in the context of his work as a whole, it resonates rather differently. Some of the more indignant responses to the essay have, perhaps, understood it in too literal a fashion.[7] "Why Has Critique Run Out of Steam?" is best seen as a provocation and thought experiment, an attempt to jolt scholars out of increasingly formulaic and predictable styles of argument rather than the stern laying down of a commandment: thou shalt never criticize! Latour himself, it needs hardly be said, has done more than his fair share of criticizing. As Blok and Jensen point out in their helpful essay, he is less interested in negating critique than in redistributing it; exploring alternative styles of thought while questioning critique's claim to be the only legitimate intellectual or political genre. And in this respect Latour is less of an outlier that he is sometimes taken to be. A reckoning with the limits of critical theory is now a significant strand within contemporary French thought, whether one thinks of Jacques Rancière or Luc Boltanski. Moreover, as Antoine Hennion shows in this volume, there are also a number of striking affinities between ANT and the tradition of American pragmatism (James and Dewey rather than Rorty). We hope, then, that this volume will contextualize and reframe Latour's work in helpful ways.

Let's begin with a thought experiment: what exactly would disappear if we lost the humanities? Such a question invites us to imagine an experience of loss and to anticipate the reactions triggered by such a loss.[8] Only in the gray early morning light, when a lover departs in a taxi for the last time, are we suddenly aware of the depth and intensity of our passion. So too, we can more fully appreciate why the humanities are irreplaceable by contemplating the prospect of their nonexistence. According to one influential line of thought, the loss of the humanities would mean, above all, the loss of critique. The idea of critique, of course, has a long history that can be spun in diverse ways; as a synonym for Socratic or Kantian modes of philosophical questioning, for example, or to denote an adversarial and agonistic style of political argument. This latter use of the term, especially, has gained increased traction in the humanities in the last half century. Critique, in this sense, typically includes the following elements: a spirit of skeptical reflection or outright condemnation; an emphasis on its own precarious position vis-à-vis overbearing social forces; the claim to be engaged in radical intellectual and/or political work; and the assumption that whatever is *not* critical must therefore be *un*critical.⁹

This association of the humanities with critique was underscored by Terry Eagleton in a widely noted essay. "Are the humanities about to disappear?" Eagleton wonders. He goes on: "What we have witnessed in our own time is the death of universities as centres of critique. Since Margaret Thatcher, the role of academia has been to service the status quo, not challenge it in the name of justice, tradition, imagination, human welfare, the free play of the mind or alternative visions of the future."¹⁰ The declining role and influence of the humanities is tied, Eagleton declares, to the evisceration of critical thinking. Thanks to an increasingly instrumental and market-driven view of knowledge, underwritten by ballooning bureaucracies that cast professors and students in the roles of managers and consumers, the concerns of the humanities are rendered ever more peripheral.

It is possible to endorse Eagleton's frustration about the sidelining of the humanities without subscribing to the terms of his defense. Indeed, his own words might give us pause, for they do not support his argument as well as he might think. Some of the ideas he invokes—imagination, perhaps; tradition, certainly—are hardly synonymous with critique; indeed, they have often been cast as its opposites. The humanities have historically defined themselves according to a broad spectrum of ideas and values, including knowledge, truth, beauty, and the building of moral character. As Helen Small writes, "The work of the humanities is frequently descriptive, or appreciative, or imaginative, or provocative, or speculative, more than it is critical."¹¹ That contemporary scholars so often invoke "critique" as a guiding ethos and principle speaks to the current entrenchment of an either/or mindset: the fear that if one is not declaring opposition to the status quo, one is being co-opted by it. Yet the practices of academic life may turn out to be messier, more variegated, and more interesting.

Is it possible to voice a defense of the humanities that is not anchored only in the merits of "critical thinking"? What other attitudes, orientations, modes of argument are in play? To what extent are humanists engaged in practices of making as well as unmaking, composing as well as questioning, creating as well as subverting? And can we talk about the social ties of the humanities in ways that avoid the dichotomy of heroic opposition or craven co-option? A multidimensional defense of the humanities would accumulate rationales rather than limiting them or narrowing them down. In this spirit, I advance four key terms—curating, conveying, criticizing, composing—hoping that the temptation of alliteration won't overly compromise the force of my argument.

These words are verbs rather than nouns; actions rather than entities. Current defenses of the humanities, as several essays in this volume note, often revive versions of the two cultures split: the sciences help us build better bridges and find cures for cancer, while the humanities make us ethical citizens or empathic individuals. The preserve of the sciences, in short, is the material and natural world, whereas the task of the humanities is to mold better persons. Such a stark division of domains of inquiry seems unwise at a time when the humanities are becoming ever more concerned with ecological questions, climate change, and the future of the planet. Thinking about the humanities as a series of actions, practices, and interventions may prove more helpful.

Curating the humanities extends well beyond the mounting of exhibitions in art galleries and museums.[12] It involves, rather, a process of caring for—the word has its origins in *caritas*—of guarding, protecting, conserving, caretaking, and looking after. Humanities scholars are, among other things, curators of a disappearing past: guardians of fragile objects, of artifacts unmoored by the blows of time, of texts slipping slowly into oblivion. What often characterizes these historical remnants, as Stephen Greenblatt writes, is their sheer precariousness; they testify to "the fragility of cultures, to the fall of sustaining institutions and noble houses, the collapse of rituals, the evacuation of myths, the destructive effects of warfare, neglect, and corrosive doubt."[13] The wounded and vulnerable artifacts of history depend on acts of caring for their survival—without which they are in danger of disappearing from view, never to reappear.

This defense of curatorship may seem like a conservative definition of the humanities, but such a view would be mistaken. Or rather, we need to disentangle the various meanings of conserving and to question the assumption that caretaking—taking care of the past—is inherently conservative in the political sense. It is now captains of industry, after all, who are eager to sweep away the old-fangled and outdated and who worship the cutting edge. Meanwhile universities are reproached for not being sufficiently attuned to an ethos of creative disruption that is eagerly embraced by deal-makers, business analysts, and CEOs. In short, the temporal schemes of modernism—which counterpose the sluggish rhythms of dominant political or economic interests against the ruptures and innovations of a marginal avant-garde—have lost their last remaining shreds of analytical purchase.

Faced with this cult of technological and capitalist-driven innovation, it is important to insist on an ethics *of preservation*—on the value of the seemingly outmoded, anachronistic, or nonrelevant. Without the humanities, how many of us would have the opportunity to loiter and linger amongst the voices of the past? Who would ever come to feel, in their very bones, the bewildering strangeness and opacity of distant forms of life? Among all the forms of knowledge in the university, writes Mark McGurl, it is the humanities that are most invested in time travel, in moving back and forth across time. We need to conserve not only the works of the past, he continues, but also to protect those institutions—such as universities and libraries—that safeguard these works and that are increasingly under threat.[14] It is time to rethink the reflexive hostility toward institutions that fuels an iconoclastic strain of literary and art criticism.

The call for a more nuanced view of institutions has recently been amplified. In a 2018 essay in the *Chronicle of Higher Education*, Lisi Schoenbach remarks that scholars in the humanities are often attached to an avant-garde rhetoric of revolt and crisis that is aesthetically compelling, but poorly attuned to the different things that institutions of higher education do. "Universities," she remarks, "are many things at once: bad actors in gentrification, protectors of individual intellectual freedoms, media influencers, producers of a humanities work force, engines of their local economies, pawns of the military-industrial complex, hotbeds of student radicalism, training grounds for local politics." Scholars trained to detect ambiguity and contradiction in literary works need to work harder at discerning these same qualities in the structures they inhabit, and in defending those aspects of institutions that they value most. Schoenbach concludes by calling for a "radical institutionalism"—an approach to change that insists on the value of the institutional spaces we need most desperately to preserve.[15]

In the opening pages of his *An Inquiry into Modes of Existence*, Latour speaks to this very issue: instead of just criticizing institutions, can we also learn to trust them? ANT has devoted much attention to tracing the networks that allow for the making of knowledge: the coalitions of human and nonhuman actors—memoranda, files, computers, administrators, articles, equipment, corridor conversations—that allow ideas and arguments to be articulated and discoveries to be made. Against a rhetoric of liberation and emancipation from ties, it insists on the inescapability and the value of ties. Meanwhile, Latour's work has long challenged modernist philosophies of history that are dazzled by the lure of the new and the now. Questioning the rhetoric of revolution and the vanguard, of the new broom and the clean slate, it underscores the extent of our historical entanglements and the ubiquity of transtemporal connections. Here we can find resources for another vision of the humanities: one that is more fully attuned to their vital role in conserving the past.

One risk of the language of conserving and preserving, to be sure, is that of conjuring up a picture of jars of homemade marmalade or pickled beetroot, arrayed in silent serried rows in a darkened pantry. History, of course, is never preserved in this way, sealed off behind glass, but can only be actualized and made meaningful in relation to the concerns of the present. "The humanistic relation to the past," writes Dominick LaCapra, "assumes it is still part of the present with implications for the future."[16] We are attached to the artefacts we study; in a way that changes both the artefacts and those who study them.

It is here that *conveying* serves as another key term for the humanities. To convey means to communicate and to transport. To argue that the humanities are conveyed is to underscore that they are transmitted across time and space into new and often unexpected arenas. And in being transported, they are also translated—into the concerns, agendas, and interests of diverse audiences and publics. Translation is a key term in

ANT, which is often referred to as a sociology of translation. In contrast to what Latour dubs "double click"—the fantasy of instant and effortless information transfer promised by computer technology—ANT underscores that translation is ubiquitous and inescapable, but never faithful or complete. "Everything is translated. We may be understood, that is surrounded, diverted, betrayed, displaced, transmitted, but we are never understood *well*. If a message is transported, then it is transformed."[17]

We are never understood well: a phrase to be kept in mind in the light of the new interest in extending and expanding academic networks. There is a growing realization that universities need to make stronger alliances with interests and communities outside their walls. At Brown, for example, students can now enroll in a master's program in the public humanities that combines intellectual content with "practical skills needed for public humanities work: to develop exhibits and websites, care for museum artifacts, conduct oral history interviews, undertake historic preservation projects, facilitate public engagement and partnership, and create and manage cultural programs."[18] Attempts to articulate the social value of the humanities are becoming more widespread, as a means of pushing back against complaints of academic aloofness and scholarship's irrelevance to public life.

Yet current thinking about public engagement and the humanities, muses Kathleen Woodward, often betrays an anti-intellectualist strain, making earnest pleas for service learning, community needs, and civic engagement, but paying little attention to intellectual development.[19] Much of this work also displays a conspicuous lack of interest in aesthetic pleasure, even though this is surely the primary factor motivating hundreds of millions of people around the globe to watch movies, read novels, or listen to music. (An important exception is Doris Sommer, who writes compellingly on questions of pleasure and play and their relation to the "engaged humanities.)[20] Here there are manifold opportunities for conveyance that have been seized by online journals such as *Public Books* and the *Los Angeles Review of Books* that are building bridges between scholars and larger publics, publishing wide-ranging commentary on art and media that combines expertise with enthusiasm and accessibility.[21]

Traditionally, humanists have kept their distance from more popular forms of aesthetic response—or queried whether they even counted as aesthetic at all. But with the rise of omnivore taste (attachments to both Samuel Beckett and *Game of Thrones*) as well as the ubiquity of cross-media translation (with most students now coming to Shakespeare or Jane Austen through film adaptations), old divisions are being breached—from both sides. Lay audiences are not only looking for entertainment but are often eager to reflect on questions of meaning; as Michael Levenson points out, there is a vibrant sphere of humanistic activity outside of formal institutions, ranging from book clubs to historical re-enactors and editors of Wikipedia. The academic and the everyday humanities are adjacent practices that currently unfold in an almost complete mutual disregard. "We need knowledge as expertise," Levenson writes, "but also knowledge as diffusion and dispersal—knowledge as discipline but also as undisciplined curiosity in a museum. Our task now is not to strengthen walls around the humanities but to roman widely beneath their banner."[22]

Much humanities research, to be sure, does not lend itself to being measured in terms of direct use-value, relevance, or public impact: indeed, it may forcefully challenge such criteria. Yet this point also needs to be conveyed, along with a case for more nuanced and capacious forms of justification. The call to demonstrate the value of the humanities cannot be waved away as just a neoliberal imposition or a grievous symptom of anti-intellectualism. Being accountable—clarifying what scholars do and why it matters—is not a task that humanists can evade, even if we strenuously object to certain forms of accounting. At a time when an older model of higher education as the leisurely self-cultivation of a cultural elite is open to question—for good reasons—we need other justifications for the costs of the humanities and more eloquent accounts of its contributions.

Ien Ang offers incisive reflections on this topic, emphasizing the need for scholars to speak to multiple constituencies while also highlighting the schism between political ambitions and political effects in her own field of cultural studies. A vanguard stance and a tendency to speak and write in academic code are likely to inhibit rather than to aid intellectuals' engagement with the messy realities of sociopolitical life. "We need," she writes, "to engage in a world where we have to communicate with others who are, to all intents and purposes, intellectual strangers—people who do not already share our approaches and assumptions."[23] These intellectual strangers are not just bureaucrats or conservative pundits, but diverse groups who may be nonplussed or mystified by what they hear about the humanities. Scholars have often been reluctant to address this wider audience—thanks, in part, to the influence of theories that assume the intrinsically mystifying functions of everyday language. Against this trend, Ang argues for a scholarship more willing to engage in positive interventions and recommendations, less quick to bridle at lay interpretations or maladroit accounts of academic ideas. In short, conveying what we do to "intellectual strangers" means being willing to go down unexpected paths and into uncomfortable places, and accepting that transportation always involves translation.

This point brings us to a third aspect of the humanities: *criticizing*. To call for humanists to engage more actively in public life is not to imply that their task is to rubberstamp what currently exists. The so-called stakeholders of higher education—whether bureaucrats, politicians, taxpayers, private donors, foundations, journalists, and public commentators, parents, or students themselves—have diverse expectations of what the humanities should do. However, one widely accepted function is that of criticizing, objecting, disputing, and taking issue. Indeed, it is hard to see how sustained thought of any kind, whether inside or outside the academy, could take place without sustained practices of disagreement.

Criticizing includes a history of philosophical and political argument that goes under the name of critique, but it also leaves room for other forms and genres of disagreement. On the one hand, the humanities cannot jettison critique, given its central and defining presence in the history of the humanities. In spite of its own questioning of tradition, critique is now a fundamental part of tradition—the intellectual tradition of modernity—and it thus falls under the curatorial function of the humanities. The history of modern thought would be incomprehensible without knowledge of the ideas of Kant and Marx, feminism and Foucault. In this respect, Eagleton's nostalgia for a lost era of critique seems misplaced; many of these ideas now play a far more visible role in the humanities than they did decades ago.

On the other hand, the broader term *criticizing* is intended to convey that there are other ways of disagreeing. Critique often insists on its difference from mere criticism, understood as ordinary forms of disagreement or objection, by underscoring its own privileged vantage point. In traditional forms of ideology critique, this is a matter of contrasting the illusions of others to the critic's own ability to decipher hidden economic or political realities. Meanwhile, in poststructuralist critique, where the very idea of truth has been "problematized," techniques of troubling or defamiliarizing now signal the critic's self-reflexive distance from the naive or literal beliefs of others. Yet in both cases, we see a methodological asymmetry: ideas that scholars object to are traced back to hidden structures of which actors themselves remain unaware, while critique remains the ultimate horizon, a synonym for rigorous and radical thought.

Latour has argued at length against this kind of asymmetry—whereby scholars sustain their own claims to authority by exposing the naive beliefs, fantasies, or fetishes of others. To portray others as being driven by hidden structures that only the critical gaze can discern is to speak about them rather than to them, from the standpoint of the vanguard. Scholars can only hope to engage non-academic audiences—rather than chastise or admonish them—if they are willing to put themselves in the shoes of their interlocutors, to combine disagreement with empathy, and to take countervailing arguments seriously as arguments, rather than treating them as symptoms. As Stefan Collini writes, we need to extend imaginative sympathy to the agents we study. "Depth of understanding involves something which is more than merely a matter of deconstructive alertness; it involves a measure of interpretative charity and at least the beginnings of a wide responsiveness."[24] Criticism that adopts such a stance is not only less likely to fall into dogmatism but also more likely to be heard by the intellectual strangers invoked by Ang.

A final verb: *composing*. In an earlier manifesto published in *New Literary History*, Latour articulates a vision of composition as an alternative to critique. Practitioners of critique, he notes, are exceptionally skilled at deconstructing and demystifying, seeking to render things less real by underscoring their social constructedness. They are very good, in short, at pulling out the rug from under people's feet, while failing to provide a place where they might stand, however temporarily or tentatively. The idea of composition, by contrast, speaks to the possibility of a common world, even if this world can only be built out of many different parts. It is about making rather than unmaking, adding rather than subtracting, translating rather than separating. Composition is relevant to both art and politics; theory and practice. The word has its roots in art, music, theater, dance, but also speaks to the creation of communities and political collectives; it directs our attention away from the uninteresting claim that something is constructed the key issue of whether it is well made or badly made. "It is time to compose," Latour writes, "in all the meanings of the word, including to compose with, that is to compromise, to care, to move slowly, with caution and precaution."[25]

The primary focus of this essay is the insufficiency of the traditional idea of Nature in grappling with the crisis of climate change. Yet many of Latour's remarks are also germane to the concerns of the humanities. Could they help inspire an alternative vision of what humanists do? One that is less invested in the iconoclasm of critique and more invested in forms of making and building? In a recent book on the future of the humanities, Yves Citton remarks that politicians are fond of invoking the "knowledge economy" and the need to equip students for the "information society." Let us do our best, says Citton, to replace such slogans with references to "cultures of interpretation"—a term that affords a stronger case for what the humanities offer. Interpretation, here, is understood not a matter of recovering original or final meanings, but of mediating and translating, as texts are slotted into ever-changing frames.[26] This view of interpretation resonates with Latour's emphasis on composition, as a forging of links between things that were previously unconnected. Interpretation becomes an act of co-making that brings new things to light rather than a deciphering of repressed meaning or an endless rumination on the deficits or opacities of language.

The language of composition can draw humanists closer to others who are invested in making, building, and constructing, whether out of joists and steel plates or musical notes and physical gestures: engineers, painters, set designers, composers, novelists, website builders, scientists, and dancers. Such rapprochements might bring unexpected fruits and unanticipated insights. For example, the departmental divisions between literary scholars and creative writers, or between art historians and studio artists, are often lamented. An embrace of the language of creating and composing might help build bridges between these adjacent yet estranged fields. Thinking of criticism as something that is composed also draws attention to its form, and to the possibility of experimenting with a wider range of genres, styles, and registers in order to reach different audiences. We teach our students to read for form—and yet pay scant attention to the forms and genres of our own writing.[27]

Kathleen Woodward has made a forceful case for the importance of the "applied humanities." Rather than focusing only on dwindling numbers of majors in English or philosophy, we should turn our attention to other potential audiences, for example by importing humanistic ideas and skills into the professional schools. Along similar lines, John Mc-Gowan surveys the growing field of narrative medicine as a collaborative enterprise that is proving beneficial to both sides. While medical students are acquiring interpretative skills that can help make them more effective doctors, science can also provide models for humanists. "Science envy, it seems to me, would serve the humanities well if it pushed us to try to assemble the received knowledge of our field, toward more collaborative modes of work, and toward a more robust ability to speak collectively as humanists about the methods and aims of our practices."[28]

To those committed to such cross-disciplinary conversations, the oft-cited adage—it is the humanities that make us human—looks like the wrong path. Sarah Churchwell, for example, claims that "the humanities are where we locate our own lives, our own meanings; they embrace thinking, curiosity, creation, psychology, emotion. . . . We need the advanced study of humanities so that we might, some day, become advanced humans."[29] Are sociologists or mathematicians to be viewed as in some sense less human than philosophers or literary critics? Is their work not inspired by intense curiosity, bursts of creativity, affect, and emotion? And is it timely to underscore the exceptional status of humans at a time when our entanglement with, and dependence upon, natural and technological forces have never been more evident? In his Tanner lectures, Latour proposes that the humanities and sciences find common ground and create new alliances in the face of shared threats to academic institutions. What connects them, he suggests, is a deep sense of puzzlement about phenomena; while scientists start with the unfamiliar, humanists seek to render things unfamiliar. In both cases, the effect is to turn self-evident substances back into contingent and often surprising constellations of actors. Whether they work in libraries or laboratories, scholars are often passionately concerned with distinctions that are invisible or uninteresting to others. Hence Latour's final rallying cry: "Hair splitters of all disciplines unite!"[30]

Perhaps, then, we can defend the humanities without relying quite so heavily on eulogies to the human, or critique. Various disciplines in the humanities are linked by a commitment to preserving, conserving, and caring for. As the ideas of these disciplines are conveyed and communicated to intellectual strangers, we should expect, and even welcome, translation, mistranslation, and transformation. An indispensable element of thinking—though one that is not limited to the humanities—is disagreement, objection, and counter-argument. Yet nay-saying also has its limits: rather than embracing a perpetual ethos of deconstructing or destabilizing, humanists should devote more effort to making, building, and connecting. Thinking along these lines may allow us to articulate a stronger case for why the humanities should matter in the twenty-first century and beyond

This book is divided into two sections. The first half consists of broad, synoptic engagements with Latour's work and its relevance for the humanities. In his opening essay, Stephen Muecke considers the humanities with the studied perplexity of an anthropologist from another planet, noting the frequent schism between what humanists claim to be doing and what they actually do. Moving across history, linguistics, and literary studies, he proposes that the techniques, practices, and modes of knowing that characterize these fields have little to do with philosophical narratives of critical reason or radical indeterminacy that hover around the humanities. How else, then, might we describe what humanists do? A Latourian perspective on literature, for example, requires us to take seriously the realness of literary objects—as residing not in their autonomy, otherness, or remoteness from the world, but in their ability to increase and multiply connections: to other texts, people, things, concepts, institutions. The field of the environmental humanities is one fertile domain of connectivity that pushes against Nature/Culture bifurcations by allowing scientists and humanists to experiment on the same terrain. Muecke proposes that we replace two major drivers of the humanities— the unmasking power of critique and the unrealistic use of theory—with an emphasis on practice-driven modes of "compositionism" and "experiment" more willing to engage with the concerns of differing publics.

Antoine Hennion offers an enlightening genealogy of the development of ANT in Latour's work and his own at the Centre de Sociologie de l'Innovation (CSI) in Paris. The Center's studies of science and of culture, he notes, pursued closely related but differing tracks, drawing on the concepts of translation and mediation respectively. In both cases, the emphasis on the mutual constitution of objects and relations remains fundamental; an emphasis that has little to do with a skeptical or demystifying language of social construction. For ANT, that things are created via relations does not make them less real but more real. Hennion illustrates the point via his own work on fans, amateurs, and aficionados: one can take the love of art seriously while also showing how this love is coproduced; by the artwork, but also by bodies and feelings, conversations with friends, knowledge of a pertinent corpus (whether we are talking of rock music or opera arias, horror movies or modernist painting), a formal or informal training in practices of perception and discrimination. Where Bourdieu went wrong, Hennion remarks, was not in situating art in a field of relations, but in inferring that the love of art was therefore an illusion to be demystified: these two ideas have no necessary connection. Meanwhile, the essay also draws out various affinities between ANT and American pragmatism; the radicalism of William James and John Dewey, Hennion suggests, has yet to be fully appreciated.

Graham Harman agrees that Latour's work offers a compelling alternative to entrenched positions within the humanities that rely on nature-culture oppositions, whether to champion the former or the latter. At odds with both the social constructivism preferred by the left and the political and philosophical conservatism of the right, it shows that "stability is neither assured nor impossible, but is achieved only at great cost, and only by way of inanimate things." In his essay, Harman traces the arc of Latour's thought, from the Hobbesian dimensions of early ANT toward a greater openness to moral concerns and "mini-transcendences" in the 1990s to Latour's most recent critique of economic thinking in AIME. If economics aspires to the status of a new master discourse, then it becomes crucial to emphasize other modes—morality, attachment, and organization, as redescribed by Latour—that are at odds with the logic of economic calculation. Via his explication of these modes, Harman draws out their relevance to a rebooted humanities willing to relinquish some of its intellectual tics and *idées fixes*.

The language of concern and care has increasingly moved to the forefront of Latour's work, underscoring that ANT should not be confused with standard accounts of social-construction-as-critique. Such a language, remarks Heather Love, has made his work more appealing to humanists, who have a longstanding investment in questions of sensibility, empathy, and subject formation. And yet, for this very reason, we can learn more from Latour's modeling of attention as a practice rather than care as an ethical disposition. In literary studies, for example, questions of method are often brushed over or decried as inherently alienating in favor of an intuitive attunement to the charismatic art work—a style of thinking that glosses over the social shaping of aesthetic response via education and cultural capital. Latour's early work, by contrast, models techniques and procedures whose value lies in being public, shareable, and learnable—that stress curiosity and careful description rather than reverence for the art work and the cultivation of ethical selfhood.

Anders Blok and Casper Bruun Jensen tackle the question of Latour's relationship to critique—the topic that has overwhelmingly dominated his reception in the humanities. They propose that his stance is not simply one of rejecting the critical but existing alongside it: "intersecting with it but not engulfed by it." In making the case, they consider his relation to Michel Serres, who refuses any analytical meta-language as well as strong divisions between fiction and fact; Latour's idea of critical proximity as enriching our repertoire of salient actors and attending to details that make a difference; his case for a "cosmo-politics" that is more fully attuned to nonhuman agency and the urgent realities of global warming (here humanists, as experts in figuration, can play a vital role); and Latour's relation to postcolonial perspectives and defenses of indigenous knowledge. Underscoring that Latour is deeply concerned with values and inspired by indignation rather than complacency about the present, they suggest that he redistributes critique, conceiving it as one strategy among others. The term "a-critical" can lift Latour out of an unproductive battle about the stakes of critical theory, while also capturing more adequately the agility and mercurial quality of his thought.

Steven Connor's title—"Decomposing the Humanities"—speaks to his skepticism about the soul-stirring claims that are often made on behalf of the humanities: as repositories of the human spirit or vigilant guardians of radical thought. The work of Latour and his mentor Michel Serres models a style of thinking that is less agonistic as well as less self-congratulatory and that can inspire an alternate, more affirmative relationship to the things of this world. As scholars in the humanities become increasingly concerned with climate change, environmental damage, and species destruction, so Latour's questioning of the philosophies of modernity becomes increasingly salient. And yet there is no automatic relation, Connor points out, between styles of thinking (such as the questioning of modern epistemology) and real-world effects. Redressing climate change is ultimately more about engineering than emancipation: if the humanities can give up their folie de grandeur and their claim to being sole custodians of the human, perhaps they will have something useful to contribute.

In his essay, Dipesh Chakrabarty underscores the drastic implications of climate change for intellectual disciplines and divisions of labor. It is no longer feasible to sustain the division of the ethical and the rational from the purely biological that has justified the separation of the humanities from the natural sciences. Chakrabarty traces out the logic of this opposition as it is articulated in Kant's distinction between animal life—as natural, given, and taken-for-granted—and human life, as the struggle to achieve a more perfect and just society. What is now often called the Anthropocene, by contrast, has forced a new awareness of the entanglement of humans with natural forces they have helped cause but cannot control, and the urgent need to adopt less anthropocentric perspectives on the future of the planet. It is in Latour's work, Chakrabarty concludes, that we can find a model of thinking that is fully attentive to the agency of the nonhuman and the related question of "deep time."

Yves Citton tackles Latour's *An Inquiry into the Modes of Existence*. The book's turn to *differing* modes of existence—religion, politics, reference, morality, etc.—responds to and revises the aforementioned tenets of ANT (where the repeated recourse to a language of actants and networks runs the risk of making everything sound very similar). This emphasis on a "pluriverse"—on multiple modes of being that are equally real—also challenges the propensity to invoke just one mode, economics, to explain or justify everything (the glories of the free market for the right; the tyrannies of capitalism on the left). How, then, does AIME speak to the concerns of literary critics? Latour's account of the mode of existence of "fictional beings"—including not just fiction, but various artistic and expressive endeavors—offers a powerful redescription of our objects of study. It speaks not only to our attachment to works of literature, Citton argues, but how such works can help us reattach—to new concerns, commitments, collectives. As forms of "meshwork," literature and art can weave fragments together, compose common agendas, binding humans to nonhumans in response to the current reality of ecological and environmental crisis. And here Citton also underscores the need for closer collaboration between literary critics and scholars of media and mediation.

In his essay, Simon During identifies a certain disjuncture between Latour's arguments and the concerns and traditions of the humanities. Latour's criticism of the "modern constitution" and its regimes of purification (oppositions of nature/culture, human/nonhuman), targets the history of the natural and social sciences. How might close attention to the "emergent humanities" of the seventeenth century complicate this account of the modern? Rather than endorsing a familiar narrative of modernity as schism and division, During wants to track the specific interests, traditions, values, and methods that have shaped humanistic thought. Here he turns to the seventeenth-century poet Abraham Cowley, whose writings embraced an innovative and ecumenical set of concerns. Cowley's "creativism" blurs distinctions between divinity, humanity, and matter; his poetry invites forms of interpretation that complicate any hard opposition of fiction and truth; his vision of translation as cultural transmission cuts across the distinction of ancient and modern; and his retreat from the world of commerce and business serves an important critical function that is not delegitimated by Latour's critique of critique. Many of these same concerns shape the subsequent history of humanistic creativity and thought—in a manner, During argues, that cannot be adequately understood via the picture of modernity in Latour's work.

Finally, Nigel Thrift takes issue with the ubiquitous "it's all got worse" lament that drives discussion of higher education. This story of decline, he points out, fails to grapple with many of the reasons why things have changed: that universities have become more bureaucratic, for example, is not unrelated to the dramatic expansion of their intellectual and public activities (including big science), as well as the much larger numbers of students they now serve. The nostalgia for an "artisanal" model of the liberal arts, meanwhile, ignores the ways in which such a model has historically shored up—and continues to sustain—class privilege. In this context, Latour's work offers a way of more fully coming to terms with the differing modes of existence that now characterize higher education. Rather than holding fast to a picture of themselves as an intellectual or spiritual elite threatened by the vulgarities of the marketplace, academics need to become more involved in reimagining and redesigning the university—in a way that can honor its varied and conflicting duties as well as its multiple intellectual and public roles.

The second part of book turns to specific fields and disciplines. David Alworth offers a wide-ranging account of Latour's ideas, arguments, and styles of thought as they pertain to literary studies. He begins by noting the heteroglossia of Latour's writing—its playful use of dialog, character, metaphor, and other novelistic techniques—as well as Latour's borrowing from Greimas's conception of actants in order to describe how phenomena interact in the empirical world. In stressing composition rather than critique and questioning the idea of "society" as a self-evident explanation or cause, Latour's way of thinking contravenes many of the current assumptions in literary studies. We are invited instead to "follow the actors" and to trace out the specific and often unpredictable ways in which literary and non-literary phenomena connect. Here Alworth underscores Latour's indebtedness to pragmatism and William James, draws connections to James J. Gibson's l notion of affordances, and details how his work is influencing both the new materialism and the new formalism. While noting the controversies and misunderstandings that swirl around Latour's work, Alworth suggests that he is poised to become increasingly influential in literary studies.

Film studies has long been conscious of distributed agency: the collectives of human and nonhuman actors, including cameras and other technologies, on which movie-making depends. Noting affinities between Latour and the phenomenological film theory of Vivian Sobchack, Claudia Breger fleshes out a layered approach to film that can trace the empirical variety of actor-networks which also attending to the force of audience attachments. That persons are networked, she points out, does not prevent us from respecting what is given in their experience. Turning to a close reading of *Western* (2017) directed by Valeska Grisebach, Breger shows how ANT's concern with tracing relations resonates with Grisebach's portrayal of German workers building a hydro-electric power station in rural Bulgaria. *Western* solicits a diffuse array of affective reactions that encourage critical proximity rather than distance—even as Grisebach's deployment of the conventions of the Western genre is respectful rather than parodic. Like other authors in the volume, Breger sees no contradiction between an emphasis on specificity (the distinctive affordances and attachments of film) and a broader commitment to transdisciplinary humanities that can be less "standoffish" in its relations to public life.

Michael Witmore discusses Latour's challenge to long-established divisions between the sciences and the humanities as it speaks to the intellectual hybrid of the "digital humanities." As an example of such digital humanities practice, he describes his own recent work on conjunctions in Shakespeare's plays: a blending of humanistic expertise and interpretation with the scanning and processing capacities of computers. The value of such an approach, he remarks, lies not just in providing statistical support for generalizations about large bodies of texts, but in the potential for highlighting distinctions invisible to the human eye—as in the case of Eadweard Muybridge's famous photos of galloping horses, we are confronted with a surprising or counterintuitive perspective on what we think we already know. Rather than seeing the sciences and the humanities as opponents, Witmore endorses Latour's recommendation that we conceive of them as allies and urges us to expand our vision of what counts as humanist scholarship.

Barbara Herrnstein Smith turns to Latour's writings on religion—as "remarkable works of lyrical philosophizing." For Latour, that scientific facts are constructed—fabricated, put together out of heterogeneous elements—does not render them less real but more real. The same is true—applying the principles of a symmetrical anthropology—of demons and divinities, whose potency is forged via numerous mediations and networks of co-actors. While both scientific knowledge and religious belief are equally composed, they are nonetheless incommensurable in their modes of veridiction and their tone and mood. In developing this line of argument, Smith remarks, Latour manages to tie together "a theoretically sophisticated account of scientific knowledge with a rhetorically deft Christian apologetics." Smith remains skeptical, however, about the strong demarcation between science and religion that she sees in *AIME*, while also noting the book's conspicuous lack of interest in religions other than Christianity. Meanwhile, Latour's vision of history turns out to be less symmetrical than his anthropology. *AIME*'s condemnation of the Moderns echoes a familiar narrative of modernity as a fall from grace that risks becoming totalizing or moralistic—against the express intentions of Latour's own method.

How might a Latourian style of thinking prompt us to rethink the question of politics—not as a final cause and explanation for all social phenomena, as critical theory would have it, but as a distinctive mode of existence with its own practices, assumptions, sites of activity, and felicity conditions? For an explanation of why politics *matters,* Gerard de Vries proposes, we should turn to Montesquieu, whose *L'esprit des lois* offers an encyclopedic survey of different societies according to their types of government. In his comparison of despotic regimes, monarchies, and republics, Montesqieu is less concerned with sources of power than how power is exercised—that is to say, how it relates in concrete ways to physical and economic conditions, ways of life, and the beliefs, attitudes, and passions of particular communities. *L'esprit des lois* thus offers a vision of politics that is very different to that of Hobbes—the primary reference point for most political theorists—in tracing the details of relations and translations, rather than setting up an opposition of the individual and the state. Montesquieu's rich vocabulary and down-to-earth analysis of how politics works turns out to be uncannily attuned to a Latourian way of thinking.

The two following essays turn to the relevance of Latour's work for the visual arts. Patrice Maniglier begins by stressing the need for diplomacy, setting out four principles to guide the interactions between supporters of *AIME* and the "militants" of art—that is, those eager to protect and defend the existence of art by emphasizing its distinctive history, forms, and interests. The mode of existence in *AIME* most relevant to art and aesthetics is Latour's discussion of "fictional beings." Yet the art critic or art historian may worry that such a term does not speak well to their concerns. After all, does not "fiction" imply a concern with representation and/or narrative that much modern and contemporary art explicitly rejects? If the term is too narrow in one sense, it seems too broad in another. As used by Latour, "fictional beings" would include virtually anything with an aesthetic dimension—from aspects of ordinary language to advertisements—and would thus fail to account for specifically artistic concerns. In response, Maniglier strives to show that these concerns can be accommodated within a Latourian framework; such a framework, meanwhile, allows us to more fully grasp the profoundly relational qualities of art and aesthetic experience.

Francis Halsall considers the parallels between Latour's ideas and contemporary art practice: not by "applying" Latour to art, but by drawing out shared techniques, approaches, and orientations. Art, he observes, has become unmoored from any relation to a specific style, object, or medium; it is now defined by a state of radical eclecticism. As a result, there is a conceptual rhyming or resonance between Latour's work and developments in the art world. The distinctive features of Latour's work—its conception of actors and networks; its questioning of nature-culture distinctions; its "flat ontology" that levels hierarchies between kinds of actors; its emphasis on the inescapability of mediation and translation—are themes that are also being picked up and explored by many contemporary artists. Meanwhile Latour is also in key respects an artist: someone who experiments with different cultural platforms and whose academic writing blurs genres by making use of literary device (dialogue, vivid description, extended metaphors, and fictional characters).

Bruno Latour takes up this theme in his concluding remarks, while remarking on the context-specific nature of "humanities discourse," with its lack of a direct French equivalent. Looking back on his own development, he traces a series of formative encounters with Nietzsche, Derrida, biblical exegesis, and semiotics—as offering not a schooling in "theory," understood as a series of vaulting philosophical claims, but a training in meticulous attention to techniques of writing. This attention to modes of expression, along with a conviction that there is no metalanguage that can subsume others, has inspired not only Latour's scholarship in science and technology studies and other fields, but also his staging of several art exhibitions (*Iconoclash, Making Things Public,* and *Reset Modernity!*) as experimental zones where different media and modes can connect and collide. Meanwhile, he concludes, the humanities have long paid attention to the extraordinary range of figurations in literature and art; they thus have ample resources with which to question, rather than to reinstate, conceptions of the human that are being radically transformed in the wake of ecological transformation and climate change.

The present volume, then, offers a substantial introduction to Latour's work, as it speaks to the concerns of literary and film critics, art historians, religious studies scholars, political theorists, and other humanists. Such an introduction, we hope, will inspire more informed conversation about the relevance of ANT for research and teaching in the humanities. And it also canvasses ways of discussing the humanities that do not fall back in the genres of the sermon or the jeremiad. All too often, it seems, humanists try to save the humanities by nurturing a sense of exceptionalism. It is as if we can only defend what we do by disparaging everyone else—those blinkered scientists holed up in their labs; our attention-challenged students enslaved to their mobile devices, a benighted public of the ignorant and the indifferent. Our contributors have opted to pursue alternative lines of thought and to experiment with other intellectual and practical possibilities. As well as defending the humanities, can we also recompose them?

Notes

The revision of this introduction was supported by a grant from the Danish National Research Foundation (DNRF 127).

1. For a discussion of the reliance of humanities discourse on sermonizing, see Simon During, "Stop Defending the Humanities," *Public Books*, March 1, 2014.

2. "Recomposing the Humanities—with Bruno Latour," *New Literary History,* 47 (2/3) 2016.

3. Bruce Robbins, "Not So Well Attached," *PMLA,* 132, 2 (2017): 375.

4. Bruno Latour, Graham Harman, Peter Erdelyi, *The Prince and the Wolf: Latour and Harman at the LSE* (Winchester: Zero Books, 2011), 27.

5. See, for example, Donna Haraway, *Modest_Witness@Second_Millennium .FemaleMan.©_Meets_OncoMouse™: Feminism and Technoscience* (New York: Routledge, 1996); Srikanth Mallavarapu and Amit Prasad, "Facts, Fetishes and the Parliament of Things: Is There Any Space for Critique?" *Social Epistemology*, 20, 2 (2006): 185–199, and the essays by Barbara Herrnstein Smith and by Anders Blok and Casper Bruun Jensen in this volume.

6. Maria Puig de la Bellacasa, *Matters of Care: Speculative Ethics in More than Human Worlds* (Minneapolis: Univ. of Minnesota Press, 2017); Kane Race, "What Possibilities Would a Queer Actor-Network Theory Generate?" in *The Routledge Companion to Actor-Network Theory*, edited by Anders Blok, Ignacio Farias and Celia Roberts (London: Routledge, 2019); and Oumelbanine Zhiri, "Practices of Early Modern Orientalism: A Latourian Perspective," in *Latour and the Nonmodern Humanities,* ed. Claire Lyu and Elisabeth Arnould-Bloomfield (London: Bloomsbury, forthcoming.)

7. See, for example, Didier Fassin's indignant response to Latour's essay, which includes a personification of critique as a long-suffering hero. His title is intended to convey, he writes, that "critique repeatedly undergoes ordeals, that it bears them with patience, and that it continues to exist beyond them." Didier Fassin, "The Endurance of Critique," *Anthropological Theory*, 17, 1 (2017): 5.

8. I owe this intriguing question to a conference at the University of Warsaw in 2014 called "Imagine there were no Humanities," where I developed an early version of these arguments.

9. See Rita Felski, *The Limits of Critique* (Chicago: Univ. of Chicago Press, 2015).

10. Terry Eagleton, "The Slow Death of the University," *Chronicle of Higher Education*, April 6, 2015.

11. Helen Small, *The Value of the Humanities* (Oxford: Oxford Univ. Press, 2013), 26. For relevant histories of the university and the role of the humanities, see e.g., Julie A. Reuben, *The Making of the Modern University: Intellectual Transformation and the Marginalization of Morality* (Chicago: Univ. of Chicago Press, 1996), and Chad Wellmon, *Organizing Enlightenment: Information Overload and the Invention of the Modern Research University* (Baltimore: Johns Hopkins Univ. Press, 2015).

12. See also Michael Meranze, "Curating the Humanities," http://utotherescue.blogspot.com/2013/11/curating-humanities.html.

13. Stephen Greenblatt, "Resonance and Wonder," in *Exhibiting Cultures: The Poetics and Politics of Museum Display*, ed. Steven D. Lavine (Washington, DC: Smithsonian, 2001), 43–44.

14. Mark McGurl, "Ordinary Doom: Literary Studies in the Waste Land of the Present," *New Literary History* 41, no. 2 (2010): 329–49. See also Nathan K. Hensley, "Curatorial Reading and Endless War," *Victorian Studies* 56, no. 1 (2013): 59–83.

15. Lisi Schoenbach, https://www.chronicle.com/article/Enough-With-the-Crisis-Talk-/243423.

16. Dominick LaCapra, "What is Essential to the Humanities?" in *Do the Humanities Have to Be Useful?* ed. G. Peter Lepage, Carolyn (Biddy) Martin, and Mohsen Mostafavi (Ithaca: Cornell University, 2006), 81.

17. Bruno Latour, *The Pasteurization of France* (Cambridge, MA: Harvard Univ. Press, 1993), 181.

18. "Program Requirements." John Nicholas Brown Center for Public Humanities and Cultural Heritage, http://www.brown.edu/academics/public-humanities.

19. Kathleen Woodward, "The Future of the Humanities—in the Present & in Public," *Daedalus* (winter 2009), 116–117.

20. Doris Sommer, *The Work of Art in the World: Civic Agency and Public Humanities* (Durham, NC: Duke Univ. Press, 2014).

21. See Wai Chee Dimock, "Experimental Humanities," *PMLA*, 132, no. 2 (2017): 241–249

22. Michael Levenson, *The Humanities and Everyday Life* (Oxford: Oxford Univ. Press, 2018): 5.

23. Ien Ang, "From Cultural Studies to Cultural Research: Engaged Scholarship in the Twenty-First Century," *Cultural Studies Review* 12, no. 2 (2006): 190.

24. Stefan Collini, *What Are Universities For?* (London: Penguin, 2012), 83.

25. Latour, "An Attempt at a 'Compositionist Manifesto,'" *New Literary History* 41, no. 3 (2010): 487.

26. Yves Citton, *L'avenir des humanités: économie de la connaissance ou cultures de l'interprétation* (Paris: La Découverte, 2010).

27. For a stimulating reflection on academic genres, see Ben Highmore, "Aesthetic Matters: Writing and Cultural Studies," *Cultural Studies*, 32, no. 2 (2018): 240–260.

28. John McGowan, "Can the Humanities Save Medicine, and Vice Versa?" in *A New Deal for the Humanities: Liberal Arts and the Future of Public Higher Education,* ed. Gordon Hutner and Feisal G. Mohamed (New Brunswick: Rutgers Univ. Press, 2016), 135; Kathleen Woodward, "We Are All Non-Traditional Learners Now: Community Colleges, Long-Life Learning, and Problem-Solving Humanities," in *A New Deal for the Humanities,* 51–71. On narrative medicine, see Rita Charon and Sayantani Dasgupta, *The Principles and Practices of Narrative Medicine* (Oxford: Oxford Univ. Press, 2016).

29. Sarah Churchwell, "Why the Humanities Matter," *Times Higher Education Supplement*, November 13, 2014.

30. Latour, "How to Better Register the Agency of Things," Tanner lecture 1: Semiotics, 12. See also Mario Biagioli, "Postdisciplinary Liaisons: Science Studies and the Humanities," *Critical Inquiry* 35, no. 4 (2009): 816–33.

PART ONE

WHAT DO THE HUMANITIES DO?

CHAPTER ONE

An Ecology of Institutions

Recomposing the Humanities

STEPHEN MUECKE

IF ONE APPROACHES the humanities with an attitude of perplexity, like the fictional anthropologist in Bruno Latour's *An Inquiry into Modes of Existence*,[1] as if from another planet, in a state of not yet knowing (which is always a good starting point for an inquiry), then it quickly becomes clear that the perplexity is justified. The humanities faculty, or the arts faculty, is made up of a disparate set of disciplines. Some sail very close to the sciences (linguistics and archaeology); others, like most branches of philosophy, are totally removed from any kind of empirical fieldwork or lab work. Others, like literary subjects, are lost if there is no text in the class.

The *Inquiry into Modes of Existence* (*AIME*) is the culmination of a number of empirical studies that Latour has written over a thirty-year period on major French institutions that, collectively, can be made to stand more generally for modern Western civilization.[2] *AIME* is a work of philosophy, of "interactive metaphysics," that uses an accompanying website to engage readers in tackling real-world problems facing the planet. It is ambitious on that scale, and no doubt doomed to fail, yet it tackles the conceptual architecture of the Moderns descriptively at the level of the organization of institutions, rather than attempting a critical overview of the whole society, as a Marxist class-based view is constrained to do. It adopts a tone of respectful and intimate dialogue that Latour calls diplomacy, and in this, it is preparatory work for future dialogues with non-Western civilizations, which would be enjoined to negotiate what they have to offer from their own heritage to see what kind of common world might be composed in the face of looming environmental and economic problems. *AIME* is thus postcolonial to the extent that it does not assume that Western modernity is the model that is everyone's destiny, especially since its serious design faults, diagnosed in this book, demonstrate its unsustainability in its current form.

The specifics of the way the modern institutions are organized sheds light on Latour's major innovation, brought to a head in this work: *ontological plurality*, that is, the ways in which different modes of existence go about their business without being able to be subsumed by one another, nor referred to some common level like "society" as a basis for their existence. For example, all the advances of science and secular modernity have not managed to eliminate a religious mode of being in the world. The churches obviously continue as flourishing institutions. For the faithful, there is a way of "being true" in a religious mode that is different from the knowledge-based truths that we find in universities. Latour's task is to describe how these truths are sustained across no less than fifteen different modes of existence, including the reproductive, the legal, the moral, the political, the habitual, the technical, and the referential, among others.

So, let's experiment by giving Latour's anthropologist another research contract, to delve into the humanities. She might still be a little nonplussed in that, having emerged hardened, with only a few bruises, from her grueling inquiry that produced the preliminary anthropological report, signed by Latour, on *all* of the Moderns, she might well say: "But I've done this, haven't I? Don't the Modes called fiction [FIC], morality [MOR], religion [REL], reference [REF], and even metamorphosis [MET] more than cover what the humanities could throw up? Are there any crucial subtleties that are not accounted for? And hasn't there been a whole process of extensive collaboration, debating, and contesting by the AIME team and the public on the project's website?"[3]

If we were to commission such a new inquiry, we would have to say that there is indeed further work to be done: the *temporal, textual,* and *theoretical* dimensions need further exploration, by grabbing the coattails of historians, literary scholars, and theorists and following them around as they move through their worlds, observing (according to good ethnographic practice) *what* they are doing, while listening with some skepticism to what they *say* they are doing. And these humanities folk often feel under siege, as if their very existence were hollowed out each time their funding is cut, while the sciences benefit by being awarded billions to find some elusive "God particle" with the CERN accelerator. Latour has not helped by brutally telling the truth as he sees it (in his famous article "Why Has Critique Run Out of Steam?"): "Is it so surprising . . . the humanities have lost the hearts of their fellow citizens, that they had to retreat year after year, entrenching themselves always further in the narrow barracks left to them by more and more stingy deans? The Zeus of Critique rules absolutely, to be sure, but over a desert."[4] So the humanities do need a pat on the back, a reassurance that they, too, deal in real worlds, and not just in representations thereof. Latour's modes of existence can surely help here, since his major innovation of plural ontologies make room for other realities. In principle, this gives them even more room to maneuver (as Ross Chambers would say) within their cherished interpretative frameworks.[5] So now the defense of poetry need no longer be based in a retreat to the usual redemption through asceticism: a view of literature as the little cave in which we can feel truly, spiritually, human, while the cold, objective world of Science and the unforgiving barbarianism of the Market imprison us, albeit providing for our material necessities. Here is reproduced that same interior-exterior opposition that is also a favorite humanities seminar trope celebrating the "freedom" to be found inside the Life of the Mind (*esprit*, *Geist*).

Instead, we now want to find a way to feed our humanities person, starving in that lonely cave of the soul (or mind), by giving full ontological importance to, in the first instance, an aesthetic mode of existence, so that it can throw its weight around, as in the work of art saying, "Screw you!" (it is the modernist ones, especially, that talk rough like this), "I can stand on my own two feet, without being reduced to a reflection of my market value, or to some sociological or biographical explanation, or to my compatibility with the interior design of your living room."

Achieving full ontological weight means being *more realistic* about works of art, about the mode of existence of the aesthetic. In practice, this does not mean leaving them alone at all, even if they are rude and abusive. Their ontological strength will not come from being isolated in their domain, but from their capacity to multiply their connections. Being realistic in this sense, for those looking at the humanities from the outside, means seeing them as *multiply real*. And as working *within networks*, something already very familiar to the Latourian investigator, who knows very well that when humanists say they are working in a particular domain, and are defensive about their own discipline or department, that following them around for a bit will reveal, very quickly, that they inhabit a heterogeneous network of associations. No surprises there—or rather, *actual* networked living is full of surprises that make life interesting in ways that the habits or ideals of living in "domains" cannot. So if we draw the typical humanists out of their cave where they reflect on interiorities ("the life of the mind" for philosophers, "forms of subjectivity" for literary scholars; forget the archaeologists and linguists, who are already foraging out there among shards and phonemes, trowel or voice-recorder in hand) . . . if they emerge from their caves and blink in the sunshine, we have to convince them that the cave was an illusion, that even for the most erudite and self-contained there was always a lot of moving around, tracking things down along knowledge-acquisition pathways and affect-acquisition pathways.[6] So let me state my initial hypothesis on the work of the humanities scholar: truth is always earned in traveling; it will not be the "reward of free spirits" nor of monastic seclusion, but "a thing of this world," as Michel Foucault famously put it.[7] Literally traveling in "the field" if you have ethnographic habits, and traveling through books and papers if you are a historian.

Let's tackle the case of history where the journey through time, and also through the career of the historian, is mundane; I would claim it is all surfaces. It does not aspire to Enlightenment in the Kantian sense, where we would be in a "state of 'immaturity' if a book should take the place of our understanding."[8] What the historian establishes as historical truth *is in* the books and papers, and in the precise textual technologies of the ordering of sources, as outlined by Anthony Grafton in his book on the footnote:

In the modern world . . . historians perform two complementary tasks. They must examine all the sources relevant to the solution of a problem and construct a new narrative or argument from them. The footnote proves that both tasks have been carried out. It identifies both the primary evidence that guarantees the story's novelty in substance and the secondary works that do not undermine its novelty in form and thesis. By doing so, moreover, it identifies the work of history in question as the creation of a professional. Like the high whine of the dentist's drill, the low rumble of the footnote on the historian's page reassures: the tedium it inflicts, like the pain inflicted by the drill, is not random but directed, part of the cost that the benefits of modern science and technology exact. As this analogy suggests, the footnote is bound up, in modern life, with the ideology and the technical practices of the profession.[9]

So, as we chase up historians and ask what they do with time (before moving on to my two other emblematic figures for the humanities, the textualists and the theorists), can we posit that the figures of history constitute a mode of existence? In the Latourian scheme, it would have to be a subset of reference [REF], a knowledge network that extends itself reliably through time, with its own felicity conditions and its own repetitive risks that are necessary for it to maintain its truths.

When does history stop being history? When its footnotes fall off the page, when it loses its referents. While a referent can just as easily be spatial as temporal, the historical text will also be peppered with dates, a simple chronological ordering, without which, again, it ceases to be history and becomes a fiction or an essay. So, there are two kinds of anchoring, one to sources (which are preferably triangulated, so that disparate sources reinforce each other) and the other to the technology of the calendar. But do these simple devices just "reassure," as Grafton says? Or is the "new narrative or argument" constructed from them more important? Where do the risks of extension lie, that is, how does history keep being history despite the risks of innovation it must take? Certainly, getting the sources or the chronology wrong is enough to invalidate the historical mode of being, but the interpretation also extends through rhetoric: argument, persuasion, ideology. I don't have time to go into the ways in which an E. P. Thompson makes Marxist critical moves while a conservative historian makes others, except to say that these moves are constitutive of the text's desire to create a reading public, which is another key aspect of the humanities disciplines: obviously, we write to be read, not just to make a record of some experience.

Now if this account of history as a chronicle rhetorically engineered to gather up a reading public seems excessively naive, that might well be because the text of history often has higher aspirations, which take the form of basic history being combined with what we call "theory." In his article "Hayden White's Philosophical History," Ian Hunter mounts a characteristically vigorous argument about how White's 1973 book *Metahistory* makes claims for historiography that favor a certain theoretical elaboration over the practice of empirical history.[10] White does this by first characterizing historians of the seventeenth century as "radically divided" into two camps: the chaotic ecclesiastical historians who were tied to their separate religious confessions, and the "historical philologists" of antiquarian erudition, who could only produce mere "chronicles," stories "geared to rhetorical effects and moral lessons" (HW 336).

Hunter disputes this account of the seventeenth-century historical field, and argues that White has set it up in this way in order subsequently to install a *modern* disposition based on the philosophy of Immanuel Kant: "[White's] divided and chaotic 'premodern' historical field is not an object of investigation and description. Rather, it is an intellectual condition that is invoked as the trigger for a particular kind of intellectual performance: the suspension of empirical attention to documents and contexts, and the reorientation of attention 'inwards,' onto the transcendental *figurae* that are supposed to lie in the practitioner's own 'mind' as the forms of 'historical consciousness.' Approached thus, as a concrete intellectual activity or 'game,' White's transcendental philosophical history can neither be false nor true" (HW 341). Hunter, across a number of articles and books, focuses on this kind of "spiritual exercise" as a pedagogical routine very much present in the humanities today, but with its origins in Kant's retention of Christian faith in the elaboration of philosophical discourse in the early modern university.[11] Hunter's analysis of White is useful for our "recomposition" purposes in that it shows White highlighting what he sees as the inadequacies and chaos of the "premodern" historians of the seventeenth century, then praising and refining the Kantian transcendental shift that later had its American reception in nineteenth-century "liberal" Protestantism (HW 342). What White ends up with, Hunter argues, is not a contribution to the craft of the historian, but a "theory" for how one should dispose oneself toward history, "a means of inducing the pursuit of transcendental structures" (HW 341), structures that the raw material of history does not demand, but that serve a pedagogical purpose: namely, the "grooming" of the moral subject in elite schools, a subject who knows how to reconcile "poetry" and "reason" dialectically, and whose mode of knowing *prefigures* empirical investigation because, as White himself states, "all historical representation is the surface effect of transcendental rhetorical tropes" (HW 337).

This kind of theory is dangerous for empirical history because it implies that the real work of history is *on human subjectivity*, on directing the evolution of historical consciousness. Now, while it might, in some instances, be a good thing for a modern secular world to have in place a pedagogical method that teaches students how to internalize moral governance (and there are vast networks of religions and psychological industries doing just that), this is surely a distraction from the real problems of what history is and what can be learned from historical texts. Hunter's rebooting of historiography, therefore, after a nostalgic glance back to the erudite practices of Renaissance humanist scholars, would involve banishing Kant's grand transcendental cognitive structures (in the forms that they survive in in the contemporary scene) in favor of empirical scholarship. Such scholarship would normally look to contextual evidence as having the capacity to inform the interpretation of texts and not as already prefigured in them (HW 352), and would not be dominated by appeals to the "prison-house of language." Henceforth, the risks taken by the historical mode of existence—its (Latourian) hiatuses—will not be the capacity for moral consciousness to survive various crises inflicted upon it and evolve towards its purer destiny, but they will be much more prosaic risks or hiatuses: misprints and misprisions, misattributions of authorship, misinterpretations of contextual ethnographic evidence.

You can see where I am heading with my recomposition program: forget internalities; you are already outside foraging. Stick to surface phenomena and resist the temptations of spiritual enlightenment. If I follow Hunter, this involves a rejection of the Kantian version of critique, and a recovery of another Enlightenment tradition that was much more civil and secular, and that he finds in Samuel von Pufendorf and Christian Thomasius. He is not alone in this kind of recovery of alternative traditions. In 1990, Stephen Toulmin wrote of a Renaissance chapter in modernity, where early humanists such as Erasmus, Rabelais, Bacon, and Michel de Montaigne had a particular style that Toulmin finds conducive: these writers—and let us emphasize their practice—wrote from experience, they contextualized ideas, and they were disciplined by rhetoric, geography, ethnography, and history. But, he argues, these early humanists were eclipsed by the next generation, platonic rationalists such as Descartes, who "limited 'rationality' to theoretical arguments that achieve quasi-geometrical certainty or necessity: for them, theoretical physics was thus a field for rational study and debate, in a way that ethics and law were not. Instead of pursuing a concern with 'reasonable' procedures of all kinds, Descartes and his successors hoped eventually to bring all subjects into the ambit of some formal theory."[12] The retreat from the empirical diversity of the world to cognition and the promise of certainty through formalization is a shift that will be familiar to any field in the humanities that has anything to do with "theory," the search for general principles or universal laws. There is nothing wrong with such a goal, except, for this argument, if such laws are seen as located in the human mind's privileged relation to "the world."

We can exemplify this point with the way modern linguistics, which took off in parallel with structuralism in the early sixties, sought to leave behind the dreary historicism of the philological disciplines that had been so central to the humanities, as claimed by James Turner in his *Philology: The Forgotten Origins of the Modern Humanities* (2014): "Until the natural sciences usurped its throne in the last third of the nineteenth century, philology supplied probably the most influential model of learning."[13] In her review of Turner's book for the *American Historical Review*, Ann Blair tells us how this worked:

> Turner identifies the persistence of the defining features of philology: concerns for authentication, dating, and interpretation through the careful scrutiny and comparison of particulars studied in their cultural context. Comparative philological thinking created an awareness of cultural distance between the Renaissance and antiquity (first classical then biblical antiquity), then between Europe and the cultures of other continents (resulting in comparative linguistics and cultural anthropology), and was transferred from texts to images and artifacts with the formation of art history and archaeology. Turner highlights the presence of this single mental toolkit across so many disciplines through the careers of individuals who were involved in multiple fields, not only in the early modern period when we expect it, but also precisely during the period when modern disciplinary distinctions became institutionalized.[14]

This is why Latour's revisiting of the Moderns, his resetting of modernity, is also crucial for the humanities. If the Moderns established the sciences as paradigmatic by exploiting the great bifurcation of Nature to give the sciences the sole power to determine its laws, this then left the humanities in a minor position, dealing with the mere relativities of human cultures and languages, however precise their scholarship. Take narrative as one example: despite the huge amount of effort that has gone into its theoretical analysis, despite the popular acceptance of the importance of "getting the story right," despite the major role of speculative fictions in creating the future, rationality is still understood as not in any way being "made up" because it is aligned with objectivity and the Laws of Nature. Henceforth, the humanities will have to show, I think, how they too have serious stakes in "the natural," not just because of so-called human nature, but because both the sciences and the humanities are always organized out of heterogeneous entanglements of facts and values, formulae and stories.

So, in noting that the modern discipline of linguistics made a resurgence in the mid-twentieth century precisely by embracing such cognitive formalism and the scientificity that it promised, we need not be cynical about this particular confluence, because all disciplinary recompositions are probably opportunistic, flourishing when the right "compost" helps them grow. No longer content to describe the finite, distributive patterns of language units, Noam Chomsky appealed to the infinite capacity of the atemporal human mind to generate language, and he and his followers could turn this metaphysical capacity (using the metaphor "deep," as in "deep" versus "surface structure") forcefully against behaviorist accounts of language, or historical ones that sought to describe languages as learned by imitation in specific times and places. Even today, at MIT, this universalist and cognitivist program continues with the outstanding discovery of "dependency length minimization," or DLM, as the latest candidate for a viable language universal, meaning that languages generally try "to place words with a close syntactic relationship as closely together as possible."[15]

I am prepared to stand corrected by my colleagues working in the digital humanities who might say that this kind of algorithmic approach to language will help hasten the cybernetic future when humans and machine will be fused. But in the meantime, I remain an advocate for the historical and geographical analysis of the distribution of language differences, strictly as finite surface phenomena.

On the issue of the spatial metaphors for reading, let me turn from language to literature, the seminar for which there must be a text. In *The Limits of Critique*, Rita Felski subtly analyzes "depth" as an archaeological metaphor owing much to Freudian repression or latency. If one uses "depth" as an analytical term in textual analysis, then it is implied that desires or ideological contradictions are not coded in the surface language, but can erupt with the help of astute readers, who might extend the archaeological metaphor in their critical accounts by using words like "gaps" and "fissures." While this kind of depth coincided with structuralism and Chomsky's use of the metaphor, the poststructuralism to follow looked more at the social distribution of discourses that "constructed" realities. This "denaturalizing" mode of critique was driven by its capacity to demonstrate the workings of power in both social institutions and texts. Thus, with Foucault, Western forms of sexuality were not natural but socially and historically constructed, and with this kind of theoretical orientation a new gender-based analysis of literary works could proceed. The implication was that an *awareness* of social and textual constructedness, with their ties to power, could lead to liberation from dominant normativity.

With Latour, these discursive modalities are upgraded to full existential modes and nature and society as baselines disappear, to be replaced by institutional workings treated as surface phenomena. Latour is even more rigorous than Foucault in his (materialist) refusal of metaphors of critique and explicitly rejects the anti-institutionalism of the generation of '68. Felski quotes Latour expressing just this kind of shift: "Where are the unconscious structures of primitive myths? In Africa? In Brazil? No! They are among the filing cards of Lévi-Strauss's office. If they extend beyond the Collège de France at the Rue des Écoles, it is through his books and disciples."[16]

A Latourian literary analysis would start with these kinds of institutional nodes and networks. Similarly, if I look at the robust size of the English Department at my own university, I have to wonder how much it is sustained by the compulsory teaching of English at secondary schools and the way our department trains teachers to pass on the values associated with the works of the canon. Despite the inherent value of Shakespeare that these teachers insist upon, interest in him is largely sustained through these interlocked organizations, including a powerful one called the Board of Studies, for whom Shakespeare has permanent tenure on the syllabus as a key figure in the literary heritage. But for the student searching for exemplary essays on *Hamlet* on the web, a different network comes into play, one that reminds us that the Author-Text-Reader triangle is a limited model for literary engagement.

With these networked institutions in mind, one can then elaborate a Latourian literary studies via the relevant modes of existence. While history and philology could reliably persist in a *referential* [REF] mode, for literature we have to pay attention to the *aesthetic*, designated as fiction [FIC]. Let me try to point to three ways in which literary aesthetics is not usually confronted head on, in its full ontological status. Firstly (and this happens less often because it is old-fashioned), literary research is carried out in a purely scholarly fashion, on historical principles that work to establish the definitive versions of texts, while footnoting the variations. Such scholars have no need to offer fanciful interpretations of the text. Secondly, works of pure literary theory continue to be written, in which case fictional beings might make only fleeting appearances. Finally, across the hall in the creative writing department, the creation of fictional beings is entirely left up to literary artists, whose talent is not seen to have any necessary foothold in scholarship or theory. When something remarkable happens, and a literary creation comes into being, neither of these literary types, critics or artists, seems to be fully present to apprehend the event. "How did I do that?" asks the creative writer when a character she has created, wrested with great effort from the raw material of language, leaves the page, and sets off on its perilous journey through the world. For the literary scholar, this being, palpable yet immaterial, is not of immediate interest as he pores over texts in the archive. And for the theorist, literary "beings" are indeed of great interest, but there is the constant danger of reducing them to our favorite "ism." Literary humanism (in its prestructuralist forms) doesn't work because these fictional beings are far from human (they move in different networks; they only "signify" human features; they can also be immortal). Meanwhile, political frameworks such as Marxism and feminism tend to reduce the aesthetic to its context-based effects, as they concentrate on tracing the characters' emancipatory potentials. So I stated earlier, it is worth trying to let the aesthetic throw its full ontological weight around by asserting that these fictional beings are really real, not characters who are "mere" representations of humans, not narrative scripts that have a trivial relation to the organization of action. To be *more realistic* about them means seeing them as *multiply real*, as not always reducible to human experience, but as having necessary and heterogeneous partnerships with concepts, things, other texts, organizations, and even the sacred.[17] Aesthetic beings, works of art, take shape in special assemblages that are distinct from those arrangements that help knowledge along (or spiritual life, or the law, or other modes of existence). A novel "takes shape" or the image of Mickey Mouse "gels" and "captures" the imagination of the public, but not before a lot of work is put into the technical and institutional means for the dissemination of this being. And once Mickey Mouse is "out there" and has achieved the kind of immortality only available to fictional beings, he is very hard to unmake, because his existence is cradled in networks of devotion.

Just as the text of history needs to create a receptive public (unlike a scientific text, which might be the summary record of an experiment), aesthetic beings need their publics even more, and they interact with them in the full force of their completed (im)material form. The impact of the genres and narratives in circulation in the massive film and literary industries is not to be denied, while schools and universities teach the humanities disciplines that have a role to play in the selection of particular cultural heritages, not to mention compositional skills in those more practical disciplines in colleges and schools where creative writing and filmmaking are taught.

The literary seminar, however, is typically taught in a fashion that treats practical "composition" or even "experiment" as inadequate. The student also has to learn to judge and to theorize in a particular manner. This means coupling an appreciation of the literary text with a suitable theoretical elaboration, a genre of writing that often demonstrates a degree of difficulty, a trial that has the effect, not of impartiality, but of an agonistic forging of an intellectual persona with "an attitude that is skeptical towards empirical experience . . . [and that] cultivates openness to breakthrough phenomena of various kinds."[18] For this exercise, the facts about Jane Austen's or J. M. Coetzee's books do need to be established, along with attention to the formal structure of the texts. But it will be a C-grade exercise if one stops at this empirical level, what in French schools is called the "explication de texte." What is important for graduate-level work is to grapple with the powerful language of theory. This exegetical work, with its arcane language, puts the student in a state of spiritual anticipation ("openness") for the moment when the moral problems that they have learned to identify will finally be overcome. The aim is a pedagogy of enlightened subjectivity, for those now liberated from "mere objectivity." This is what Hunter calls a "spiritual exercise," as we saw, a manifestation of Kant's transposition of Christianity to the early modern confessional university.[19] Today, the humanities are often defended as the set of disciplines that has the special capacity to deliver critical reason to *society as a whole*; these knowledges are not treated as instrumental or diverse, but unified in their conception of the life of the mind (or, if you like, the soul) as something that can be introspectively improved. This singularity, argues Hunter, was made possible through Kant's "preserving the idea of the divine mind as the source of all thinkable things."[20] It is this striving for unity that conjoins "ever higher forms of philosophical abstraction and principle with ethical dispositions";[21] hence the emphasis on "theory" in English departments, alongside the continued phenomenological orthodoxy, what Quentin Meillassoux calls "correlationism," of how the mind relates to the world, that familiar subject-object dualism that obscures all kinds of other active relations that can make knowledge.[22]

This particular deployment of theory is not confined to the humanities, I hasten to add. Latour has also identified economics with "spiritual exercises": "If the 'cameral sciences' are not exactly sciences, it is because they belong to a wholly different register from physics, chemistry, or pedagogy. They are more like spiritual exercises, like the challenging discipline of yoga, like self-control and the control of others."[23] And black-letter lawyers have been dismayed to find theory infiltrating legal studies in the form of "critical legal studies,"[24] which involves moving from the empirical description of cases in their social or historical context to a theory of the law that subordinates legal practice to critique in a discourse with a higher moral register, a morality that anticipates the breakthrough to "a more humane, egalitarian, and democratic society."[25] This, clearly, is legal studies for those for whom legislation (step by step through the usual institutions) can never move fast enough.

Now, suppose we bracket out this kind of theoretical exercise in our recomposition of the humanities? What would be the result? As I indicated in passing, some disciplines or subdisciplines would remain untouched, because they do not rely on this particular kind of critical move. Linguistics could "return" to its meticulous descriptive tasks without the need to posit generative deep structures in human cognition. What does change is that the humanities will refuse to claim, as I said, any *special capacity to deliver critical reason to society as a whole*, and one might thus be able to avoid the unpopular tendency of sounding as if one is standing in judgment. Much has been done to complicate processes of perception, feeling, and judgment, and the relations among them, in much recent work in queer, feminist, and political theory. But even as affect and judgment are subtly explored by, for instance, Linda M. G. Zerilli, she still uses a formula like "the central question of what it means to-be-in-the-world and to-be-open-to-the-world."[26] Latour is useful for refusing not only the syntax of the subject-object dualism, but also the conceptual singularity of "the world" and its opposition to the equally singular "society."

The alternative to being "truly open to the world" (Zerilli again) is not, obviously, dogmatism, but a confidence in the technological capacities of the various humanities disciplines to do what they can do well. Instead of inflating theory into the capacity to prepare oneself (critical grooming) for an eventuality (liberation, revolution), the professionalization of the humanities would involve the deployment of various kinds of expertise that stand up as such in the public *fora* in which they might appear. This might well involve critical judgments (rather than "critical judgment") as a set of more deflated and technological kinds of calculation.

Conclusion

According to Hunter's account, some important characteristics of the modern humanities were forged out of a crisis that posed a problem for rival confessions during the drawn-out religious wars in seventeenth-century Europe: how to live in peace. After thirty years of bloodshed, that particular set of wars was continued by philosophical means, in academic cloisters, and the metaphysical disposition promoted by Leibniz and Kant won out over the civil secular philosophy of Pufendorf and Thomasius. The modern humanities were thus "recomposed," at that time, out of an "ecological" upset in European heritage, and parts of Christian doctrine were incorporated in a neoplatonic metaphysical anthropology that, says Hunter, "envisaged a person whose self-perfecting ascent to the domain of transcendent concepts ('perfections') qualified them to exercise an integral moral-civil authority."[27]

Today, with global warming and various environmental crises, the contemporary humanities are responding to the task of changing the planetary mind, not the more local task of finding a way for Protestants and Catholics to coexist. This has called for a much more realist, dare I say, *scientific* response. But the call to take a "more realistic" approach no longer means a *reduction* to Science or to the Laws of Nature, as Latour has repeatedly shown. Not only do the sciences comprise heterogeneous elements and take zigzag courses of action, but in practice, and by necessity, they meet up with their friends in the humanities on their equally nonlinear paths.

It is in this spirit that we find our anthropological investigator knocking on the door of a noted academic in the new field of environmental humanities. As she probes him to find out what his discipline cares about most, he readily agrees that his field is beset by *concerns*, by all sorts of crises: extinctions, global warming, pollution. But these are not used as pretexts for moral blackmail, as if "we, the humans, are responsible, therefore we must act now." Surprisingly, he values a new humility on the part of humans whose arguments are entangled with those of other nonhuman agents, perhaps with different styles of rationality.[28] So human rationality or critical reason no longer occupies the central, elevated position it once had in the quasi-theological corner of the humanities academy described by Hunter. The various forms of expertise of the humanists can now be specified as the sciences and the humanities working together to make the arguments put by whales or Amazonian rainforests *sound reasonable in particular contexts*, but not necessarily sound universal, not yet.

Thanks to the reliable projections of science, the planet's troubled future is now known as never before, and time, alarmingly, seems to be going backward, crashing back from a *known* future onto a chaotic present to reconfigure our ways of composing our human and nonhuman communities. The arguments put by the Anthropocene do indeed sound universalist and radical because that weird phenomenon is a *hyperobject*, to use Timothy Morton's phrase: everywhere at once and infiltrating everything.[29] Even the kinds of stories that are told about human destiny now have interesting contradictions: we have an age named after the human, yet one in which the human loses its sovereignty, while key organizing concepts such as "freedom" have hit an ecological wall.[30] "Nature," in the context of this new interdiscipline, can no longer be a rallying point for puritanism or apocalypse, but there continue to be all sorts of surprises in store for those investigating the emerging natures that "we" will have never mastered. Human reason will now be localized and limited to the range of its methods, scientific as well as humanist, and in many cases working together on the same terrain, on the same project. Few would now advocate "facts first, values later," as was the case when nuclear weapons were being invented last century by "disinterested" scientists.

So the environmental humanities academic is often doing philosophy "in the field," collaborating with real scientists, while inflecting their joint results with understandings that are not based on a singularized and objectified Nature.[31] Sometimes the results come out a bit queer, as in the work of Donna Haraway or Karen Barad, in which case the environmental humanities may indeed be experimental and compositional rather than judgmental. Given the plurality of communities, experiments are necessary in that we don't know a priori just what composition of agents will enhance the life of the ecosystem and propel it forwards.

I should like to finish with the question of what will drive the humanities in the future (in the context of what Toulmin called at the end of his book "an ecology of institutions"[32]) if we remove a couple of major drivers from among the critical moves the humanities makes: the unmasking power of critique and the unrealistic use of theory. The first is powerful because it claims to reveal the truth hidden under illusion or false consciousness. Meanwhile, the second claims the power of prophecy, as its discourse is released from any empirical shackles to envelop the speaker in its aura; it is a spiritual pedagogy that is forever preparing the subject for the event-to-come. For answers, for new drivers, we could turn to Latour and ask if "compositionism" and "experiment" are sufficiently attractive to engage the labor of researchers and students. I have, for many years, been convinced that experimentalism in history can offer a new structure of argument that entails a switch of tonality. Women's history, for example, instead of critiquing patriarchy for its repression of the truth that was always there, can ask positive questions about what the archive actually contains to build up this new discipline. Now the critique, strengthened by the evidence, becomes part of a *recomposition* project. The field of history is rebooted for its new niche: somewhat different, less combative, and more pluralistic.[33] Recomposition, in other words, is not the kind of dialectic that depends on opposed positions. Conceiving of the humanities as an "ecology of institutions" prevents us from being inward-looking. Knowledge is really only tested by being tried out in publics that are not one's own. Why did psychology rapidly become the discipline with the truth about human subjectivity, such that its experts' words could often be called upon in courts of law? Whereas the poor literary critics are rarely called as experts, because they don't seem to have proved that serious conduct is guided by fictional scripts, and others are even derided because all they can do is "deconstruct." Let's take this expertise to be exemplary, rather than literal. For it is not the aim of the humanities to produce folk who are fully armed and puffed up with expertise (and we know how it was psychology's clever deployment of science that got it its transferable expertise). The aim is rather to test our propositions or techniques via a detour, an excursion into a public sphere. Truth would thus necessarily be produced and made visible in such Latourian detours, never through direct application (as in the internet "double-click" version of accessing the easy answers). And let's not forget that your "day in court" (or your day as a philosopher in the field with environmental scientists) makes you a changed person through the experience. The humanities you return to will henceforth not be quite the same domain either because, through your mediation, it has been touched by its public exposure. The test was a reality endured and paid for in an expended effort to translate one set of concerns into another. And it can be a political act to the extent that allies have been gained or lost, depending on the outcome of the experiment.

It is easy to see, then, that this Latourian idea of necessary detours and transformations pertains to the basic elaboration of chains of knowledge acquisition in the humanities. You don't have to wait to be called as an expert witness, although such a trial would help focus your effort. I might go so far as to say, however, that if humanistic studies are not experimenting, trying things out against alterity, then they are at rest. I would modestly advocate that we build up and follow some practical procedures to strive toward various reasonable truths as well-wrought constructions, rather than, for instance, submitting to the depressing deconstructive lesson of learning to live with the disappointment of knowing one never really gets to make anything real.[34]

Can literature be taught without enjoining students to "open themselves" to the text, to celebrate the ambivalent and the in-between, as if "in the interstices" they might glimpse a hidden meaning or coherence that literary theory might predict? We know that good teachers in the humanities can skillfully negotiate the pathways that relate the given heritage, the canonical texts, to their pupils' lives. If these students are taught to look beyond the Author-Text-Reader triangle and to pay attention to the complex ecologies that keep texts alive from place to place (for example, looking at how *The Tempest* migrates from the cacophony of the Elizabethan theater, zigzagging along various networks to the present), they might realize that descriptions of contexts can be fascinatingly extended, and not just in opposition to "the text itself."[35] Matters of current environmental concern might also have the effect of making the storm itself a more central actor in *The Tempest*. If so, this is one of the shifting contingencies that sustain the life of the work of art. So, too, is the editing and interpreting of Shakespeare's language to render it a little more accessible for contemporary audiences. There is a *multiplicity* of such aesthetic effects that are adjusted for each representation of Shakespeare, on stage or in the classroom, where it is on trial for its survival as something *real*. The stakes are ontological, not representational, because we have given up the bifurcation that makes the text into a (mere) reflection of the world. Our students might rather be invited to join in the experiment of *ongoing composition* of vital old/new, text/ world, nature/culture, human/nonhuman assemblages.

Notes

Thanks to Gary Wickham, Ian Hunter, Frances Muecke, Rita Felski, and the *NLH* readers for their help in the preparation of this essay.

1. Bruno Latour, *An Inquiry into Modes of Existence: An Anthropology of the Moderns*, trans. Catherine Porter (Cambridge, MA: Harvard Univ. Press, 2013).

2. The following books by Latour are examples of works in translation that analyze, respectively, the institutions of the law, science, religion, and the social sciences: *The Making of Law: An Ethnography of the Conseil d'État* (2002); *Laboratory Life: The Construction of Scientific Facts* (1979), with Steve Woolgar; *Rejoicing: or the Torments of Religious Speech* (2013); and *Reassembling the Social: An Introduction to Actor-Network-Theory* (2005). Known as an originator of Science and Technology Studies (STS), Latour's work is steeped in European philosophical traditions and takes its cues from the more progressive strands of continental philosophy (Gilles Deleuze and Félix Guattari) or less mainstream social science (Gabriel Tarde). It addresses "matters of concern" more than purely academic debates, as in *The Politics of Nature* (2004).

3. See http://www.modesofexistence.org/.

4. Latour, "Why Has Critique Run Out of Steam? From Matters of Fact to Matters of Concern," *Critical Inquiry* 30, no. 2 (2004): 239.

5. Ross Chambers, *Room for Maneuver: Reading (the) Oppositional (in) Narrative* (Chicago: Univ. of Chicago Press, 1991).

6. Latour, "A Textbook Case Revisited—Knowledge as a Mode of Existence," in *The Handbook of Science and Technology Studies*, ed. Edward J. Hackett, Olga Amsterdamska, Michael E. Lynch, and Judy Wajeman, 3rd ed. (Cambridge, MA: MIT Press, 2007), 83–112; Stephen Muecke, "Public Thinking, Public Feeling: Research Tools for Creative Writing," *TEXT* 14, no. 1 (2010), http://www.textjournal.com.au/april10/muecke.htm.

7. Michel Foucault, "Truth and Power," in *The Foucault Reader*, ed. Paul Rabinow (New York: Pantheon Books, 1980), 72.

8. Foucault, "What is Enlightenment?" in *The Foucault Reader*, 34.

9. Anthony Grafton, *The Footnote: A Curious History*, rev. ed. (Cambridge, MA: Harvard Univ. Press, 1997), 4–5.

10. Ian Hunter, "Hayden White's Philosophical History," *New Literary History* 45, no. 3 (2014): 331–58 (hereafter cited as HW).

11. Hunter, "The Mythos, Ethos, and Pathos of the Humanities," *History of European Ideas* 40, no. 1 (2014): 11–36.

12. Stephen Toulmin, *Cosmopolis: The Hidden Agenda of Modernity* (Chicago: Univ. of Chicago Press, 1990), 20.

13. James Turner, *Philology: The Forgotten Origins of the Modern Humanities* (Princeton, NJ: Princeton Univ. Press, 2014), x.

14. Ann Blair, *American Historical Review* 120, no. 2 (2015): 557–58.

15. Cathleen O'Grady, "MIT claims to have found a 'language universal' that ties all languages together," *Ars Technica*, August 6, 2015, http://arstechnica.com/science/2015/08/mit-claims-to-have-found-a-language-universal-that-ties-all-languages-together/.

16. Rita Felski, *The Limits of Critique* (Chicago: Univ. of Chicago Press), 202n. Latour, *The Pasteurization of France*, trans. Alan Sheridan and John Law (Cambridge, MA: Harvard Univ. Press, 1988), 179.

17. Muecke, "Reproductive Aesthetics: Multiple Realities in a Seamus Heaney Poem," in *Mindful Aesthetics: Literature and the Science of Mind*, ed. Chris Danta and Helen Groth (London: Bloomsbury Academic, 2013), 164.

18. Hunter, "The History of Theory: Notes for a Seminar," *Critical Inquiry* 33, no. 1 (2006): 81.

19. Hunter, "The Mythos, Ethos, and Pathos of the Humanities," 11–36.

20. Hunter, "The Mythos, Ethos, and Pathos of the Humanities," 16–17.

21. Hunter, "The Mythos, Ethos, and Pathos of the Humanities," 17.

22. Quentin Meillassoux, *After Finitude: An Essay on the Necessity of Contingency*, trans. Ray Brassier (London: Continuum, 2008).

23. Latour, *AIME*, 465.

24. David Saunders, *Anti-Lawyers: Religion and the Critics of Law and State* (London: Routledge, 1997).

25. Duncan Kennedy and Karl E. Klare, "A Bibliography of Critical Legal Studies," *The Yale Law Journal* 94, no. 2 (1984): 461–90.

26. Linda M. G. Zerilli, "The Turn to Affect and the Problem of Judgment," *New Literary History* 46, no. 2 (2015): 271.

27. Hunter, *Rival Enlightenments: Civil and Metaphysical Philosophy in Early Modern Germany* (Cambridge: Cambridge Univ. Press, 2001), xii.

28. "The Anthropocene unmakes the idea of the unlimited, autonomous human and calls for a radical reworking of a great deal of what we thought we knew about ourselves and the humanities as fields of enquiry." Deborah Bird Rose et al., "Thinking Through the Environment, Unsettling the Humanities," *Environmental Humanities* 1 (2012): 3.

29. Timothy Morton, *Hyperobjects: Philosophy and Ecology after the End of the World* (Minneapolis: Univ. of Minnesota Press, 2013).

30. Dipesh Chakrabarty, "The Climate of History: Four Theses," *Critical Inquiry* 35 (2009): 197–222.

31. Bruce V. Foltz and Robert Frodeman, eds., *Rethinking Nature: Essays In Environmental Philosophy* (Bloomington, IN: Indiana Univ. Press, 2004).

32. Toulmin, *Cosmopolis*, 209.

33. Muecke, "Experimental History? The 'Space' of History in Recent Histories of Kimberley Colonialism," *The UTS Review* 2, no. 1 (1996): 1–11.

34. Latour, *AIME*, 156.

35. See Felski, "Context Stinks!" *New Literary History* 42, no. 4 (2011): 573–91.

CHAPTER TWO

From ANT to Pragmatism

A Journey with Bruno Latour at the CSI

ANTOINE HENNION

BRUNO LATOUR and I have been traveling companions for a long time, starting with his long period at the CSI (Centre de Sociologie de l'Innovation) at the École des Mines in Paris, through to the SPEAP program (Experimental Programme in Political Arts) that he instituted in 2010 at Sciences Po. In the latter, students start by inquiring into emerging problems, such as environmental ones, then begin rebuilding the links between political, scientific, and artistic representations that the traditional disciplines have so systematically disconnected. Rather than discuss in an abstract fashion the possible relations between the humanities and Latour's *An Inquiry into Modes of Existence*, I am going to retrace these kinds of relations by looking at exchanges that actually occurred, in particular at the CSI. These links and trajectories have been numerous, as much among domains (sciences and technology, law, culture, economy, health) as among concepts (translation, mediation, regimes of enunciation, agency, attachment, etc.). The *Inquiry* is the culmination of a long-running project,[1] from the first work of Latour and Steve Woolgar on laboratories, then to STS and the beginnings of actor-network-theory (ANT) worked out together with Michel Callon and John Law (and quickly subjected to their own critique, as in Law and John Hassard's 1999 *Actor Network Theory and After*), through to the interrogation of the Moderns launched in a spectacular fashion by *We Have Never Been Modern* in 1991, and the less diplomatic proposition to sociologists that they rebuild sociology without Émile Durkheim (*Reassembling the Social*, 2005). The *Inquiry* is both a recapitulation of the results of ANT and an explicitly pluralist and pragmatist reformulation of ANT methodology, as well as a response to the questions that the latter had left open—the price to pay for a radicalism that did its job well. One should also mention that the *Inquiry* has installed Latour and his current colleagues, especially Isabelle Stengers, on the pathway of those she calls the philosophers of difference: Gabriel Tarde, William James, Alfred North Whitehead, Étienne Souriau, Gilles Deleuze, Michel Serres, Donna Haraway, etc.

I am going to put this overall path under the heading of a "return to the object," enabling me to draw out the threads supporting thought, like Latour's, that is always *in the process of making*. I will base this approach on the reciprocal exchanges that my own work on music, amateurs, and attachments has been able to weave between problems to do with science and technology on the one hand, and culture on the other.[2] Music is a network of crisscrossing mediations, the production of a disappearing object, and is always being remade through the artist's performance or amateur activity; it has been most suggestive in forging unusual pathways with technical objects that ANT has allowed us see differently. Rather than a historical review, I would like to carry out a kind of partial archeology of the *Inquiry*, taking questions that I am particularly interested in as points of departure, such as mediation and its contrast with translation, the question of attachments, our critical relations to Bourdieu's critical sociology, and a revival of pragmatism in social inquiries.

Networks, Association, Translation: Getting Sociology to Be Object-Friendly

From the end of the seventies, the CSI became known as a pioneer in the sociology of science and technology. Here, with Latour and his Dutch and English colleagues, Callon created the sociology of translation, then actor-network-theory, which, using the acronym ANT, took off in the anglophone world.[3] In the wake of David Bloor's powerful approach, *Knowledge and Social Imagery*, and of the *Sociology of Scientific Knowledge* (SSK) that set itself the task of treating supposedly true or false statements on the same footing, ANT was able to generalize this principle of symmetry in relation to a project's technical successes or failures, to the explanatory factors (social or technical) that were brought to bear, and to accountable actors (human or nonhuman).[4] In order to better describe science as it was happening and to analyze technological innovation, ANT proposed a series of concepts: network, association, *intéressement*, translation, and obligatory passage point.[5] These were notions that brought about a radical reversal between objects and relations: action made actors, *intéressement* made interest, the relation made the object, and not the other way around.

It's a bit suspect to claim radicalism for oneself, but, as it happens, this is part of the history of ANT, which burst onto a very diversified field with a polemical tone. Requiring sociology to take into account scientific and technical objects created a tension right away—a tension that I have placed at the heart of this text—first in relation to those dedicated to their autonomy, like epistemologists, but also, hot on their heels and from the opposite direction as we got better at explaining our project, in relation to those wanting to reduce objects to "social constructions."[6] Callon, a sociologist and engineer, had worked on the politics of private and public research in order to rethink the place of the sciences in society. His foil was the Mertonian school of sociology of science. Latour, as a philosophy graduate, was for his part convinced that one could no longer do philosophy without carrying out inquiries in social science. Working with Françoise Bastide, he took the notions of shifter and actant, at the heart of his theory of translation, from Algirdas Greimas's semiotics. Apart from philosophy (in particular Gottfried Wilhelm Leibniz, Tarde, and Whitehead), his first references were primarily to anthropology, to the ethnology of technique and of culture, and to ethnomethodology (Aaron Cicourel, Woolgar, Michael Lynch).[7] He was more fired up by his intense battles with epistemology and the history of science than he was by sociology. The theoretical audacity that marked Callon and Latour's collaboration owes a lot to this complementarity of disciplinary sources.[8]

We were not the only ones in France advocating a return of the object to the social sciences. Among the first in these strands of sociology to take the pragmatic turn, which gave rise to very diverse interpretations and uses, were the GSPM (the Groupe de Sociologie Politique et Morale founded by Luc Boltanski, Michael Pollak, and Laurent Thévenot, which came up with the call for a "pragmatic sociology"); the CEMS (Centre d'Études des Mouvements Sociaux, where ethnomethodology, analysis of conversation, and situated action were read and discussed in terms of strengthening the theoretical bases of an American sociological research that the Bourdieusian tradition had reduced, a little too quickly, to symbolic interaction); and other places where the treatment of objects was made central to sociology.[9] It was a moment when all of these things were emerging.

At the CSI, over and above our various research projects, there was discussion of things that were strongly related at the level of the grammar of our debates, at their *hypertextual* level: the concept of association, the fact that it was not necessary to maintain a strict division between subjects and objects, or between humans and the things they manipulated. As work in STS advanced, the areas of research spread out toward the environment, law, health, the body, public debate, and politics.[10] For its part, the CSI, starting with relations between objects and their uses, became involved in domains such as the environment, health, or markets,[11] and developed other themes like agency and performativity, new concepts cutting across the domains of investigation (regimes of enunciation, modes of existence, attachments, "agencements," framing and overflowing) that updated previous and tenuous divisions between users and producers, culture and technique, and politics and economy.[12] In this context, the fact that I studied music is not just of anecdotal interest. These ideas—objects seen as provisional results of a heterogeneous tissue of relations being continually tried out, tested, reshaped, in order to produce other objects, without being able to reliably distinguish content and support, network and actors, products and users—were as valid for music as they were for technical projects. This association between culture and technology was simply reaffirming a classical idea of anthropology. I had come to the CSI to pursue a sociology of music that was not to be carried out against it, but with it. On the question of the place given by sociology to the object according to how "actors themselves" define it and become attached to it, this musical detour allows me to retrospectively link up projects that have dealt with technology or projects on culture without always making explicit their mutual borrowings or their differences.[13]

Can Science and Culture Be Treated in the Same Way?

That the CSI had acquired its international reputation in the field of STS gave rise to the belief that the projects on culture carried out under its auspices were a broadening of its first research on technology. Historically, things went in the other direction. As soon as the Centre was founded in 1967 by Lucien Karpik, the aim was not only to take an interest in science and technology, markets, and users (natural subjects given our location in the École des Mines), but also, in a comparative sense, in several domains such as law and culture, drawing on the same idea, which was very new at that time in the framework of traditional sociology: content counts. Whether dealing with sociology of science, technology, culture, or law, it was not just a matter of doing institutional or professional histories, of speaking about organizations, social networks, fields, or reception—in other words, of enclosing a domain by reference to sociological realities, coming to terms with its functioning independently of its particular object—but, on the contrary, of realizing that it was impossible to understand what was going on without taking into account the fruits of the activity. This was already a way of recognizing that such fruits had a capacity for action, an *agency*, even if this idea was expressed more trivially at the time. So, the study of actors, organizations, etc., of course, but also of objects themselves with their specific assemblies, often very sophisticated, by means of which a domain is progressively elaborated through controversies and challenges. And in the other direction, understanding that these objects of collective action create their actors and organizations, especially through their capacity to install their relative autonomy by interiorizing their own effects. Instead of shunning objects like the plague, sociology, in order to come to terms with the force of science and technology (or the law, or, for me, music), had to dare to face up to these objects. Far from being socially inert, they resist, they "work," they make things happen, they transform their users.

But things don't present themselves in the same way; it depends on whether one is speaking of science and technology, or of culture. Saying that law and culture are human things, made by people, "instituted," as the Romans would say, is common sense. Saying that physical laws, natural or formal, are socially constructed realities is, on the other hand, immediately shocking. In the STS debates, this thesis was understood as being constructivist, and for good reason. In the first instance, it was indeed a constructivist battle against the idea of an absolute truth, independent of the proofs that allow it to be known. This was also the moment when the CSI came to international attention.[14] In relation to dominant sociology, the initial constructivism of SSK—socially explaining *all* of science and not just its mistakes—and then that of STS (still fairly undifferentiated in its different versions, as was the case later, with ANT happening to be folded in) were both attacked for their radicalism and accused of being relativist.[15]

This explains the subtle differences among the CSI members according to their fields of research, something we were not entirely conscious of at the time. What was the same project came down, in the case of science, to making more social what was seen as objective, whereas the opposite was aimed for in the case of culture: respecting the objectivity of what sociology had reduced to social signs, to markers of differentiation among groups. This was, of course, on the condition of not understanding objectivity as an aesthetic absolute, or as the autonomy of a pure object; but rather of redefining the object as a knot of relations, as a tissue of associations and links that test each other and are more or less resistant, this object in turn transforming the collectives that take hold of it.

Translation or Mediation

In short, if we shared a concern to distance ourselves from a sociology believing in the autonomy of the social—leaving each of its specific objects to their respective sciences, while just studying their social aspects—this concern needed to express itself in opposing ways. Hence the preference for either the word "translation" or "mediation"; it's a good example of seeing a problematic in the process of emerging. In the case of music, the artwork had to be reconceived as a heterogeneous tissue (human, material, corporal, collective . . .), with its resistances and cumulative effects (a keyboard, a sound, a scale, the body of the instrumentalist, a dedicated space and time . . .). To express this resistance of music to sociological reduction without going so far as to turn it into an autonomous object, and to show that these tissues of association "hang together," without dissolving in a codification of social differences, I had foregrounded the word "mediation."[16] The word translation was well chosen for science and technology. While both words suggest that this necessitates a betrayal, translation also emphasizes passage or movement, the fact that the instauration of a truth requires links, work, trials. Mediation, for its part, is a better word for music because, if it supports the same general idea, it insists on the other side: that is, not just establishing but also interrupting the relation, making it overflow. A passage is not reduced to the transmission of an object; it does something else. It does not refer back to causes; it is a performance, with unforeseeable effects, that are not deducible from the sum total of causal factors.

We have to realize retrospectively the strength of the model that this line of argument was challenging and its pervasiveness in the social sciences. For anthropologists and sociologists, culture is defined as humans collectively projecting their social relations onto arbitrary objects. Durkheim propped up this view with a positive definition of society, Bourdieu repeated it by reversing the idea, turning it into a foundational mechanism for negating social domination: cultural objects are totems, pure signs pertaining to a code, which, on top of everything, doesn't know itself to be one. We rejected this commonsensical notion.

Music does something other than what the humans gathered around it would like it to do, something other than what they have programmed. This is why they listen to it; it is not their double, nor the mirror of their vanity. "Made" the way it is, it has its own capacity to act. It forges identities and sensibilities; it does not obey them. It does act (*fait oeuvre*) in this sense. To say this is neither to fall back into turning it into a social sign, nor to take it as an absolute object.[17] Souriau speaks eloquently of a "work to be made," which is calling us. So, on the culture side, the problem was posed in completely the opposite way to science, where, in the eyes of *social scientists*, it was impossible to question the absolute status of truth. The idea that the object is everything (science) or it is nothing (music)? Absolutely not! By seeing the musical thing as something that emerges, a presence, yet without having an object sitting there in front of us that one can isolate, mediation breaks this sterile dualism. By insisting on association and on passages, translation does the same work when faced with a diesel motor or with a mathematical truth, which, for their part, give the impression that the object is untouchable. In sum, if the word mediation really belongs to what I have called our hypertext in that period, and if it shares many of the features of translation, it deals with the other side of the problem that things pose for sociology: not just that they associate, but that they also stop doing so. In other words, we were developing the same idea, but on two opposite slopes. Latour was battling with epistemology, while at the same time hoping to "save" the specific regime of the determination of things that science has implemented. This made it possible for both camps to turn him into the enemy; those who held with the absolute character of scientific truth were scandalized, and he was treated as a traitor by those constructivists in the "linguistic turn" club who saw science as a story just like any other. On the topic of art, I recognized myself in this battle. Sociology had justifiably challenged the aesthetic of the absolute and its infinite quest for the autonomy of the Work with a capital W, while I was looking, for my part, for ways to recognize in works, with a small w, their capacity to act.

It took us some time to understand these articulations, and the need to use different words. Latour and I did so in two coauthored works, while Latour was writing on "factishes," developing the same idea.[18] We crafted a single rich definition of mediation or translation, as a way of dealing with the problem that things posed for sociology: as what resists (a common critical position, no doubt) but also that which goes beyond, that overflows, comes back; objects that you make, but that in return make you; that are made, but escape. In other words, they have their *agency*, their capacity to act. After a first article that compared mediations in science and in art, a second and more provocative essay inverted, term by term, Walter Benjamin's critique of modernism via his theory of the *aura* of the work of art and the destructive effects of mediations.[19] There are certainly many other ideas to be culled from Benjamin, but as it happens we wanted most of all to question the celebrity of this article centered on the idea of mediation, taking its very success as a symptom of the attraction (both critical and complicit) that is contained in every appeal to the idea of the modern.[20]

The Reflexivity of Things: Were We Really Constructivists?

Speaking of mediations means taking music away from the kind of analysis that uses external explanations and rule-bound effects; the kind a sociologist would come to measure according to her own concerns, or a musicologist or a psychologist according to theirs. There are only partial and heterogeneous causes that cannot be tabled clearly. They are necessary; they make things emerge. From these assembled causes, effects erupt in an unforeseeable way, always being remade, and being themselves irreducible to the causes that brought them about. No doubt this is more difficult to express than putting things in a simple cause and effect relation, but at the same time I am not saying anything esoteric here. Mediation understood in this way is obvious to any musician sitting down to the keyboard. He knows there are his scales, his score, his touch, and the skills he has acquired, that without them he is nothing, and yet, even if he starts with these mediations, nothing is settled, the music will have to emerge; there is nothing that is automatic or guaranteed. Incidentally, as often happens, theory is lagging behind common sense here, and not the other way around. The surprise that peels away from the flux of things is the most ordinary of experiences, for an audience member, a painter, a footballer, a drinker. It is an experience shared by professionals and amateurs alike.

Out of the fabric of familiar things, a small but decisive deviation has effects that can be enormous, but they arise from the things themselves as they present themselves. It is the jazz improviser who plays the same piece a hundred times, and yet . . . wait, this time it's going this way, insisting on a quite new pathway. He follows it, tries it again . . . comes back, it has opened up a space. It's a difficult but important question, this "bringing about." What part does each element play in such a creative offering? Certainly the musician takes things up again, but the offering is realized in the movement itself, in the thrust of things, which hold out in some way a possibility to be seized. I have suggested that we speak of a reflexivity that things have in themselves, which are not given, but give themselves.[21]

There is a crucial factor here, that of recognizing in objects this "making" of things: both the fact that they are made and the fact that they make their making. A making that cares for things and does not oppose them (does not denaturalize or deconstruct them) because they are fabricated—the latter being a quite different aim, that of social constructivism. This moment of divergence (from social constructionism) and explication was very important for us. At the beginning the theme was confused, the question difficult and open to all sorts of misunderstandings. Latour suggested various solutions to extricate ourselves: contrast constructivism as such to social constructivism, call it constructionism and not constructivism, talk instead of fabrication, then recall, via his word "factishes," that these *facts*, as hard as wood, with which positivists challenge him, are saying by their very name that, yes, they are made![22] The very word pragmatism no doubt helped us to realize that in reality we were not constructivists *at all*, in the sense of "socially constructed," which had in the meantime become an automatic slogan in sociology. Of course, initially, every sociological move is constructivist in the broad sense of the term. Faced with its object, whether it is art, religion, truth, morality, or culture, one shows that it is historical, that it depends on a time and a place, that it conveyed via corporeal practices and that it varies according to context, that it has procedures, that it is underpinned by convention, that it is supported by institutions . . . showing the believer how belief is produced, as Bourdieu said.[23]

Doing sociology necessarily means, to some extent, partaking of the original constructivism of the discipline. On the topics of science and culture, we traveled with sociologies very different from our own, as long as it was a matter of being opposed to the absolutism of truth in itself, or the beauty of Art for Art's sake. But as we went further, the same word designated two divergent paths: showing that things are constructed, *and that therefore they are nothing*, or, on the contrary, giving things themselves a role to play in these matters.[24] We were following the second path, and we first had to understand ourselves, and then to make understood, at what point it radically departed from what is generally understood as constructivism, whether it is Bourdieu's version or that of the linguistic turn, SSK, a good part of STS, or cultural studies. From a common starting point, the paths go in completely opposite directions.[25]

Obviously, things don't have an inherent nature; the work of the social sciences is to show their instauration. But once this is done, the next question that arises is even more arduous, a decisive bifurcation that Latour expressed admirably with his "factishes." Does this fabrication of things have to be played out *against* them or *with* them? The social sciences will remain at the threshold of this new territory as long as they maintain at any price their two founding intangible distinctions, between human action and the agency of objects, and between social interpretation and natural realities—the very distinctions that ANT challenged.[26]

Beyond Bourdieu?

On the culture side, moreover, this required me to situate myself in relation to critical sociology, the necessary background to any problematic on this subject—shall we say in relation to Bourdieu, who at that time was carrying out a massive task of anthropologization to get his discipline out of its positivist self-conception.[27] One forgets to what extent sociology, in the 1970s—its writing, its reasoning, even its concepts—had a fundamentally realist character, in the most banal way, that did nothing to diminish its lively expressiveness or its political importance in public life—quite the contrary. Organizations, power, social class, interactions—all this was what was in front of us "for real," as much for the actors as for the observer. Thanks to Bourdieu, his philosophical view of culture, a reflexive writing style, a distinctive way of developing his arguments via circular phrasing, his attention to practices, capacities, and apparatuses, to the weight of the body, to collective and embodied history, there crystalized, in contrast to sociological common sense, what it seems appropriate to call an anthropological revival. On the topic of art, faced with the dualism of the work and its admirer, he brought about a welcome desubjectivation of the relation to works of art, a collective, instituted, incorporated redeployment—one that is situated, as one would say now. Of course, at the end of the day, and in line with the scientist and critical sociology that he was defending, Bourdieu took hold of the reins again and spirited back to the collective the same object that he had just put in the saddle, giving it back to the sociologist—your objects are not what they seem; they are the hidden play of your relations, which is what builds your common belief. The social is nothing other than your effort to install it, while at the same time hiding this installation from you.

It's a brilliantly devastating thesis, but even though its premises and consequences have been much discussed, I think what's been missed is that it is not required by the preceding anthropologization and can't be deduced from it. There is no need, starting from the latter, to launch into a sociological disqualification of the object, to change it into an *illusio*, *inlusionem*, into the stakes that make up the social. The object, which at the beginning of the analysis was reinserted into a tissue of relations, bodies, apparatuses, and histories, now ends up being a totem. This is a magical trick, a sleight of hand.[28] The sociologist is himself creating what he believes he is describing among his actors; he conjures away the object of common action to replace it with the inert symbol of a purely social collective; in this movement he turns the social into a scientific object and attributes the study of it to himself. One way of presenting the work conducted at the time at the CSI, as much in STS as on the topics of culture, health, or markets, is to say that it preserved the "pragmatization" of our activities, the anthropologization that Bourdieu was carrying out in putting history, institutions, habitus, body, apparatuses, and dispositions into play.[29] But while Bourdieu did all this work *against* the object, in a traditional, very dualist procedure, opposing himself to those he believed were believers in the object,[30] our own projects tried to respect the objects and the actors attached to them. Not in a "nevertheless" or "still . . ." fashion, a call for moderation ("objects have a certain autonomy, after all"), but, on the contrary, by emphasizing this general pragmatics without reservations, that is, by carrying it out *with* objects and not against them. Why not treat the objects in question in the same way that Bourdieu constantly does with regard to bodies, collectives, or apparatuses, but which he refuses to do for objects: why not see them as beings in formation, open, resistant, that make each other, in a reciprocal fashion, acting reflexively on those who cause them to come into being?

Perhaps it should be less a matter of criticizing Bourdieu than of taking up what he has done and applying it *also* to objects, instead of using it to quash them. It is true that this procedure turns everything around. To use contemporary language, it goes from a theory of practice to a real pragmatism. Objects have their *agency* that we make and that makes us. The key point here is the status given to objects. This means not taking them as external, fixed givens (conceding their natural reality in the case of science; making them into simple signs in the case of culture), but rather seeing them as indeterminately composed, made of the links that are knotted or unraveling as they undergo their trials, thereby creating unique and composed worlds. Certainly it means "socializing" objects, but not by emptying out their content. They can be allowed to fill up and fill us up, form diverse and connected worlds, and then layer by layer spread out and spread us out—at this point, no doubt, the paths diverge. To go on means leaving the trail blazed by Bourdieu: the bracketing of objects that leads to a sociological dead-end. The challenge they throw up does not mean denouncing them, nor welcoming them with open arms, but asking what they are doing and what they are making happen. How do we speak of the love of art, or of wine, or of any object or practice, taking this question seriously, without being satisfied by showing that it is really a matter of something else than what it thinks it is? No one reading Bourdieu's 1966 *The Love of Art* would have thought for a moment that the book would *actually* speak about the love of art: come on, you are not going to take the artwork "itself" seriously, are you? That would mean falling back into aesthetics, or letting actors seduce you with their talk, getting sucked into belief rather than showing its mechanism. Well, as it happens, taking the love of art seriously is exactly what I'm working on. The price, as we shall see, is that pragmatism is taken back to its founding principles, *pragmata*, the *agency* of things, and is not used as an alias to shift out of critical sociology without reconsidering the very narrow arena it had reserved for objects. And as I was saying, pragmatism thus conceived is far from being an esoteric new fashion; it helps reconnect with common sense (a criterion of relevance essential in social science). In art, as well as in sciences and technology, objects count. What the sociologist lacks is perhaps something like a respect for the thing in itself.

Affordances, Situated Action, Extended Mind: Another Opening Toward Pragmatism

At the time, other traditions helped in the reformulation of these questions, this time coming from American extensions of action theory, a path that emerged from pragmatism and in turn helped us rediscover the latter. Authors such as James J. Gibson, Don Norman, Edwin Hutchins, or Lucy Suchman broke the model of instrumental action, with its intentions, means, and ends, in favor of a vision of situated action, extended mind, and distributed cognition. In contrast to our experience of Bourdieu, we had not been immersed from the beginning of our training in the language of *affordances*, those orientations of objects toward new uses whose possibility they nevertheless suggest.[31] Nor in situated action. Suchman's 1987 book was a revelation, really helping us to focus our project.[32] Similarly, Hutchins's 1995 *Cognition in the Wild* amazed us. This work came out of the American "history of technology" tradition, along with splendid books like Thomas Hughes's 1983 *Networks of Power* on the electricity grid, or works on material culture by impressive authors such as Chandra Mukerji.[33] Hutchins highlighted technical apparatuses and their linkages in a quite new way. These writers had considerable influence on STS in general, and on us in particular.

Here again, the fact that we were working at the CSI on both technology and culture helped us. The crossover to the questions posed by the knowledge of amateurs was explicit. For example, in an article co-written with Émilie Gomart, we compared the attachments of amateur musicians with those of the consumers of drugs with whom Gomart worked.[34] The aim was to question the limits of ANT and of those intellectual trends that had been able to redeploy action outside of a linear and instrumental model. We wanted to extend the logic of their challenge, but by departing from the framework of action in which they were still situating themselves, in order to recognize the active role of objects and grasp other forms of agency beyond the active/passive dualism. We wanted to do so by, for instance, finding room for an active passivity or, in relation to musicians and drug "amateurs," for an action aimed at making oneself passive. That is the reason that, rather than referring to cause and effect, we went back to the word "attachments" that Callon had used in his 1992 analysis of markets, and which Latour, in the style of his example of the puppeteer and the puppet (who controls whom?), applied to a Mafalda comic strip in a very hard-hitting article on cigarettes and liberty.[35]

This American detour also allowed us to be more explicit about the different meanings of the word "pragmat-*ic*." There is scarcely a sociologist in France today who doesn't claim this qualifier, but pragmatics understood simply as a theory of action is not really pragmat-*ist*. You can see the divergences when you look at the place that is given to objects. Now, according to this criterion, Hutchins, for instance, was already directly echoing philosophical pragmatism, in James's sense, by presenting the piloting of a boat as a collective task to which many kinds of machinery contributed, from the instruments to the layout of the control room to the water resistance and the radio—in short, by making use of the idea of the extended mind. I said James, because among the founding fathers, it was he who took *pragmata* most seriously in the battle against dualism; it was he who formulated the symmetry principle, *avant la lettre*, in the most radical way. The symmetry between the knowing subject and the world to be known is his problem as a philosopher, but he also defended this symmetry in relation to beings and things. I'll come back to this in my concluding example of amateurs. It is *pragmata*—thing-relations, plural and extended—that are at the heart of pragmatism, not practice, which doesn't require anyone to challenge the grand divide between human actions and the things they act upon.[36]

In this context, the discovery of James left us stunned. For example, he refuses to distinguish between things and their effects; he considers things and their relations to be made "of the same stuff."[37] Here we were reading one of the key themes of our own research—and one of the most controversial—written in someone else's hand.[38] After reaching its peak, pragmatism, that American philosophy, was looked down upon, even in the US, and was stifled by analytic philosophy. As far as the growing interest in this line of thought in France in the 1980s went, it focused on a kind of pragmatism that was largely rewritten to fit the occasion. Combining elements of enunciative pragmatics, analytical philosophy, and theories of action, it made very little reference to the arguments of pragmatism's founding fathers. So in the end, thanks to debates nourished by interested researchers, this shared framework nominated a new direction: pluralism, the rejection of exteriority, trials and investigations, public controversy and debate, the competence of actors.[39] At the CSI, it was above all the radicality of James's and John Dewey's propositions that surprised us, as if, even though these authors made little reference to the technology and objects on which STS had been expending all its efforts, they had expressed, in advance, a vision of the world and of objects amazingly compatible with our own research ideas: objects that are pragmata, i.e., things "in their plurality"; and concerns, these things in common that emerge from public debate by being put to the test, without one being able to list a priori the stakes involved, the actors, or the arenas of discussion.[40] It was as if Dewey were confronted by contemporary problems, such as the environment, development, energy, sexuality . . . all of this in a world without exteriority, but plural and open, an expanding tissue of heterogeneous realities, but connected loosely, "still in process of making," as James nicely puts it.[41] This feeling that everything had already been said back then was no doubt partly illusory, giving the impression of a miraculous coincidence between the contours of James's "pluriverse" and the tentative articulation of ANT's ontology, especially with regard to the ideas of association and general symmetry. All that was missing were the ideas of inquiry and the concerned public, this time imported from Dewey, and we were in our STS universe: association, mediation, testing, *agencement*.

There was nothing of a backward-looking attitude in this return to pragmatism. It helped us make the shift that all of us at the CSI, with our different objects, were looking for: going from a theory of practice to an agency distributed across a multitude of links. No longer working with dualisms, with an instauration of things by humans against things, but toggling between the *assemblages* that Latour borrowed from Deleuze;[42] an *agency* dispersed in an "actor-network." The expression aimed to stress that the network itself is acting, and, far from the binary opposition between humans and nonhumans, actors of very different natures form each other. As Latour says, this is to take the word "socio-" in the etymological sense of association, of link.[43] There are only relations, and this "there are only" is not understood in a critical and sociological mode (in fact these are *only* social relations), but in a full and ontological mode. Yes, things are themselves relations. This is the lesson of pragmatism.

Conclusion: Objects That Make Demands

Whether it is a popular song, a contemporary art installation, an opera aria, or a painting, once a work is created, it escapes from its author, it resists, it has effects or it doesn't. These effects change according to circumstance; the work lives its life.[44] This is precisely what attaches aficionados to such objects; the object has its own presence; it makes itself by making us. Works that we create, that we fabricate, that escape us and come back changed: what a mysterious relationship!

Amateurs are experts in this consequential testing of objects they are passionate about. They confront them; they do whatever is necessary to test and feel them (in French, *éprouver* has the two meanings), and they thus accumulate an experience that is always challenged by the way in which these objects deploy their effects. Rather than experts, as I said, they are experimenters; *éprouveurs* would be better, if there were such a word. It is with this perspective that I would like to conclude, by coming back to taste, taste as an appreciation of things that come about via the act of appreciation itself. Less an object of study, in other words, and more an experience to be approached.[45]

Amateurs are not believers caught up in the illusion of their belief, indifferent to the conditions under which their taste came about. On the contrary, their most ordinary experience is that of doubt and of hope.

They are well placed to know, experiencing one disappointment after another, that there is nothing automatic about the appearance of the work or of their emotion. They are on the hunt; the experience of taste continually forces them to question its origin: is it my milieu, my habits, a quirk of fashion; am I being taken in by a too-easy procedure; could I be too much under the influence of so-and-so, or the plaything of some projection that makes me see something that isn't there? This question of the determination of taste is at the very heart of the formation of the amateur subject; it is a long way from being the sociologist's discovery of a truth that everyone has repressed. No one feels more than amateurs the open, indeterminate (and hence disputable, *contestable*) character of their object of passion. *De gustibus* est *disputandum*.

Amateurism is the worship of what makes a difference. It is the opposite of indifference, in the two timely meanings of the word. That is why I treat amateurs as little teachers of pragmatism. They know better than anyone (by truly living it) that there is no opposition between the need to "construct" an object—having permanently relied, to that end, on a body trained by past experience and the techniques and tastes of others—and the fact that, from the entanglement of crisscrossing experiences out of which the object arises, it is just as capable of surprising, escaping, or doing something else entirely. If the smallest brick is missing from this fragile construction, it all collapses. But they also know, like Souriau's sculptor, that, far from implying a reduction of the object to "only being" a reflection of those that make it, this is the very condition for it to emerge in all its alterity, and that in return it alters its "constructors." The passion of the amateur is not a state or an accomplishment; it is a self-dislocating movement that starts with the self, via a deliberate abandonment to the object. The word passion expresses it beautifully, even if one has to be careful of its grandiloquence. If it is the right word, it is not because it adds a supplement of soul to our relations with things, but because it is the exact autochthonous expression of our specific relations to those things that *seize us*.

No one thinks of "passion" as passivity. If something is to seize you, then you have to "make yourself love" it. But we are no longer talking about mastery, action, or a theory of action. Passion is not this kind of calculation; it is being transported, transformed, or taken, and despite all these passive turns of phrase, it is anything but passive. For things to appear, something has to be made of them! One has to actively abandon oneself, as it were, to do everything so things can take their course. This was the gist of my article with Gomart. A strange grammatical construction, no doubt, but the very one that lays out the rules, and that the word "passion" refers to: to be taken/to allow oneself to be taken by whatever arises in the midst of experiencing things. This uncovers another, less expected aspect of the activity of amateurs: the ethical dimension of an obligation, of a sustained engagement with the things one loves, with oneself, with the quality of the ongoing experience. There is clearly a dimension of obligation in taste. An obligation to run the course, to respond to the object holding out its hand, to rise to the demand that its very qualities call forth. Souriau puts this beautifully when he talks about creators being obliged to do what their own work demands of them. This also implies that this obligation in relation to oneself and to things is an ethical task that certainly extends to the *social scientist* as well, when he values and makes more widely known the experience of amateurs. For my own part, I find that this spurs my interest in pursuing a sociology of taste. It is not just the amateur that the object puts under an obligation, but also the philosopher or the sociologist.

Notes

Translated by Stephen Muecke

This text is based on an online interview with web philosopher Alexandre Monnin published in *SociologieS* (Hennion, "D'une sociologie de la médiation à une pragmatique des attachements," in Théories et recherches, *SociologieS*, June 25, 2013, available at http://sociologies.revues.org/4353).

1. See Antoine Hennion, "Review Essay: 2012. *Enquête sur les modes d'existence: Une anthropologie des Modernes*, de Bruno Latour," *Science, Technology, and Human Values* 38, no. 4 (2013): 588–94.

2. For my work on music, see Hennion, *The Passion for Music: A Sociology of Mediation*, trans. Margaret Rigaud and Peter Collins (1993; Farnham: Ashgate, 2015) and Hennion, "Playing, Performing, Listening: Making Music—or Making Music Act?" in *Popular Music Matters*, ed. Lee Marshall and Dave Laing, trans. Rigaud-Drayton (Farnham: Ashgate, 2014), 165–18. For my work

on amateurs, see Hennion, "Pragmatics of Taste," in *The Blackwell Companion to the Sociology of Culture*, ed. Mark D. Jacobs and Nancy Weiss Hanrahan (Oxford: Blackwell, 2005), 131–44; Hennion, "Those Things That Hold Us Together: Taste and Sociology," *Cultural Sociology* 1, no. 1 (2007): 97–114; Hennion, "Paying Attention: What is Tasting Wine About?" in *Moments of Valuation: Exploring Sites of Dissonance*, ed. Ariane Berthoin Antal, Michael Hutter, and David Stark (Oxford: Oxford Univ. Press, 2015), 37–56; Hennion, "Music Lovers. Taste as Performance," *Theory, Culture, Society* 18, no. 5 (2001): 1–22. For my work on attachments, see Hennion and Émilie Gomart, "A Sociology of Attachment: Music Amateurs, Drug Users," in *Actor Network Theory and After*, ed. John Law and John Hassard (Oxford: Blackwell, 1999), 220–47; Hennion, "Vous avez dit attachements? . . ." in *Débordements: Mélanges offerts à Michel Callon*, ed. Madeleine Akrich et al. (Paris: Presse de l'École des Mines, 2010), 179–90. Hennion, "Enquêter sur nos attachements: Comment hériter de William James?" in "Pragmatisme et sciences sociales: explorations, enquêtes, expérimentations," ed. Daniel Cefaï et al., *SociologieS*, February 23, 2015, available at http://sociologies.revues.org/4953.

3. See Akrich, Michel Callon, and Bruno Latour, eds., *Sociologie de la traduction: Textes fondateurs* (Paris: Presses de l'École des Mines, 2006).

4. See David Bloor, *Knowledge and Social Imagery* (London: Routledge and Kegan Paul, 1976).

5. Callon and Latour, "Unscrewing the Big Leviathan: How Actors Macro-structure Reality and How Sociologists Help Them To Do So," in *Advances in Social Theory and Methodology: Toward an Integration of Micro- and Macro-Sociologies*, ed. Karin Knorr-Cetina and A. V. Cicourel (New York: Routledge, 1981), 277–303; Latour, *The Pasteurization of France* (Cambridge, MA: Harvard Univ. Press, 1988); Latour, *Science in Action: How to Follow Scientists and Engineers through Society* (Cambridge, MA: Harvard Univ. Press, 1987); Latour, *Reassembling the Social: An Introduction to Actor-Network-Theory* (Oxford: Oxford Univ. Press, 2005); Callon, "Some Elements For A Sociology of Translation: Domestication of the Scallops and the Fishermen of St-Brieuc Bay," in *Power, Action and Belief: A New Sociology of Knowledge?*, ed. Law (London: Routledge and Kegan Paul, 1986), 196–233; Akrich, "Comment décrire les objets techniques?" *Technique et culture* 9 (1987): 49–64.

6. Knorr-Cetina, *The Manufacture of Knowledge: An Essay on the Constructivist and Contextual Nature of Science* (Oxford: Pergamon, 1981); Harry Collins, *Changing Order: Replication and Induction in Scientific Practice* (1985; Chicago: Univ. of Chicago Press, 1992); and Bloor, "Anti-Latour," *Studies in History and Philosophy of Science* 30, no. 1 (1999): 113–29.

7. In his first fieldwork, between 1973 and 1977, Latour was already following a "symmetrical" program, studying Ivory Coast researchers as if in a modern laboratory, studying the Salk Institute in San Diego without assuming its scientific character. Basically just like savages! Latour and Steve Woolgar,

Laboratory Life: the Social Construction of Scientific Facts (London: Sage, 1979). See also his references to André Leroi-Gourhan or Gilbert Simondon, and to the journals, *Technique et Culture* or *Technology and Culture*.

8. As well as, from the 1970s, one of them meeting with Law, the other with Woolgar.

9. The journal *Raisons pratiques* published an issue entitled "Les Objets dans l'action: De la maison au laboratoire," ed. Bernard Conein, Nicolas Dodier, and Laurent Thévenot, 4 (1993), while, drawing on the activities of experts and counterfeiters, Christian Bessy and Francis Chateauraynaud set up the question of the qualification of objects in terms of holds and markers [*prises et repères*]. Bessy and Chateauraynaud, *Experts et faussaires: Pour une sociologie de la perception* (Paris: Métailié, 1995). See also the positive reception in France of Arjun Appadurai's 1986 collection, containing in particular Igor Kopytoff's "cultural biography of things." Appadurai, ed., *The Social Life of Things: Commodities in Cultural Perspective* (Cambridge: Cambridge Univ. Press, 1986).

10. For a take on the evolution of this field, see the volumes edited after the 4S conferences, the *Society for Social Studies of Science*, http://4sonline.org/, http://stshandbook. com/. Founded by a small group of colleagues in 1975, today the membership comprises approximately 1,200 individuals and covers vast domains that are closer to *critical studies*, and sometimes a long way from the initial field studies on technology.

11. Latour, *Politics of Nature: How to Bring the Sciences into Democracy* (1999; Cambridge, MA: Harvard Univ. Press, 2004); Vololona Rabeharisoa and Callon, *Le Pouvoir des malades: L'Association française contre les myopathies et la recherche* (Paris: Presses de l'École des Mines, 1999); Callon, Yuval Millo, and Fabian Muniesa, eds., *Market Devices* (Oxford: Blackwell, 2007).

12. Callon, Millo, Muniesa, *Market Devices*; Latour, "Factures/Fractures: From the Concept of Network to the Concept of Attachment," *RES: Anthropology and Aesthetics* 36 (1999): 20–31; Hennion, "Vous avez dit attachements?"; Hennion, "Enquêter sur nos attachements."

13. I will take up later the example of the related but distinct notions of translation and mediation.

14. Latour, who had already worked for some years with Callon, came to the CSI in 1982.

15. Raymon Boudon and Maurice Clavelin, eds., *Le relativisme est-il résistible? Regards sur la sociologie des sciences* (Paris: PUF, 1994); Pierre Bourdieu, *Science of Science and Reflexivity*, trans. Richard Nice (Chicago: Univ. of Chicago Press, 2004).

16. Hennion, *The Passion for Music.*

17. The double quarrel between the sociology of culture and aesthetics locked them into a sterile debate that they can scarcely get out of.

18. Latour, *On the Modern Cult of the Factish Gods*, trans. Catherine Porter and Heather MacLean (1996; Durham NC: Duke Univ. Press, 2010).

19. Hennion and Latour, "Objet de science, objet d'art: Note sur les limites de l'antifétichisme," *Sociologie de l'Art* 6 (1993): 5–24; and Hennion and Latour, "How to Make Mistakes on so Many Things at Once—and Become Famous For It," in *Mapping Benjamin: The Work of Art in the Digital Age*, ed. Hans Ulrich Gumbrecht and Michael Marrinan (1996; Stanford, CA: Stanford Univ. Press, 2003), 91–97.

20. Walter Benjamin, "The Work of Art in the Age of Mechanical Reproduction," in *Illuminations: Essays and Reflections*, ed. Hannah Arendt (1955; New York: Schocken, 1968), 217–51.

21. Hennion, "Those Things That Hold Us Together," 106.

22. In French, "les faits sont faits." See *pragma*, Greek for thing. Latour reminds us that it is the same in all languages: *res, Ding*, thing, *chose/cause* in French, the word that, designating assembly, the public thing, the judicial case, the common cause, in other words collective discussion, also names things in their most material, nonhuman sense.

23. Bourdieu, "The Production of Belief: Contribution to an Economy of Symbolic Goods," trans. Nice, *Media, Culture, Society* 2, no. 3 (1980): 261–93.

24. More precisely, for critical sociology, they are everything, absolute, when they relate to science, and nothing, pure arbitrary signs, when they relate to culture.

25. If we gauge the fruitfulness of a field by the way it can set up harmful debates, then the virtue of STS is in having put the "epistemological chicken" of Harry Collins and Steven Yearley on the table. Collins and Yearley, "Epistemological Chicken," in *Science as Practice and Culture*, ed. Andrew Pickering (Chicago: Univ. of Chicago Press, 1992), 301–26. See also the reference of Callon and Latour to the baby and the bathwater (alluding to the Bath school) in their response to Bloor's "Anti-Latour" (the same journal issue presented Bloor's attack, Latour's response, and the counter-response); and the positions subsequently taken by Barry Barnes, Collins, Pickering, Susan Leigh Star, Malcolm Ashmore, Lynch, and of course, Ian Hacking, who put the question bluntly in *The Social Construction of What?* (Cambridge, MA: Harvard Univ. Press, 1999).

26. Latour, *Reassembling the Social*.

27. Bourdieu, Alain Darbel, and Dominique Schnapper, *The Love of Art: European Art Museums and their Public* (1969; Cambridge: Polity, 1997); Bourdieu, *Outline of a Theory of Practice*, trans. Caroline Beattie and Nick Merriman (Cambridge: Cambridge Univ. Press, 1977); Bourdieu, *Distinction: A Social Critique of the Judgment of Taste*, trans. Nice (London: Routledge, 1984).

28. Hennion, "The Price of the People: Sociology, Performance, and Reflexivity," in *Cultural Analysis and Bourdieu's Legacy*, ed. Elizabeth Silva and Alan Warde (1985; London: Routledge/CRESC, 2010), 117–27.

29. Rather, let us speak for the moment of reinsertion into practices. Pragmatism (I will come back to this question) is not a theory of practice, but a taking into account of things, which is a different matter entirely.

30. "Believing is believing that others believe," as Michel de Certeau wrote very astutely. Michel de Certeau, "Croire: Une pratique de la différence," *Documents de travail et prépublications* no. 106 (Centro Internazionale di Semiotica e di Lingüística: Università di Urbino, 1981), series A, 1–21.

31. James J. Gibson, *The Ecological Approach to Visual Perception* (Boston: Houghton Mifflin, 1977); Donald A. Norman, *The Design of Everyday Things* (New York: Doubleday/Currency, 1988).

32. Lucy A. Suchman, *Plans and Situated Actions: The Problem of Human Machine Communication* (Cambridge: Cambridge Univ. Press, 1987).

33. Chandra Mukerji, *From Graven Images: Patterns of Modern Materialism* (New York: Columbia Univ. Press, 1983); Mukerji, *Territorial Ambitions and the Gardens of Versailles* (Cambridge: Cambridge Univ. Press, 1997).

34. Gomart and Hennion, "A Sociology of Attachment," 220–47.

35. Hennion, "Vous avez dit attachements?"; Callon, "Sociologie des sciences et économie du changement technique," in *Ces réseaux que la raison ignore*, ed. Akrich et al. (Paris: L'Harmattan, 1992), 53–78; Latour, "Factures/ Fractures," 20–31.

36. Hennion, "Enquêter sur nos attachements."

37. In French, "de la même étoffe," from the famous Essay 8, "La Notion de conscience." William James, *Essays in Radical Empiricism* (New York: Longmans, Green and Co., 1912), 233.

38. If James is not the first name to come to mind when one thinks of STS (or even of ANT or the thought of Latour or Callon), I am happy to say that during the ANT epoch, when we were talking about translation and mediation, we were Jamesian without knowing it. To say this is not to insult James, an author who asks to judge a philosophy less by its positions than by the effects and uses it can give rise to!

39. Luc Boltanski and Thévenot, *On Justification: Economies of Worth* (1991; Princeton, NJ: Princeton Univ. Press, 2006); Dodier, *Les Hommes et les machines: la conscience collective dans les sociétiés technicisées* (Paris: Métailié, 1995); Bessy and Chateauraynaud, *Experts et faussaires*; Céfaï and Isaac Joseph, eds., *L'Héritage du pragmatisme: Conflits d'urbanité et épreuves de civisme* (La Tour d'Aigues: Éditions de l'Aube, 2002); Céfaï, *Pourquoi se mobilise-t-on? Les Théories de l'action collective* (Paris: La Découverte, 2007).

40. James, *The Meaning of Truth: A Sequel to Pragmatism* (New York: Longmans, Green, and Co., 1909), 210; John Dewey, *The Public and Its Problems* (New York: Holt, 1927).

41. James, *A Pluralistic Universe: Hibbert Lectures at Manchester College on the Present Situation in Philosophy* (New York: Longmans, Green, and Co., 1909), 76; James, *The Meaning of Truth*, 263.

42. Callon prefers, for his part, to go back to *agencement*, the original French word for the English term assemblage, sometimes picked up in French. Callon, "Qu'est qu'un agencement marchand?" in *Sociologie des agencements*

marchands, ed. Callon et al. (Paris: Les Presses de l'École des Mines, 2013), 325–40. Assemblage being too supple, it diminishes the active dimension of *agencement* and the subtlety of the adjustments required between humans and things.

43. Latour, *Reassembling the Social.*

44. Alfred Gell, *Art and Agency: An Anthropological Theory* (Oxford: Oxford Univ. Press, 1998).

45. Hennion, "Those Things That Hold Us Together," 97–114; Hennion, "Vous avez dit attachements?"

CHAPTER THREE

Demodernizing the Humanities with Latour

GRAHAM HARMAN

AT FIRST GLANCE, the primary lesson of Bruno Latour for the humanities appears to be simple: the humanities are not just about humans. Though the ultimate picture is more complicated than this, it remains a useful guiding principle, since few contemporary thinkers have had more success than Latour at incorporating nonhuman entities into their writings. Such popular human-centered terms as "language," "society," "power," or even "capitalism" are reassigned by Latour to a derivative position. None of these things is made up of purely human material; all are shown to be composed of hybrid networks that feature viruses, earthworms, computers, and ozone holes no less than police stations and other cynical panoptica. Indeed, one of the reasons that Latour is starting to look like Michel Foucault's eventual replacement as the default citation in the humanities—he is quickly approaching that point in the social sciences—is that whereas Foucault treats inanimate entities primarily as means by which the human subject is historically molded, Latour's ever-expanding oeuvre better equips us to take such entities on their own terms, rather than merely as human accessories.

I. The Current State of the Humanities

Latour's work provides us with a unique opportunity to outflank the existing trench war in the humanities, which can be helpfully caricatured as follows. A typical "conservative" position holds that human nature never grows better or worse. The sobering conclusions of Thucydides and Tacitus still mark the limits of human endeavors today; the same basic gallery of characters appears in every era despite the distracting tempests of historical color that surround them. We read Shakespeare and Jane Austen because of their wisdom about "timeless human emotions." The resulting prescription is usually a Great Books education based on canonical authors (I am grateful to have received such an education myself, though I reject the conservative interpretation of its merits) and with it we often find an excessive suspicion of theoretical novelty. Meanwhile, a typical "progressive" position holds that humans are constantly changing and infinitely improvable, as mutable as time itself. Hence there is no excuse for allowing the continued oppression of marginalized groups: injustice is not a tragic recurring feature of human existence to be resignedly managed, but an ongoing moral affront to all that is best in us. The Great Books are said to represent only the limited viewpoint of dead white European males, and our oppression by this privileged group can be remedied, in part, by adding more diverse voices to the conversation. The more marginalized the group, the greater the purported intellectual revolution wrought by its fresh inclusion. Intellectual life tends to become a permanent sit-in, a constant reenactment of "speaking the truth to power," on the Rousseauian assumptions that (a) power means corruption and (b) the truth should already be clear to the uncorrupted. I would say that the vast majority of academics in the humanities today belong recognizably to one of these two camps, though with varying degrees of intensity.

Yet there is something naggingly inadequate in both of these positions: namely, their shared assumption that human nature must be the central focus, whether to affirm or deny its constancy over time. This anthropocentrism remains in place despite increased talk of "materialism" in the progressive camp: for on closer examination the so-called "new materialism" merely affirms the social constructionist and anti-essentialist dogmas to which progressive academia has already adhered for decades, though with more references to science and nature than were found in the heyday of postmodernism.[1] One of the attractions of Latour is the alternative he provides to both social essentialism and social constructivism. His alternative view is that stability is neither assured nor impossible, but is achieved only at a certain cost, and only by way of inanimate things. Consider for instance his joint research in the

1980s with the primatologist Shirley Strum, which concluded that baboons are even more socially oriented than humans are.² The life of these Old World monkeys is a state of constant vigilance in the face of a constantly shifting social hierarchy. Which baboon is now grooming or sharing food with another? Which animal stuns the community by defeating an apparently stronger baboon? Can a baboon continue to eat where it currently stands, or must it leave this food source to stay close to its moving comrades? For baboons there is constant social negotiation and renegotiation. By way of contrast, consider human beings. Though we humans are hardly free from struggles over rank and symbolic capital, we do awaken most mornings with a relatively secure position. On most days our money has not vanished, our spouse has not left us, our job has not ended, our name has not changed, and our country has not collapsed into civil war. Unlike baboons, we only need to negotiate our status intermittently. This is not because humans themselves are inherently stable, but because we lean on nonhuman objects— more durable than we are—to provide the stability for us. Uniforms, contracts, wedding rings, passports, bridges, tunnels, and private dwellings are among the thousands of entities that prevent the slide of humans into a baboon-like cosmos of permanent social anxiety.³

In his breakthrough book *We Have Never Been Modern*, Latour's method was to attack modernism as the futile attempt to purify nature and culture as two separate and pristine realms that must not be mixed.⁴ The modernist either appeals to a nature of things that exists whether we like it or not, or appeals instead to a human society that is nothing but an arbitrary projection of values onto a cold grey world of physical matter. More often, the modernist shifts back and forth between calling some things "natural" and others "cultural" depending on the political interests of the moment. For instance, American liberals tend to proclaim the current status of women as being socially constructed, while insisting at the same time that homosexuality is the result of nature rather than a deviant social adaptation. To say that women are naturally very different from men sounds like a reactionary claim, but to say that homosexual behavior is "not natural, but contingent and reversible through therapy" also sounds like a reactionary claim, though contingency and reversibility are usually progressive tropes. By contrast, American conservatives like to view war as an ineffaceable fact of human nature, while depicting domestic gun violence as the socially constructed product of immoral films and video games and the poor treatment of mental illness. To say that war might one day be abolished sounds like a progressive claim, while saying that handguns in the possession of the citizenry will naturally increase violence also sounds like a progressive claim, even though "nature" is usually a conservative trope. Whichever side of this divide we occupy, "nature" and "culture" have become the two basic terms of our political vocabulary. In my view, *We Have Never Been Modern* is the most important book of the 1990s precisely because it remains the best source for overcoming the nature/culture banality that still pressures most of our thinking.

Nonetheless, Latour is not always recognized as having found a novel third path between appeals to nature and appeals to culture. In France the young Latour was treated with all possible rudeness by Pierre Bourdieu, who viewed him as a reactionary essentialist for challenging the omnipotence of society, as partly summarized by Anders Blok and Torben Elgaard Jensen: "In his book on the sociology of knowledge . . . Bourdieu spends ten pages accusing his younger fellow countryman . . . of having created 'a sociological work with absolutely no merit whatsoever.'"[5] The opposite fate awaited Latour in America, where he was mocked by the likes of Alan Sokal for supposedly reducing science to a contingent power struggle.[6] Though I do not agree with the old maxim that if one is being criticized then one must be doing something right— idiots and incompetents are also criticized, after all—it is probably true that one must be doing something right if one is criticized simultaneously for opposite reasons. What this phenomenon suggests is that one is being equally misunderstood from both directions.

This article will make several interrelated points, en route to claiming that Latour's work contains great possible benefits for the future of the humanities. First, I will discuss the initial position of Latour's actor-network-theory (hereafter, ANT), which I hold to be rooted in the political philosophy of Thomas Hobbes.[7] Second, I will discuss Latour's partial rejection of Hobbes in *We Have Never Been Modern*, and consider the new trajectory on which this led him. Third and finally, I will consider Latour's latest incarnation of his work, as published in his monumental but still little-understood book, *An Inquiry into Modes of Existence*.[8] Since this recent book is too vast to be summarized in a single article, I will focus on Latour's decomposition of economics into three modes of existence that he calls attachment [ATT], organization [ORG], and morality [MOR]. My hypothesis is that his account of these three modes heralds a new model for the humanities.

II. Early ANT

Actor-network-theory has grown into one of the dominant methods in the social sciences, with perhaps tens of thousands of advocates. The development of the method is usually credited to three people in particular: Latour, Michel Callon, and John Law, but since Latour is the subject of this article, I will speak here only of his version of ANT. It is easiest to understand Latourian ANT, at least up until 2012, as an example of what Manuel DeLanda calls "flat ontology."[9] Many philosophies assume some sort of basic distinction between *kinds* of entities: the human subject and the object outside the mind, nature and culture, eternal forms and mere appearance, the scientific and the unscientific, the phenomenal and the noumenal. By contrast, a flat ontology is one that begins by placing all objects on an equal footing. In Latour's case this means that all entities are *actors*, and are only real insofar as they have some sort of effect on something else. Rocks and chemicals have effects, as do the human mind, cartoon characters, daydreams, and unicorns. As he would later put it concisely in *Pandora's Hope*, an actor is nothing more than whatever it "modifies, transforms, perturbs, or creates."[10] The actor is not an autonomous substance that preexists its actions, but exists only *through* those actions. There are no nouns in the world, only verbs. This is a radically relationist philosophy, and that is why my 2009 book *Prince of Networks* compared Latour primarily with one of his intellectual heroes, Alfred North Whitehead.[11] In his great book *Process and Reality*, Whitehead analyzed entities into what he called their "prehensions," or relations, while rejecting any Kantian claim that there is a vast ontological difference between the relation of thinking rational subjects to the world and the relation between two mindless pieces of rock.[12] For Whitehead all relations are prehensions, and all entities must be analyzed into their prehensions rather than posited as enduring substances that undergo adventures in space and time.

The comparison of Latour with Whitehead remains an excellent way to understand Latour, despite some crucial differences between them. But Latour's interest in Whitehead came somewhat later in his career, through the influence of a colleague he has always greatly admired: the Belgian philosopher Isabelle Stengers.[13] It was only while attempting to reconstruct Latour's political philosophy in my 2014 book *Bruno Latour: Reassembling the Political* that I realized an even better comparison for Latour than Whitehead would be Hobbes, who is mentioned only three times in *Prince of Networks*.[14] The crucial point of connection is Hobbes's hostility to any form of transcendence, which is sufficiently intense as to make him one of the greatest philosophers of "immanence" the world has ever known. It is well known that Hobbes, who experienced the brutal English Civil War firsthand, held that civil peace requires that nothing be allowed to transcend the sovereign authority of the Leviathan.[15] Anyone claiming direct access to religious truth beyond the authority of the civil state risks subverting the power of that state, thereby plunging us back into the horrific state of nature, in which murder, rape, and theft would be the norm. Thus, any independent religious movement must be crushed as a threat to social peace. It is less widely known that Hobbes thought the same about science, which also poses the danger that someone might claim direct access to a truth beyond the Leviathan. Steven Shapin and Simon Schaffer remind us of this in their classic work on the dispute between Hobbes and Robert Boyle over the meaning of Boyle's air-pump, a book—as we will see—that provided a negative stimulus for the second phase of Latour's intellectual career.[16] Hobbes not only made intellectual objections to Boyle's claim to access the truth of nature through his air-pump experiments, but even reported Boyle to the British authorities as a threat to public order.

Now, nothing is more typical of the early Latour than his ingrained hostility to transcendence. There is no reality "out there," beyond the strife of actors as they compose their various networks. Scientists do not just use dispensable implements to gain access to an external reality uncontaminated by various actors. Instead, science for Latour fully *consists* in the networks we use to stabilize our sense of reality. His version of ANT gives us a Hobbesian model of science as a power struggle: but *not* as a human power struggle in which "the most powerful scientist wins." Instead, it is a broader ontological power struggle in which chemicals and neutrons are also involved, since they often have sufficient power to tell the "powerful" scientist that she is wrong. In a sense it is true that Latour can be viewed as a "social constructionist," but only if we expand society well beyond human beings and their institutions. Latourian society includes the "natural" as well as the "social." The scientist is no longer a high priest who gains access to eternal truths lying beyond society, but just another social entrepreneur who links chains of actors in which gossip and trash cans may be just as decisive as the most expensive pieces of technical equipment. Though it was written before Latour's career began, James Watson's controversial book *The Double Helix* gives us a good picture of science viewed in the early Latourian manner, in which the scientific and the "unscientific" mix in often shocking and hilarious ways, though still yielding the golden result of the structure of DNA.[17] Latour's explicit political remarks from this early period display the same spirit. Along with the references to Hobbes in a joint article on the topic with Callon, the young Latour gives us a straight-faced defense of Machiavelli, though a Machiavelli expanded to include the inanimate world.[18] Latour shows little respect for political moralists, since there is simply no use being right without also having might. In what transcendent world would it do any good to be "right" while failing to have an effect? Nothing exists beyond the perpetual struggle of human and nonhuman actors to increase their place in the sun of some network. Morality, just like science before Latour reinterpreted it as the forging of hybrid networks, is not only treated as a villain in the properly Hobbesian framework. Rather, the moralist is also portrayed as a misguided loser for only caring about what is right, while failing to inscribe this rightness in the immanent world of networks.

III. The Turn Away from Hobbes

It is near the beginning of *We Have Never Been Modern*, first published in French in 1991, that Latour makes his sudden break with Hobbes. This event has governed the direction of Latour's career ever since. The context for the break is Latour's discussion of Shapin and Shaffer's classic book on Hobbes and Boyle, a work that Latour greatly admires despite rejecting its conclusion. We recall that the central topic of *We Have Never Been Modern* was the modern duality of nature and culture. Boyle, as one of the first modern scientific researchers, is an exemplar of those who prioritize the truth of nature over the supposed arbitrariness of human culture. Conversely, Hobbes is one of the canonical stars for those who wish to subject any concept of nature to the dictates of society. Though Shapin and Schaffer spend much of their book trying to balance the competing claims of both, they end by taking sides rather decisively: "Knowledge, as much as the state, is the product of human actions. Hobbes was right."[19] Latour glosses their statement as follows: "If all questions of epistemology are questions of social order, this is because, when all is said and done, the social context contains as one of its subsets the definition of what counts as good science."[20] But Latour reaches the opposite conclusion: "No, Hobbes was wrong." Or rather, both Hobbes and Boyle were wrong. Latour continues: "[Shapin and Shaffer] offer a masterful deconstruction of the evolution, diffusion and popularization of the air pump. Why, then, do they not deconstruct the evolution, diffusion and popularization of 'power' or 'force'? Is 'force' less problematic than the air's spring? . . . unless one adopts some authors' asymmetrical posture and agrees to be simultaneously constructivist where nature is concerned and realist where society is concerned (Collins and Yearley, 1992)! But it is not very probable that the air's spring has a more political basis than English society itself. "[21]

As we have seen, the early Latour has no qualms with taking power, in the sense of "effect," as the root form of all reality. Thus he was every bit as Hobbesian as Shapin and Schaffer's own book. But in 1991, for what seems to be the first time, Latour notices a problem with Hobbes, and by extension with his own ontology. If power or effect is said to be the root form of reality, then this itself is a truth-claim every bit as dog-matic as that of Boyle or any scientist, priest, or philosopher. More than a decade later, this is precisely the critique that Latour will make of ma-terialism. For though reality itself is always somewhat mysterious, and partially resists our various attempts to come to grips with it, scientific materialism wants to replace this mystery with the all-knowing dictate that nothing is real except hard material particles moving through a vac-uum.[22] Latour in 1991 discovers that the Hobbesian model of power makes the same basic mistake as the Boylean model of truth, in that both are too sure of themselves, too insensitive to the permanent surplus of reality beyond all efforts to theorize it. Mind you, this never leads Latour to endorse full-blown transcendence. He is never tempted to em-brace Heideggerian concealment or Aristotelian substance as if they were able to elude normal interaction in networks. Latour remains a thinker of immanence, though from 1991 onward he supplements im-manence with the need for what he calls a *mini*-transcendence, and in AIME this is joined by the synonymous term *hiatus*.

Whereas transcendence in Kant's philosophy referred to an inaccessible-thing-in-itself, Latour's mini-transcendence means that which *can* be accessed but is simply *not yet* accessed. This has conse-quences for both his theory of knowledge and his political philosophy. Or rather, the consequences for his theory of knowledge were already there from the beginning of his career. Instead of a model of knowledge in which a human subject leaps outside itself and makes direct contact with an objective world, Latour's flat ontology entails a model in which all entities are on the same plane and progressively linked by mediators. In a brilliant early chapter of *Pandora's Hope*, for instance, Latour de-scribes all the numerous grand and crude translations required to dis-cover scientifically that earthworms are enabling the advance of the Amazon rainforest by changing the composition of nearby soil. The cor-respondence theory of truth, featuring the "adequation between mind and reality," is replaced by an "industrial" model of truth as successive unbroken transformations between one stage and the next. "When . . . a student of industry insists that there have been a multitude of trans-formations and mediations between the oil trapped deep in the geological seams of Saudi Arabia and the gas I put into the tank of my car from the old pump in the little village of Jaligny in France, the claim to reality of the gas is in no way decreased."[23] Each stage of translation is a mini-transcendence with respect to the next: not something unknowable in Kant's sense, but something still unknown yet knowable. In this way Latour makes his unique attempt to synthesize immanence with that which lies beyond it, without yielding to the mysterious thing-in-itself that he forever distrusts.

But here it is the change in Latour's *political* thinking that best registers the change brought about by his 1991 break with Hobbes. Though his political opinions have always remained those of a liberal with an occasionally progressive streak, his political *discourse* is quite different after 1991 from what it was before. In the 1980s Latour reveled in adopting the stance of a remorseless Hobbesian, Machiavellian, or Nietzschean admirer of "power." Though he extended the theories of these thinkers to cover the inanimate sphere (though arguably Nietzsche was already there), Latour was always willing to say "tough luck" to the losers: the burden was on each actor to express its own reality effectively in networks. This changes markedly in his *Politics of Nature*, published in French in 1999 and in English five years later.[24] Here, far from the Leviathan crushing any assertion of transcendence, it is said that part of the function of democracy is to detect mini-transcendences that are not currently recognized by the *polis*. And whose political task is it to conduct this night-watch for the new and unexpected? It is none other than scientists and moralists. This is a change in role for the scientists, who were previously asked to link actors in a chain of translations leading from an experimental Point A to a resultant Point B of heavily mediated scientific knowledge that never reached any nonexistent things-in-themselves. But now it is the job of scientists to keep a close eye on what may be eluding all known networks, even if the nature of these elusive beings is only a mini-transcendence rather than a full-blown unknowable otherness. But it is the moralists whose station is raised the most in *Politics of Nature*: "In the old framework, the moralists cut rather a sorry figure, since the world was full of amoral nature and society was full of immoral violence."[25] But now that mini-transcendence has been added to Latour's previously Hobbesian framework, moralism is a legitimate enterprise, aimed at detecting that which has been previously excluded. We should note, however, that the excluded are not only the "voices" of marginalized social groups, since it might also include the forgotten work of a dead white male reactionary European heterosexual physicist who has possibly earned a new hearing due to changes in science itself. On top of this, Latour remains skeptical toward any claim to "truth" by academic moralism. For unlike the Rousseauians, Latour finds truth to be something rather elusive, as we saw in his critique of materialism.

From here Latour moved even further along the path of political uncertainty and away from a pure Hobbesian political immanence, due in large part to the influence of his Dutch doctoral student Noortje Marres, now on the faculty at Warwick University in the United Kingdom.[26] What Marres undertook in her thesis was a reinterpretation of the relation between the political theories of two prominent American authors: the journalist Walter Lippmann and the philosopher John Dewey. Lippmann's book *The Phantom Public* took a rather pessimistic view of American democracy.[27] Though our national American ideology holds that education is essential since the citizens must rule themselves, Lippmann's portrait of America features legions of ill-informed dupes completely unable to master the information needed for self-governance. Thus, America is destined to evolve into a technocracy led by chilly teams of professional experts. Though Dewey was highly stimulated by Lippmann's assessment, he found a characteristically more optimistic solution.[28] Why, wonders Dewey, is it necessary for anyone to have mastery of all or even most political issues in order for democracy to function? Lippmann himself had admitted that even he, the most well-connected journalist of his era, had not had sufficient leisure to formulate clear opinions about some of the most important issues of his time. But rather than seeing this inevitable ignorance as grounds for technocracy, Dewey took it as a reason to reconceive the relation between politics and the public. Instead of a single mass public called upon to speak intelligently on all possible issues, Dewey conceived of democracy as triggered piecemeal by specific *issues*. Each issue attracts its own public, and there is no politics in the absence of an issue. Moreover, issue politics never involves the expert revelation of the truth of an issue thanks to some professionalized class of technocrats. The truth of an issue will never be clear, and the most that democracy—or any form of government—can hope for is to reach a fruitful deal or temporary settlement on that issue.

For some readers of Latour, this was the high point of his political philosophy. Traces of this position can still be found even in 2012's *An Enquiry into Modes of Existence*, where he borrows my term "object-oriented" to speak of Dewey and Lippmann as theorists of an "object-oriented politics" (*AIME* 337). Before moving to a discussion of AIME itself, a word is in order about where Latour's political theory has been heading most recently. Since Latour's break with Hobbes in 1991, his explicit references to Hobbesian political theory have decreased. He now seems to prefer Hobbes's fellow firebreather Carl Schmitt, as in his astonishing call for Schmittian warfare against climate change skeptics in his 2013 Gifford Lectures in Edinburgh.[29] The appeal to Schmitt does not signify a reactionary turn on Latour's part, any more than his early fondness for Hobbes was reflected in reactionary views. For one thing, the Left has always been fond of Schmitt due to his hostility to liberalism and his antibourgeois assertion that violence and political life go hand-in-hand. For another, Latour made a very good point when he said, during our 2008 dialogue at the London School of Economics: "Because usually it's true, I mean this is a common thing in political philosophy, that reactionary thinkers are more interesting than the progressive ones ... in that you learn more about politics from people like Machiavelli and Schmitt than from Rousseau."[30] Though both Hobbes and Schmitt can fairly be described as reactionaries, in one important sense they are polar opposites. For Hobbes, the goal is to close off the horrors of the state of nature: the iron hand of the state is meant only to secure safety for all, so that civilians are insulated from politics and can safely go about their nonpolitical pursuits. In this respect, Hobbes can be viewed as the founder of political liberalism, despite his most unliberal-sounding advocacy of the crushing power of the state.[31] But for Schmitt, as for many Marxists, this Hobbesian closure of politics is the very nadir of bland bourgeois existence, and we must face up to the state of nature—the conflict with the enemy—in all its ugliness if we are to regain a properly political existence.[32] In his own defense, Schmitt also argues that liberals are the truly nasty ones, since it is they who define their enemies as the morally evil who must be annihilated for the public good (think only of al-Qaeda or the Islamic State in our own time), while Schmitt holds that the enemy need only be defeated rather than annihilated.[33] Granted, for Schmitt this requires something approaching dictatorship, since it is the sovereign alone who can decide on the state of exception, in which the enemy is faced in an existential struggle for one's own survival. It is easy to see how this attitude led to Schmitt's own Nazism, though others have tried to displace Schmittian ideas to the Left by viewing the enemy in terms of internal class struggle.[34]

In any case, it is fascinating that Schmitt now seems to have more appeal for Latour than does Hobbes. Some readers of Latour have suggested off the record that this fondness for Schmitt may point to a "closet authoritarianism" on Latour's part. But having known Latour personally long enough to be fairly certain that he is not a political authoritarian, I think that his Schmittian turn is better explained by Schmitt's greater compatibility than Hobbes's with the search for mini-transcendence that has guided Latour's work since 1991. And in any case, Latour's Schmittianism is a strange sort of "Green Schmittianism." A struggle for survival with climate change skeptics is not a project that many leftists or liberals are likely to oppose, or one that many fascists will be quick to endorse. But the time has come to turn from the history of Latourian ANT to the present reality of AIME, Latour's new philosophical system.

IV. Modes of Existence

If Latour's philosophical career were a normal one, we would call AIME his "later" philosophy, a distinct break with the older theory of ANT, which he had partly retracted as early as 1999 in these famously stirring words: "There are four things that do not work with actor-network-theory: the word actor, the word network, the word theory, and the hyphen!"[35] But there are two crucial problems with calling AIME a post-

ANT theory. The first is that Latour was already working in secret on AIME by the late 1980s, as he explains in an interesting history of the project.[36] Thus, ANT and AIME do not have the usual straightforward chronological relation of "after" and "before" that we find in the relation between Heidegger's 1949 *Insight into That Which Is* and his 1927 *Being and Time*.[37] Imagine if we discovered that Heidegger's later poetic discussions of earth, sky, gods, and mortals were already present in some manuscripts from the 1920s: that is roughly how surprising it was to learn that Latour had been hard at work for two decades on his new project before even a draft version was circulated to colleagues at the Cerisy colloquium in 2007. The second difficulty with viewing ANT and AIME in the usual chronological terms is that—whatever the problems may be with "actor," "network," "theory," and "the hyphen"—AIME does not *revoke* ANT. Quite the opposite, since network [NET] is still one of the most important modes of existence found in AIME, responsible for half the work of the new theory. A better understanding of AIME would see it as an ANT "expansion pack," as if the already vast universe of the World of Warcraft videogame were revealed to be just one of more than a dozen parallel worlds now equally playable by the gamer.

We have seen that the flat ontology of ANT allows it to treat all actors of all types in precisely the same way. If we were to analyze the long debate between Niels Bohr and Albert Einstein in ANT terms, the quantum phenomena they disputed would have no inherent priority over important personal anecdotes or the social atmospherics of Princeton, New Jersey. While this does make ANT one of the most flexible methods ever devised, that flexibility is at times almost *too* broad, as Latour recognizes. Early in AIME, he introduces a fictional character who remains with us throughout the book: an imaginary ethnographer who is just learning the basics of ANT for herself. Though we imagine that the ethnographer gets off to a good start with ANT and finds the method very useful, "to her great confusion, as she studies segments from Law, Science, The Economy, or Religion she begins to feel that she is saying almost *the same thing* about all of them: namely, that they are 'composed in a heterogeneous fashion of unexpected elements revealed by the investigation.'" Though she moves "from one surprise to another . . . somewhat to her surprise, this stops being surprising . . . as each element becomes surprising *in the same way*" (*AIME* 35).

The remedy to this problem is not to abandon ANT's relational model of actors and restore some classical theory of substance; Latour never considers doing such a thing. Instead, he finds it necessary to say that there are multiple "modes of existence," a term borrowed from the French philosopher Étienne Souriau, once again under the influence of Stengers.[38] The central idea is that there is not just one giant network encompassing everything. Instead, networks come in many modes, just as the radio in one's car picks up many, mutually impenetrable frequencies, each playing its own song at any given moment. Along with network [NET], Latour thus introduces a mode called preposition [PRE], which inflects a network into one of the other modes. Perhaps the simplest example is to compare the scientific mode of reference [REF] with the familiar modes of politics [POL], law [LAW], and religion [REL]. Though Latour still adheres to the ANT teaching that science consists of a chain of translations rather than of direct access to a reality outside the mind, at least [REF] aspires to make contact with some sort of distant reality. This turns out not to be the case for [POL], [LAW], and [REL]. Modern Western civilization—which according to Latour has never really existed—has a tendency to view [REF] as the only legitimate mode of truth, and thus it ridicules religion as a fraud, and condemns politics for its crooked deals, phony smiles, and mediocre summits. But Latour is one of the few prominent contemporary philosophers to appreciate religion, and one of the very few to admire politics. He does so in part because he recognizes [REL] and [POL] as having different conditions of success from [REF]. Far from attempting to access an objective, transcendent world, religion simply tries to mobilize God in various assemblies and processions: Latour's God is not an objective, hidden entity outside the mind, but comes alive only in acts of worship. As for politics, he sees the politician's betrayal of her constituent's wishes and the reciprocal betrayal of the politician's commands not as some inherent tragedy of the political realm, but as the very condition of political success. Politics is indeed a circle of betrayals, but only this betrayal of translations back and forth makes political success possible. Law is an interesting case, since modern civilization always retained some respect for the autonomy of the law, and never tried very hard to turn it into a science. No one is shocked if a crucial piece of evidence is ruled inadmissible by a judge on a technicality, and neither do we criticize a court for "taking something to be true" simply because it went undisputed at trial.

Perhaps these differences between [REF], [POL], [REL], and [LAW] suffice to give some idea of Latour's goal in distinguishing between fifteen different modes of existence, arranged into five triads. But in fact there are really only four triads of modes. One of Latour's trinities consists of [NET], [PRE], and another mode he wittily terms "double-click" [DC], in reference to our expectation that double-clicking on a computer mouse will take us directly to the desired information without need of further mediation. But this means that [DC] is not a mode, but rather a "category mistake" that effaces the differences between the other modes, and for this reason the inclusion of [DC] seems like a forced attempt at triadic symmetry. What is really going on in Latour's new book is that [NET] and [PRE] are the two overarching modes whose overlap generates the others.

The four genuine triads in the book are differentiated by the varying permutations of how they relate to what Latour calls "quasi-subjects" and "quasi-objects," though we must leave this theme for another occasion. We have already considered one of these triads: [POL] [REL] [LAW], which are said to pertain to quasi-subjects but not quasi-objects. But the one we will focus on here is the triad said to *link* quasi-subjects with quasi-objects. This is the only triad in the book that Latour generates by examining the failure of an existing discipline: economics, sometimes called "the dismal science." If Science with a capital "S" is the primary enemy of ANT, then Economics with a capital "E" is the mortal foe of AIME. While Science claimed access to a nature that would be able to make everyone shut up and just look at the facts, Economics also hopes to make everyone shut up by simply calculating the best interests of all. In this respect, we could say that Latour's importance for the humanities lies less in ANT's alternative model of science than in AIME's decompositional approach to Economics. The three modes that Latour derives from the purported failure of economics are [ORG]anization, [ATT]achment, and [MOR]ality. But since we have already been discussing morality as one of the entryways to mini-transcendence, we will reverse Latour's order of presentation and begin with morality here.

V. [MOR] [ATT] [ORG]

As mentioned, Latour views Economics as a new master discourse that aspires to replace natural science in this role. The only way to escape this looming new disaster is to focus directly on what economics misses: "The second Nature [economics] resists quite differently from the first, which makes it difficult to circumvent The Economy unless we identify some gaps between The Economy and ordinary experience" (*AIME* 381). Otherwise, the resultant disaster will lead to two problems. First, allowing the Economy to become the master discourse would stop the work of building a new human collective that Latour cherishes more than anything, by wrongly taking it to be a *fait accompli*. "With The Economy, there would always be mutual understanding, because it would suffice to calculate . . . since, from now on, the entire Earth would share the same ways of attributing value in the same terms" (*AIME* 383). To prevent misunderstanding, it should be noted that Latour has no paranoid fear of calculation. In a neglected 2010 article on Gabriel Tarde, Latour notes approvingly that Tarde sees calculation as belonging *more* to the social than the natural sciences, since the latter see their objects from so far off that they can only make statistical abstractions, where a close-up view can see the many quantities at work in any given individual.[39] But to believe too much in the power of calculation would be giving too much credit to what is usually called "Capitalism." As Latour sarcastically puts it: "If The Economy is universal, it is because of the deadly ailment, the unpardonable crime known as CAPITALISM, which continues to be a monstrous product of history that has infected nearly all the cells of a body unequipped to resist" (*AIME* 384). Here again we encounter Latour's objection to materialism, which abstracts so entirely from what it discusses that it becomes a form of idealism

(*AIME* 381). But let's go through each of the three modes that supposedly emerges from the failure of this new economic master discourse, and see what the implications might be for the humanities.

It would be inaccurate to say that the economy has nothing to do with morality. For one thing, most businesspeople and consumers accept clear moral limits as to what they are willing to pursue economically. Child prostitution remains well beyond the pale of most people's economic lives, as does the purchase of slaves or freshly harvested ivory; most of us have already boycotted a business or two for sound moral reasons. For another, the fiercest critics of the economy treat it in moral terms in the inverse form of *immorality*: with capitalism as the antimoral force par excellence, as that which makes it conceivable to put a dollar value on childhood innocence or rhinoceros horns. But in neither of these cases is the question of morality *interwoven* with that of the economy. In the former case it is simply an outer limit where each of us would end his or her calculations and cease our business in a mood of moral outrage. In the latter, it simply denounces all commodity exchange as a moral evil. For Latour, by contrast, morality is involved in every decision, and indeed in the application of every mode of existence: "the question of morality has already been raised for each mode" (*AIME* 443), since each mode has its proper felicity conditions. It would be just as morally wrong to try to end a political dispute with scientific canons of truth as it would be to bully astronomers with machetes into returning Pluto to the rank of a legitimate planet. Latour claims that for his new project, morality is not some "supplement of soul, a treat, a sweet note, like dessert after a copious meal" (*AIME* 452). Instead, "there is not a single mode that is not capable of distinguishing truth from falsity, good from evil *in its own way*" (*AIME* 452). Sounding a bit like Nietzsche, just as he did in his early work "Irreductions,"[40] Latour sees moral valuation at work even in the nonhuman sphere: "Everything in the world *evaluates*, from von Uexküll's tick to Pope Benedict XVI—and even Magritte's pipe. Instead of opposing 'is' to 'ought to be,' count instead how many beings an existent needs to pass through and how many alterations it must learn to adapt to in order to continue to exist" (*AIME* 453). Moreover, "those who don't understand that glaciers, too, have acquired a 'moral dimension' are depriving themselves of any chance to accede to morality" (*AIME* 457). And finally: "Just as a geologist can hear the clicks of radioactivity, but only if he is equipped with a Geiger counter, we can register the presence of morality in the world provided that we concentrate on that particular *emission*" (*AIME* 456).

This might make it sound as if Latour were simply ontologizing "morality," turning it into a metaphysical name for the cosmos as a whole. Yet he adds two important caveats to his concept of [MOR] to distance it from that sort of superexpansive metaphysical concept. One of these caveats merely repeats Latour's career-long suspicion of a full-blown transcendence that would trump the surprising empirical exploration of the world. As he puts it: "Extracting oneself from situations to seek the 'external' viewpoint that alone makes it possible to 'judge' situations that otherwise would remain merely 'factual' is the classic example of bad transcendence that tips you outside the experience, or, better, outside moral experimentation, and leads to MORALISM" (*AIME* 462). This is Latour's familiar call for a good or "mini-" transcendence that does not claim direct access to truth in a way that silences all debate, but simply remains open to that which is excluded from, but not inaccessible to, our current understanding. The second caveat is that even if morality is scattered throughout the universe well beyond the human realm, not *everything* that happens is moral. Morality applies only to one specific aspect of the existence of human or nonhuman things:

> To seek to found [morality] on the human or on substance or on a tautological law appears senseless when every mode of existence manages excellently to express one of the differences between good and bad. And yet, in addition to the moralities scattered throughout the other modes, there is indeed another handhold. This handhold, as we now understand it, is the *reprise of scruples about the optimal distribution of ends and means*. If every existent remakes the world in its own way and according to its own viewpoint, its supreme value is of course that of existing on its own, as Whitehead says, but it can in no case shed the anxiety of having left in the shadows, like so many mere means, the multitude of those, *the others*, that have

allowed it to exist and about which it is never very sure that they are not its *finality*. (*AIME* 455)

Since everything must be counted and measured, even as nothing can be reduced to calculation and measurement, every arrangement of beings is haunted by scruples, since every being always partly reduces to a means what is also an end in itself—a fact that holds not just for humans, as Kant wrongly held. While the economic "optimum" may pass for morality for a while (*AIME* 450), numerous individual beings will be crushed by the optimum, just as countless victims are crushed by the Lisbon earthquake in Leibniz's "best of all possible worlds." Though entities must be woven together in networks rather than each remaining in its own precious and pristine dignity, we must always wonder whether we are exploiting or oppressing the maid, the house cat, or the worms beneath our lawn. This is why we are never "quits," never out of debt (*AIME* 448–49), as some non-Western economies recognize in purposely leaving a bit of indebtedness in the wake of each transaction. The dream of being debt-free is the dream of adequately compensating for one thing with another, an impossibility in a world of incommensurable moral entities. A further implication of our permanent interwovenness with debt is the fact that we are always morally *committed* through our actions (*AIME* 460), and must look at these commitments rather than merely speaking about morality (*AIME* 459). Just as we cannot rely on a Big or Bad transcendence to judge for us, we cannot absolve ourselves of responsibility by passing the buck to some economic providence, whether it be the Invisible Hand of the free market, or the Visible Hand of the communist state. "One cannot both deny Providence and reintroduce the supernatural of The Economy. If one cannot serve *both* God and Mammon, it is because one must not serve *any* Gods while believing that they are transcendent" (*AIME* 467).

We move now to the second posteconomic mode: [ATT]achment. If [MOR] is concerned with what lies beyond the horizon of our current experience, [ATT] is the adhesive that glues us to that experience. For Latour, economics misses the *passion* that drives the acquisitive interest.[41] In his own words:

You observe goods that are starting to move around all over the planet; poor devils who drown while crossing oceans to come earn their bread; giant enterprises that appear from one day to the next or that disappear into red ink; entire nations that become rich or poor; markets that close or open; monstrous demonstrations that disperse over improvised barricades in clouds of teargas; radical innovations that suddenly make whole sectors of industry obsolete, or that spread like a dust cloud; sudden fashions that draw millions of passionate clients or that, just as suddenly, pile up shopworn stocks that nobody wants any longer Everything here is hot, violent, active, rhythmic, contradictory, rapid, discontinuous, pounded out—but these immense boiling cauldrons are described to you as the ice-cold, rational, coherent, and continuous manifestation of the calculation of interests alone. (*AIME* 386)

Detachment is perhaps the modern virtue par excellence. The cold and aloof attitude of Star Trek's Mr. Spock is taken as paradigmatic for intellectual life as a whole. Refusal to adopt a definite political attitude in favor of a cynical dismissal of all politicians as greedy and corrupted counts in some circles for the most sophisticated attitude. In personality terms, the present-day Brooklyn barista who halfheartedly shrugs or points vaguely while smirking in response to questions from customers counts as the height of social excellence. The common root of all these positions lies in the modern supposition that subject and object are two alien things, and that the best course of action lies in disentangling one from the other as cleanly as possible, thereby freeing us from "alienation" at the hands of nonhuman things.

Latour's mode of [ATT]achment, however, suggests that we do precisely the opposite. Perhaps the most annoying symptom of the modern cynic is the claim not to be surprised by anything anymore: "I've seen it all." By contrast, I have always admired Paul Berman's claim that "wisdom consists of the ability to be shocked."[42] Shock implies two things: (a) the acknowledgment that a hidden reality lurks beyond one's own purported knowledge of reality, and (b) the admission that one is entangled enough in this reality to care about it. Both are denied by Latour's "modern," who claims to understand everything in advance thanks to a systematic ideology, and also claims to be unperturbed by anything occurring in the breast of mere *res extensa*. In this spirit, Latour credits [ATT] with "exteriority . . . surprise before abrupt transformations . . . brutal alternation before the enthusiasm of being carried away by energizing forces and the depression of being subjected to forces that exceed us in all directions" (*AIME* 425). Another product of the modern ideology, though viewed by its adherents as the sole non-ideology, is the notion that any interest in things amounts to "fetishism" (*AIME* 431). In response, Latour tries to show us the silver lining in the tediously maligned "consumer society," which in fact multiplies passionate interests to perhaps the greatest level ever known: "You were walking by that store without thinking about it; how does it happen that you can no longer get along without this perfume when ten minutes earlier you didn't know it existed? You bought that plastic-wrapped chicken without attaching any importance to it; what unexpected discovery has left you really disgusted with it now?" (*AIME* 429). Or even better: "What is the discernment of a confessor or a psychoanalyst worth next to the the 'shopping know-how' of a featherbrained twenty-year-old who is capable of comparing two underwear fabrics and distinguishing the difference that will make some notions store in China rich or bankrupt?" (*AIME* 436). Though in principle anything *can* be calculated, "nothing here is defined by calculation, or at least not by the part of calculation that would presuppose equivalences and transports without transformation" (*AIME* 437). The phrase "transport without transformation" is one of the most grievously Latourian insults, referring as it does to an attempt to equate two forces directly without mediators, which is precisely what the whole of his philosophical work opposes.

We come, finally, to the third mode that emerges from Latour's decomposition of economics, which is known as [ORG]anization:

> When the talk turns to The Economy, [our imaginary ethnographer's] informants assert with respect, one has to approach vast sets of people and things that form organizations of astounding complexity and influence, covering the planet with their reticulations.

And she sets out to approach enterprises, organizations qualified under the law as "corporate bodies." She extends her hand and what does she find? Almost nothing solid or durable. A sequence, an accumulation, endless layers of successive disorganizations: people come and go, they transport all sorts of documents, complain, meet, separate, grumble, protest, meet again, organize again, disperse, reconnect, all this in constant disorder; there is no way she could ever define the borders of these entities that keep on expanding or contracting like accordions. (*AIME* 387–88)

Whereas [MOR] added a mini-transcendent exterior to the realm of calculations, and [ATT] entwined us with objects that do not altogether transcend us, the role of [ORG] is to account for the rather loose way in which individuals combine and separate from each other to form those larger organizations that are misunderstood as maxi-transcendent categories. I speak of "society," "corporation," or "capitalism," all of them objects of Latour's career-long contempt, since they posit in advance that which they should be explaining: the way in which larger entities emerge from a surprisingly loose confederation of smaller ones.

Latour's most extensive example of [ORG] is that of two friends, Peter and Paul, who speak on the phone and agree to meet tomorrow afternoon under the big clock in the Gare de Lyon in Paris. Given that their planned scenario is still just imagined and has not yet occurred, it almost fits under Latour's mode of [FIC]tion. But the difference from fiction is found in the level of existential commitment implied by their promise, which is known in the philosophy of language as a "performative" statement. "Between the end of today's phone call and 5:45 p.m. tomorrow, Peter and Paul are going to be held, organized, defined—in part—by this story, which *engages* them" (*AIME* 390). And as we have seen, "these narratives are fairly close to what sociolinguists call PERFORMATIVES, since the stories, too, do what they say and engage those who are their authors" (*AIME* 391). In some ways, what Latour offers here is a less heroic and more quotidian vision of the commitment implied in Kierkegaard's theory of choice or Alain Badiou's concept of fidelity or wager, in which truth is not truth if human devotion to it is missing (*AIME* 391).[43] As a technical term for these performative agreements about future actions, Latour chooses the name "scripts." Sometimes we are the authors of scripts, but more often we are characters in scripts created by others. To some extent we have the power to change or withdraw from these scripts, but only by paying a certain price. And in any case, we always participate in many scripts at once:

> It can never be said that [Peter and Paul] had nothing else to do but meet each other at the Gare de Lyon. At any given moment in time . . . there [are] dozens of "Paul" and "Peter" characters who [reside] in *other scripts*, scripts that gave them *other roles* and anticipated *other due dates*. "Paul" no. 2 was expected at the dentist; "Peter" no. 2 was seeing his girlfriend; "Paul" no. 3's boss was waiting for him to turn up at an impromptu meeting, and "Peter" no. 3's mother wanted him to bring a special gift for her granddaughter when he came to Paris. (*AIME* 392)

Latour's point with [ORG] is already partly known to us. There is no maxi-transcendence that organizes everything in advance (*AIME* 394, 399). An organization does not have a "stable essence" but is constantly reproduced by its components, until perhaps one day they are no longer able to hold together (*AIME* 396). A new observation in this context, if not in the sciences, is that reducing entropy in one place only increases it in another: "To organize is not, cannot be, the opposite of disorganizing. To organize is to pick up, along the way and on the fly, scripts with staggered outcomes that are going to *disorganize* others" (*AIME* 393). There is no way to avoid this, since life cannot be perfectly systematic any more than economic calculation can be perfectly amoral: "Paul has rushed to the station from the dentist, Peter from his girlfriend; Paul loses his job because he missed the meeting with his boss; Peter has his mother mad at him again. 'I can't be everywhere at once.' 'I only have two hands.' 'I can't be in two places at the same time.' Impossible for any human to unify in a coherent whole the roles that the scripts have assigned to him or her" (*AIME* 393). Though Latour speaks of [ORG] in one passage as the source of many important concepts in the history of philosophy (*AIME* 397), I know of no other philosopher who draws metaphysical implications from "such modest, trivial activities" (*AIME* 397). Latour is also certainly right that "no other mode ensures *borders* to the entities that it leaves in its wake" (*AIME* 397). But despite this setting of borders, one of the implications of [ORG] for economics is that the difference between individuals and societies is not so clear: are Peter and Paul individuals? Latour's answer is clear enough: "As if there were INDIVIDUALS! As if individuals had not been dispersed long since in mutually incompatible scripts; as if they were not all indefinitely *divisible*, despite their etymology, into hundreds of 'Pauls' and 'Peters' whose spatial, temporal, and actantial continuity is not assured by any isotopy" (*AIME* 401).

It is the mode of [ORG] that perhaps resonates most with what we think of as classic ANT. [ORG] reminds us that there are no vast social structures conditioning everything else, but only local assemblies of loosely correlated actors that are always fragile, and rarely as "hegemonic" as might be believed. This has been lacking in the recent humanities, which generally shifts from larger actors to smaller ones only when it wishes to launch a protest in the name of some "subaltern" entity. But one of Latour's primary themes is the fragility of the supposedly strong. And though I once questioned how far one can push the supposed reversibility of power, it is a useful corrective to that bleak trend in the humanities that describes everything *first* as either "powerful" or "powerless."[44] For as Latour discovered in the course of his own career, the dethroning of "power" as a master category is a crucial step toward demodernizing philosophy.

VI. Conclusion: A Latourian Humanities

It is safe to say that the modern humanities were born from a division of labor with the natural sciences. The humanities were supposed to deal with human beings, but not to encroach on the terrain of the sciences: inanimate entities, for the most part. Animals and plants were left in an ill-defined neutral zone, with philosophers not shedding much light on that zone and rarely hazarding a guess as to where animals and plants were supposed to fit in this taxonomical schema. Philosophy took the human-centered path more or less permanently after the time of Kant, who turned philosophy from a "dogmatic" meditation on things-in themselves into a reflection on the internal conditions of human access to those things. To be sure, the humanities and natural sciences have not always respected this division of labor, and at different times both have made attempts to attain a position of crushing hegemony over the other. Most recently this has occurred in attempts by various neurophiloso- phers to subsume the humanities under the all-encompassing umbrella of brain science. It is obvious that Latour, from his earliest years, does not try to confine the humanities taxonomically to the study of human entities, since his flat ontology of actors and their actions effaces this distinction from the start. But it is also obvious that, at least from *We Have Never Been Modern* forward, he does not try to subsume both nature and culture under the "culture" side of the divide. From the mo- ment in 1991 when he rejects the Hobbesian conclusion of Shapin and Shaffer, Latour rejects "power" for good as an all-embracing explana- tory term. One might well think that *action* still plays that role, since Latour continues to work in the ANT idiom, which recognizes nothing other than actions. While this is correct, it also misses the growing anti- Hobbesian importance of *transcendence* or *hiatus* in Latour's thinking, even if only in the form of a "mini"-transcendence. Whereas the young Latour viewed moralists with contempt, the Latour of 1999 grants them an important political function, and by the time of AIME in 2012 a case can be made that [MOR] is the leading mode of them all. And whereas scientists were treated by the early Latour as nothing but actors linking hybrid entities together in networks, both in *Politics of Nature* and even in AIME, they are allowed some breathing room to face the outside of all known networks. These are signals of the crucial new role of mini- transcendence for Latour.

In any case, we cannot adequately conceive of a Latourian humani- ties simply by adding nonhumans to the humanities as currently consti- tuted. Instead, we must *demodernize* the humanities, not in the sense of paving the way for a reactionary approach to the Greek, Latin, and Catholic classics, but in the Latourian sense of removing the modern conceptions of nature and culture altogether. But ANT already does this, and the lingering problem with ANT as a basis for the humanities is its excessive promotion of "immanence" in the sense of everything being solely the equivalent of its actions. This would turn the humanities into no less a form of "calculation" than the natural sciences as conceived by Heidegger, or economics as conceived by Latour. For now, we need not ask whether Heidegger and Latour give fair disciplinary accounts of science and economics, respectively. More important for us is the assertion that calculation must be excluded from the humanities, and not just in the trivially obvious sense that the humanities cannot be mathematized, despite the attempts of Badiou and Quentin Meillassoux to do just that.[45] Instead, we should look beyond ANT to the three modes of AIME that Latour explicitly opposes to calculation.

Let's begin with morality. To call for increased morality in the humanities might sound like a deeply conservative call for finding timeless virtues in the Western canon, virtues that may not even be there. Is Chaucer a font of virtue? Is Dante, grabbing sinners roughly by the hair and happily reporting the disembowelment in Hell of the Prophet of Islam? What about Plato, and his more-than-suggestive praise of beautiful boys? If there is any morality to be retrieved from the great Western tradition, it is not in the sense of explicit moral content, which for Latour could count only as "moralism." Indeed, the morality of which Latour speaks is one for which we are always on the hunt, much like the wisdom that *philosophia* can only love but never attain. On the other hand, some might say that morality is already with us in the humanities, in the form of "speaking the truth to power." Unfortunately, this morality too is only a moralism, since it is quite confident that it already has the truth, and seeks only to defeat those who purportedly do not have it: difference is always better than identity; the queer is less enchained than the straight; the weak are morally superior to the strong; power and truth are inherently in opposition; power inevitably corrupts; the developing world is morally superior to the wealthy Western nations, whose sins are legion. This too is moralism rather than morality, because it too thinks that the truth is already in its grasp, and thus it has lost the ability to be shocked or surprised. Morality might also entail the recognition of certain things that will rarely be appreciated on the Left: the courage of the soldier in defending the homeland; the prudence of austerity in the face of mounting debt. Such actions are moralistically inscribed on the side of the oppressor, and there alone. To bring Latourian [MOR] into the humanities would require not so much "speaking the truth to power," as "speaking uncertainty to truth." This is nothing especially new, since it is already what Socrates did, but it would certainly be refreshing in a time when the humanities have been moralized from two directions at once.

The mode of attachment is Latour's attempt to combat the modern virtue of aloof or cynical distance from whatever one observes. This has nothing to do with a supposed "re-enchantment of the world," since Latour's point is that the world was never disenchanted in the first place, even if modernity thought it was. The inadvertent sincerity of even the most egregious attempt at cynicism is best exemplified by a fictional 2005 editorial in the comic newspaper *The Onion* by the nonexistent Noah Frankovitch, "Why Can't Anyone Tell I'm Wearing This Business Suit Ironically?"[46] The general concept of the article is already clear from the title. Frankovitch reports that he bought a nice business suit and wore it "ironically" in places where he knew it would be disdained. "Fresh from the tailor's in my new suit, I hit all the hippest spots, just waiting for the scenesters' jaws to drop at my sheer audacity. To make sure the irony was pitch-perfect, I got the matching shoes, the cuff links, everything—I even matched my silk socks to my eye color and the accents in my tie! I could barely keep a straight face!" When his friends fail to grasp the irony of his new sartorial tastes, Frankovitch only pushes things further: "I got a leather Hermes attaché case, and I filled it with— you guessed it—actual legal briefs! And my watch? Lame-ass TAG Heuer. Most expensive one I could find. Is that the avant-garde of hipness, or what?" He goes on to "ironically" apply for, and receive, a job at a law firm. Though he initially frames his first paycheck as an "irony trophy," he eventually decides to cash all future checks in order to pay for future exercises in irony. As his legal career progresses even as he continues to smirk, Frankovitch ends by pushing his "irony" all the way: "I even married this clueless girl from Connecticut—loves shopping and everything—and we have two ironic kids. I swear, they look like something out of a creepy 1950s Dick and Jane reader—I even have these hilarious silver-framed pictures of them in my cheesy corner office."

There is a sense in which postmodernity was the Frankovitch Era of the humanities. Though the postmodern "scenesters" did not ironically pretend to be conservatives, at least not to my knowledge, they did often pretend to be above every possible straightforward discourse or position. Yet in doing so, they simply collapsed into a new and energy-draining discourse of their own: one marked by jargon-laden prose, ludicrous strings of punning wordplay, and an affected lack of enthusiasm about nearly every possible topic. Latour's mode of [ATT], much like Badiou's conception of fidelity—though without Badiou's Maoist dogmatism—is well suited to restore passion and commitment to the humanities. Latour is an anti-Frankovitch, as if he were writing a career-long editorial entitled "Why Can't Anyone Tell I'm Critiquing Modernism Unironically?" Among other things, a Latourian humanities would entail the writing of vivid and persuasive prose, not the endless hedging of fifty-page exercises in pretending to subvert everything while actually moving nothing a single inch.

Notes

1. The most admirably systematic book from this camp is surely Karen Barad, *Meeting the Universe Halfway: Quantum Physics and the Entanglement of Matter and Meaning* (Durham, NC: Duke Univ. Press, 2007).

2. S. S. Strum and Bruno Latour, "Redefining the Social Link: From Baboons to Humans," *Social Science Information* 26, no. 4 (1987): 783–802.

3. See also Graham Harman, *Bruno Latour: Reassembling the Political* (London: Pluto, 2014).

4. Latour, *We Have Never Been Modern*, trans. Catherine Porter (Cambridge, MA: Harvard Univ. Press, 1993).

5. Pierre Bourdieu, cited in Anders Blok and Torben Elgaard Jensen, *Bruno Latour: Hybrid Thoughts in a Hybrid World* (London: Routledge, 2011), 143.

6. Alan D. Sokal and Jean Bricmont, *Fashionable Nonsense: Postmodern Intellectuals' Abuse of Science* (New York: Picador, 1999).

7. Thomas Hobbes, *Leviathan* (1651; Oxford: Oxford Univ. Press, 2009).

8. Latour, *An Enquiry into Modes of Existence: An Anthropology of the Moderns*, trans. Porter (Cambridge, MA: Harvard Univ. Press, 2013) (hereafter cited as *AIME*).

9. Manuel DeLanda, *A New Philosophy of Society: Assemblage Theory and Social Complexity* (London: Continuum, 2006).

10. Latour, *Pandora's Hope: Essays in the Reality of Science Studies* (Cambridge, MA: Harvard Univ. Press, 1999), 122.

11. Harman, *Prince of Networks: Bruno Latour and Metaphysics* (Melbourne: re.press, 2009).

12. Alfred North Whitehead, *Process and Reality: An Essay in Cosmology* (New York: Free Press, 1979).

13. Isabelle Stengers, *Thinking with Whitehead: A Free and Wild Creation of Concepts*, trans. Michael Chase (Cambridge, MA: Harvard Univ. Press, 2011).

14. Harman, *Bruno Latour: Reassembling the Political*.

15. Hobbes, *Leviathan*.

16. Steven Shapin and Simon Schaffer, *Leviathan and the Air-Pump: Hobbes, Boyle, and the Experimental Life* (Princeton, NJ: Princeton Univ. Press, 1985).

17. James D. Watson, *The Double Helix: A Personal Account of the Discovery of the Structure of DNA* (New York: Norton, 1980).

18. Michel Callon and Latour, "Unscrewing the Big Leviathan: How Actors Macro-Structure Reality and How Sociologists Help Them to Do So," in *Advances in Social Theory and Metholodology: Toward an Integration of Micro- and Macro-Sociologies*, ed. Karin D. Knorr-Cetina and Aaron V. Cicourel (London: Routledge and Kegan Paul, 1981), 277–303; Latour, "*The Prince* for Machines as Well as for Machinations," in *Technology and Social Process*, ed. Brian Elliott (Edinburgh: Edinburgh Univ. Press, 1988).

19. Shapin and Schaffer, *Leviathan and the Air-Pump*, 344.

20. Latour, *We Have Never Been Modern*, 25–26.

21. Latour, *We Have Never Been Modern*, 27.

22. Latour, "Can We Get our Materialism Back, Please?" *Isis* 98, no. 1 (2007): 138–42.

23. Latour, *Pandora's Hope*, 137.

24. Latour, *Politics of Nature: How to Bring the Sciences into Democracy*, trans. Porter (Cambridge, MA: Harvard Univ. Press, 2004).

25. Latour, *Politics of Nature*, 160.

26. Noortje Marres, "No Issue, No Public: Democratic Deficits After the Displacement of Politics" (PhD diss., Univ. of Amsterdam, The Netherlands, 2005), http://dare.uva.nl/record/165542.

27. Walter Lippmann, *The Phantom Public* (New Brunswick, NJ: Transaction Publishers, 1993).

28. John Dewey, *The Public and its Problems: An Essay in Political Inquiry* (University Park: Pennsylvania State Univ. Press, 2012).

29. Bruno Latour, *Facing Gaia: Eight Lectures on the New Climatic Regime*, trans. Catherine Porter (Cambridge: Polity Press, 2017).

30. Latour, Harman, and Peter Erdélyi, *The Prince and the Wolf: Latour and Harman at the LSE* (Winchester, UK: Zero Books, 2011), 96.

31. Leo Strauss, "Notes on Carl Schmitt, *The Concept of the Political*," appendix to Carl Schmitt, *The Concept of the Political*, trans. George Schwab (Chicago: Univ. of Chicago Press, 2007), 99–122.

32. Schmitt, *The Leviathan in the State Theory of Thomas Hobbes: Meaning and Failure of a Political Symbol*, trans. Schwab and Erna Hilfstein (Chicago: Univ. of Chicago Press, 2008).

33. Schmitt, *The Concept of the Political*, trans. Schwab (Chicago: Univ. of Chicago Press, 2007).

34. Slavoj Žižek, "Carl Schmitt in the Age of Post-Politics," in *The Challenge of Carl Schmitt*, ed. Chantal Mouffe (London: Verso, 1999), 18–37.

35. Latour, "On Recalling ANT," in *Actor Network Theory and After*, ed. John Law and Jon Hassard (London: Wiley-Blackwell, 1999), 15.

36. Latour, "Biography of an Inquiry: On a Book About Modes of Existence," *Social Studies of Science* 43, no. 2 (2013): 287–301.

37. Martin Heidegger, *Bremen and Freiburg Lectures: Insight into That Which Is and Basic Principles of Thinking*, trans. Andrew J. Mitchell (Bloomington: Indiana Univ. Press, 2012); Heidegger, *Being and Time*, trans. John Macquarrie and Edward Robinson (New York: Harper, 2008).

38. Étienne Souriau, *The Different Modes of Existence*, trans. Erik Beranek (Minneapolis, MN: Univocal, 2015).

39. Latour, "Tarde's Idea of Quantification," in *The Social After Gabriel Tarde: Debates and Assessments*, ed. Matei Candea (New York: Routledge, 2010), 145–62.

40. See the second half of Latour, *The Pasteurization of France*, trans. Alan Sheridan and Law (Cambridge, MA: Harvard Univ. Press, 1988).

41. For more on this topic, see Latour and Vincent Antonin Lépinay, *The Science of Passionate Interests: An Introduction to Gabriel Tarde's Economic Anthropology* (Chicago: Prickly Paradigm, 2010).

42. Paul Berman, *Terror and Liberalism* (New York: Norton, 2003), 9.

43. Søren Kierkegaard, *Fear and Trembling*, trans. Sylvia Walsh (Cambridge: Cambridge Univ. Press, 2006); Alain Badiou, *Being and Event*, trans. Oliver Feltham (London: Continuum, 2007).

44. Harman, "Entanglement and Relation: A Response to Bruno Latour and Ian Hodder," *New Literary History* 45, no. 1 (2014): 37–49.

45. Badiou, *Being and Event*; Quentin Meillassoux, *The Number and the Siren: A Decipherment of Mallarmé's* Coup de Dés, trans. Robin Mackay (Falmouth, UK: Urbanomic, 2012).

46. Noah Frankovitch, "Why Can't Anyone Tell I'm Wearing This Business Suit Ironically?" *The Onion*, November 30, 2005, http://www.theonion.com/blogpost/why-cant-anyone-tell-im-wearing-this-business-suit-11185.

CHAPTER FOUR

Care, Concern, and the Ethics of Description

HEATHER LOVE

IN *MATTERS OF CARE: Speculative Ethics in More than Human Worlds*, María Puig de la Bellacasa describes Bruno Latour's turn to "matters of concern" in the early 2000s as inaugurating a shift in the "affective charge" of his work.[1] In writing on the production of scientific knowledge, Latour had long recognized the complexity of things and their imbrication in human systems of value and meaning. But if in his earlier scholarship he attended to the *interests* that compete in and through scientific practices and objects, he later broadened his focus to include investments—attachments, commitments, and beliefs—implied by the term *concern*. Latour coined the phrase "matters of concern" in his 2004 essay, "Why Has Critique Run Out of Steam?" In the context of widespread skepticism about scientific facts, Latour called on critics to get closer to their objects of study; to strive to add reality to them; to explain them rather than to explain them away. Bellacasa frames Latour's intervention in the field of science, technology, and society studies (STS). Because scholars in this field regularly produce constructive, detailed, and richly material descriptions of sociotechnical systems, Latour's turn to matters of concern (MoC) was a "renaming" rather than a wholesale invention: it emerged from an established disciplinary practice, a "collectively learned lesson" by an "academic community" (Bellacasa 38). What was new, according to Bellacasa, was the translation of shared methodological commitments into a categorical imperative. Writing in the pages of *Critical Inquiry*, Latour "propose[d] to critical thinkers more generally to do what STS had, in his view, already learned to do: to respect things as MoC" (Bellacasa 39). Cashing in the methodological specificity of STS, Latour extracted values from the "life of things" and put them back into circulation as free-floating or "normative added values" (Bellacasa 31).[2] He thus reached a wider audience in the humanities, and gained a larger theoretical point: it is good to care about your objects, no matter what they are.

Latour frames his argument about matters of concern in the context of climate change denial, suggesting the political and existential urgency of building more robust and persuasive accounts of reality. "Why Has Critique Run Out of Steam?" pairs this urgent call with an indictment of contemporary scholarship, in particular the excesses of the "critical spirit" associated with STS in the wake of the Science Wars.[3] Arguing that the dominance of large-scale structural explanations and slash-and-burn ideology critique makes it difficult for scholars to account for their objects, Latour suggests that such approaches are not only ineffective but dangerous in a world that badly needs objectivity. "Why Has Critique Run Out of Steam?" offers a rueful retrospective on the reception of Latour's own work, noting his failure to emphasize the constructive aspects of his method. He writes, "The mistake we made, the mistake I made, was to believe that there was no efficient way to criticize matters of fact except by moving *away* from them and directing one's attention *toward* the conditions that made them possible" (231, italics in original). As a countermeasure, Latour suggests the cultivation of a "*stubbornly realist attitude*" (231, italics in original). In offering a riposte to corrosive skepticism about scientific and other facts, Latour elaborates his material-semiotic network analysis as a positive methodological program, a "renewed empiricism" (231) distinct from positivism but nonetheless attuned to the materiality of objects and practices. But at the same time, he offers something more than a methodology: noting the violence of structural explanation, Latour expresses dismay at "our critical arsenal" stocked "with the neutron bombs of deconstruction, with the missiles of discourse analysis" (230). Latour's intervention echoed arguments made earlier by his colleague in science studies, Donna Haraway; writing in a feminist and decolonial framework, Haraway had nonetheless questioned the value of "the acid tools of critical discourse," noting their incompatibility with constructive aspects of scientific practice and with elements of a transformative life praxis.[4] But "Why Has Critique Run Out of Steam?" also resonated with an influential essay by the literary critic Eve Kosofsky Sedgwick, "Paranoid Reading and Reparative Reading; or, You're So Paranoid, You Probably Think That This Essay Is About You" (1997), in which she argued for the importance of cultivating more generous and less punitive critical moods or dispositions.[5]

Latour's explicit invocation of politics, in the form of climate change denial, and of ethics, in his attention to critical disposition, gained him unprecedented attention among scholars in the humanities. While an earlier book, *Aramis, or the Love of Technology* (1993/1996), included an explicit address to humanists, its ambitions were still contained within the ambit of STS. In the Preface to *Aramis*, Latour writes, "I have sought to offer humanists a detailed analysis of a technology sufficiently magnificent and spiritual to convince them that the machines by which they are surrounded are cultural objects worthy of their attention and respect. They will find that if they add interpretation of machines to interpretation of texts, their culture will not fall to pieces; instead, it will take on added density."[6] In his multigeneric description (or "scientifiction" [VII]) of the failure of an individualized mass transit system, Latour draws on the resources of the literary, but still understands his task as "the interpretation of machines"; Latour argues that the interpretation of machines should be accorded the same prestige traditionally granted to the interpretation of texts, but he understands this kind of analysis as a supplement to rather than a displacement of humanist inquiry. However, in broadening his claims to address questions of truth and skepticism generally in "Why Has Critique Run Out of Steam?," Latour expanded the reach of his project, speaking to scholars with little investment in STS as a field or a method. By reaching beyond questions of methodology to invoke not just research ethics but the question of critical ethos or disposition, Latour gained the attention of humanities scholars for whom questions of affect, recognition, and attunement, rather than knowledge production, are primary.

Latour further distilled the ethical implications of STS methodology by allowing *concern* as a research protocol to shade into *care* as an ethical imperative. Care, a keyword in feminist science studies, flickers up intermittently across Latour's work: as a form of relation marked by vigilance and empathy, it is linked to the figure of the feminine and, frequently, to the name of a female critic. Within this framework of feminist science studies, Latour's attention to the ethico-political dimension of scholarly practice seems belated and, as Bellacasa argues, incomplete. In turning from interest to concern, Latour was drawing, but not deeply, on philosophical disputes regarding the feminist ethics of care. In "Why Has Critique Run Out of Steam?" and *Reassembling the Social* (2005), Latour folded many positive qualities of care into an account of how facts are made and sustained, and how collectivities gather. But he disregarded the underside of care—unavoidable in a feminist framework— and its links to histories of domination, labor, and to systems of racial, neocolonial, and gender oppression. By ignoring the material conditions under which care is provided, for instance in the reproduction of life or in the global economy of care work, Latour is able to treat care as a fixed good, with predictable and beneficent consequences. The key context for Latour's thinking about both concern and care is knowledge production, a framework in which labor practices and intersubjective ethics are typically sidelined. As Bellacasa notes, knowledge production is typically understood as a "labor of love," but, she adds, such accounts "silence not only the nastiness accomplished in love's name but also the work it takes to be maintained" (78).

Bellacasa suggests that Latour's account of matters of concern does not go far enough, which is why she names her book *Matters of Care* and centers the work of scholars in feminist science studies. Although I am deeply influenced by Bellacasa's analysis of Latour's ethico-political turn, my emphasis in this essay is different. I consider the place of care and concern in Latour's work, suggesting that with his address to fields beyond STS, he began to blur the distinction between them. Furthermore, I argue, this elision of the ethical and the methodological made Latour ripe for uptake in the critical humanities, particularly in literary studies. By addressing knowledge production as a matter of concern, Latour appealed to humanities scholars' longstanding commitment to pedagogical and exemplary practices of subject-formation, self-examination, and pastoral care.[7] Latour's injunction to care for your objects, as well as his challenge to the distinction between human and nonhuman actors, resonates in literary studies, where attending to texts can be understood as a proxy for attending to human others. Scholars in the field of literary ethics address this homology directly, arguing for the value of reading in cultivating empathy, openness to difference, and a capacious sense of justice. The problem in this context is that Latour's idealization of care fits *too well* within the disciplinary framework of literary studies. Literary critics have responded to the ambiguity in Latour's call to scholars to treat their objects as matters of concern by heeding it as an ethical injunction to treat their objects with care. But it is Latour's methodological injunction that stands out above ethical his one. As such, literary critics will gain the most from Latour's modeling of close attention as a practice, rather than his modeling of care as an ethical disposition. Not only does care have the status of doxa in literary studies, but it also tends to float free as a "normative added value." Returning to Latour's early work in STS demonstrates the value of a methodological ethics that attends to the values in things, and that relies on attention rather than empathy as a critical tool.

The last two decades have witnessed a rapprochement between Latour's semiotic-materialist network analysis and literary studies. The connection is somewhat surprising. Latour's signature method of actor-network-theory (ANT) was developed in the field of science and technology studies, and its most robust applications have been in social scientific contexts. ANT, with its focus on objects and infrastructure, is much more obviously suited for the analysis of laboratory practices, transportation systems, and campaigns for public hygiene than it is for the analysis of novels or poems. And yet a range of critics have taken up Latour's methods, considering their utility in the analysis of aesthetics and culture. This interest in Latour can be understood in part as a reaction against the textualism of post-structuralism. As if in recovery from the rejection of the world outside the text, many critics have been aggressively reinserting the text back into the world, and the world back into the text. Recent years have witnessed a rise in approaches that treat the text as a material object; in systemic approaches to questions of literary value, circulation, and exchange; in digital and quantitative approaches to noncanonical literature; and in ethnographic accounts of reading as a social practice. As empiricism has taken hold in literary studies, scholars have turned to ANT as a materialist framework that is capacious enough to deal with aesthetic objects. Latour's work has been taken up to rethink literary networks and reader relations; to address "setting" as a point of transfer between fictional and real worlds; to recalibrate interpretation and description in textual analysis; to reimagine character, agency, and cognition; and in a host of other experiments aiming to treat literature as a worldly phenomenon, connected to other forms of textual and social practice. At the same time, his work has had an impact in interdisciplinary fields beyond literary studies; in fields such as disability studies, queer studies, and animal studies, increasing interest in questions of ontology, ecology, and the liveliness of matter has made Latour's work on the interrelation between human and nonhuman worlds increasingly salient.

During this time, imaginative literature has played a prominent role in Latour's own writing and thinking. Latour's engagement with literature has proceeded on two fronts. Latour attends to the materiality of texts, suggesting that books, though they may point to other worlds, are embedded in this one. He is also invested in the reality of the worlds represented in texts, and he calls on scholars to lend "ontological dignity" to imaginative literature.[8] Because of the worldliness of literature and its descriptive power, Latour has suggested that it might serve as a model for the social sciences. In *Reassembling the Social*, he touts the capacity of literature to produce accounts that are "disciplined, and enslaved by reality": fictional narratives paradoxically model the values of "accuracy and truthfulness," through their ability to reflect the complexity and liveliness of reality itself.[9] Latour's experiments in reading literary texts, and his uptake by literary scholars, has allowed a spirit of experiment to flourish. This willingness to try new approaches, to engage in counter-factual thinking, and to open, once again, the question of what literature is and is for, has all been salutary. But the grounds on which Latour embraces literature are in some tension with the epistemological commitments of the interpretive humanities. If Latour's call for the demilitarization of criticism struck a chord with many literary critics, his call for a renewed empiricism was more puzzling to scholars devoted to the analysis of aesthetic artifacts. What is the utility of Latour's post-hermeneutic, quasi-empiricist, incrementally objective analysis outside of the context of science studies? What is the status of the fact in a fictional text? Are these tools even vaguely suited for the analysis of literary texts? And, if so, what can they accomplish that literary critical methods cannot?

Reassembling the Social, framed as an introduction to ANT, addresses concrete questions of methodology that resonate with the practices of literary scholars. Latour's insistence on going slow and keeping things flat; his suggestion that scholars must "pay the price, in small change" for their claims (35); and his image of scholars tracing "the delicate trails of the regions we wish to map" (35), all invoke the kind of small-scale, incremental, and recursive forms of attention deployed in close reading. Latour insists on the significance of close attention, but disregards the distinction between the sciences and the humanities that codes such attention either as objective scrutiny or as subjective responsiveness. Research in the social sciences is presented as, ideally, obsessed with details, sensitive to unfolding context, delicate. Latour also resists the idea that research in the humanities discovers a world that is more concrete, rich, or warm than other realms of social and material practice. Latour does not object to the language of phenomenology per se, but insists that it should be extended beyond the sphere of the human. Particularly useful for literary criticism is Latour's advice to "describe, write, describe, write" (149). In a section called "Deployment not critique," he writes, "No scholar should find humiliating the task of sticking to description. This is, on the contrary, the highest and rarest achievement" (136–137). Latour argues that a good description is not in need of an explanation— explanation should instead take the form of "extending" a description "one step further" (137). This emphasis on description as a privileged form of analysis reflects the way that, in literary criticism, metalanguage often emerges from the text itself.

At other times, the friction between Latour's methodological injunctions and the practice of literary criticism is more evident. If the idea that a textual reading is a kind of higher order description resonates with literary scholars, they may have some trouble taking on Latour's claim in *Reassembling the Social* that an effective scholarly text is not a portrait but rather an "artificial experiment" to trace the pathways of its objects.[10] It is not clear in what sense novels, plays or poems have pathways: we may consider books as objects that circulate in networks but it is puzzling how to adapt step-by-step network analysis to the analysis of complex literary forms. The necessity of considering texts as objects can also limit the usefulness of Latour's methodological directives. In *Reassembling the Social,* Latour insists on the importance of tracking one's research activity. He writes, "The best way to proceed at this point . . . is simply to keep track of all our moves, even those that deal with the very production of the account. This is neither for the sake of epistemic self-reflexivity nor for some narcissist indulgence into one's own work but because from now on *everything is data*" (133). Latour discusses the necessity of keeping a log of activities, sorting information chronologically and by topic, and engaging in ongoing "writing trials" (134), arguing that taking notes well is the "only way there is to become slightly more objective" (135). While humanists may stumble over the disparaging treatment of self-reflexivity or the approving mention of objectivity (even if it is only "slight"), they may be stopped in their tracks by the assertion that "from now on *everything is data*." Can everything be converted into data? What count as moves for the literary scholar? And what space does the ideal of comprehensive accounting of research practice leave for opacities in the text, or for gaps in the researcher's self-knowledge?

Latour engages such doubts in *Reassembling the Social* by addressing several "uncertainties," one of which is the question of the "precise sense social sciences can be said to be empirical" (22). It is clear that this question is much more vexing for those who deal with entities that are fictional and aesthetic. Latour offers a specific example of the empirical value of literature in his 2008 essay, "Powers of the Facsimile: A Turing Test on Science and Literature" (2008). As is often the case, Latour insists on the hyper-exemplarity of the essay's central figure, the science fiction author Richard Powers (reminiscent of his lionizing of Gabriel

Tarde, or Alfred Whitehead): for those interested in the general relevance of novels to the social sciences, the emphasis in the essay on the singularity of Powers's contributions can be a bit confounding. Furthermore, Latour's interest is even more specific. Across his work, Powers models how to "provide a realistic description of science" using "a rich literary repertoire" to bring the reader "closer to the matters at hand."[11] His contributions in *Plowing the Dark* are even more specific: he deploys "different levels of realism" (7) to investigate and unsettle what it means to be a character. Latour champions Powers because he is, for him, a novelist who treats things as matters of concern, and contrasts him to Zola, whose forensic naturalism offers the reader only matters of fact. In this sense, Latour embraces novelistic description, but without acceding to the traditional critical denigration of description—for instance in Georg Lukács essay, "Narrate or Describe?," as capable only of a lifeless reproduction of a commodified reality. In addition to championing a specific form of literature (which may only be a set of one), Latour promotes a form of literary criticism: "All the usual resources of criticism, fiction, and illusion which usually go into chic commentaries of Escher-like 'abyme' effects, are here all telescoped by Powers to provide more reality, not less. Constructivism is made to be the exact opposite of deconstruction while, at the same time, using many of the same resources" (12). Contrasting deconstruction (chic, knowing, skeptical) with constructivism (curious, credulous, even dorky), Latour imagines a reading practice attuned to "what language can do in terms of model-building" (13).

These reflections on the value of literature precede by a few years Latour's epic 2013 *An Inquiry into Modes of Existence: An Anthropology of the Moderns (AIME)*. In this book, Latour elaborates a method that would take each of the major "modes of existence" (such as law, religion, politics, and aesthetics) on its own terms, attending to their characteristic forms of veridiction, aims, procedures, and felicity conditions. *AIME* builds on Latour's application of ethnographic method to the world of the natural and physical sciences. Here, in elaborating an "experimental metaphysics," a "comparative anthropology," or a "practical ontology" (481), Latour turns this method on a wider set of practices and domains, elaborating a multiverse made up of worlds that propose and sustain their truths in distinctive ways. This scaling up constitutes another move in Latour's translation of STS methodology into a wider domain of knowledge and practice. As in the case of his reframing of concern as an ethics with implications far beyond the analysis of sociotechnical systems, this translation involves both trade-offs and ironies for Latour.

One of the key modes in *AIME* is the fictional, which he treats in the chapter "Situating the Beings of Fiction." Latour accords density and presence to the "beings of fiction," treating them as real actors rather than as textual figments. It is not obvious what Latour means by the "Beings of Fiction": in the idiosyncratic table in the back of the book, Latour lists the "Beings to Institute" in the mode of fiction as "dispatches, figurations, forms, works of art." While the emphasis, as in his earlier work, is on realist and science fictional narrative, his discussion in *AIME* takes in a broader range of aesthetic objects. *AIME* is organized around the concept of mode as a kind of frame or key (borrowing from Erving Goffman) for organizing experience. Latour also compares modes to genres, and, in a telling explanation of the concept of "preposition" (a position-taking that comes before a proposition is stated, determining how the proposition is to be grasped and thus constituting its interpretive key" [57]), he invokes the scene of the bookstore. He writes,

> If you find yourself in a bookstore and you browse through books identified in the front matter as "novels," "documents," "inquiries," "docufiction," "memoirs," or "essays," these notices play the role of prepositions. They don't amount to much, just one or two words compared to the thousands of words in the book that you may be about to buy, and yet they engage the rest of your reading in a decisive way . . . Like the definition of a literary genre, or like a key signature on a musical score, at the beginning an indication of this sort is nothing more than a signpost, but it will weigh on the entire course of your interpretation. (57)

Latour later offers another comparison to elaborate the concept of preposition or interpretive key, which is the first page of the novel (*page de garde* or warning page in French [271]), which does not conceal what is to follow in the book but "founds its reality" (271). Rather than see these invocations of the book as arbitrary metaphors, I would argue that they reveal how central the concept of genre is to the modes of existence. While Latour shies away from the specter of "aestheticizing" reality whenever it looms, in making reality generic he has imported the material apparatus of the book into his comparative anthropology.

In *AIME*, Latour reframes discourses usually either dismissed as irrational (religion) or seen as irrefutably true (economics) ethnographically, suggesting that they are a distinct mode of existence, characterized by non-fungible assumptions, practices, and criteria. In a tribute to work on the "worlds of art," Latour suggests that historians and sociologists of art have already learned to treat aesthetic objects ethnographically, adding to the significance of works by situating them in dense material-semiotic networks. He writes,

> History and sociology have made themselves capable of deploying the trajectories of a work . . . in which one has to take the whims of princes and sponsors into account as well as the quality of a keystroke on a piano, the critical fortune of a score, the reactions of a public to an opening night performance, the scratches on a vinyl recording, or the heartaches of a diva. On following these networks, it is impossible to separate out what belongs to the work "properly speaking" from its reception, the material conditions of its production, or its "social context." Even more than the anthropology of the sciences and technology, that of art has succeeded, by stunning erudition, in continuing to add segments without ever taking any away: in the work, truly, to sustain it, everything seems to count. All the details count, and the details, all together, pixel by pixel, are what sketch out the composite trajectory of the work. (243)

Latour suggests that a model for inquiry into the modes of existence has already been undertaken in such work on aesthetic objects (undertaken, I would add, not only by historians and sociologists, but also by literary critics, art historians, film scholars, and others).[12] Latour's emphasis on the importance of details resonates with the practice of literary critics; his argument about the way that the treatment of practice, materials, reception, and economics might bring us closer to our objects rather than distance us from them suggests the promise of his ethnography of the modes.

In *AIME*, Latour extends his call to respect the objectivity of the worlds described in works of fiction. He links the dismissal of the "ontological dignity" (247) of the aesthetic to its prestige as evidence of the imagination. He writes, "If it is true that the beings of fiction have been swamped by honors, they have paid a big price for the central place they have been given in the collective. . . . They have been valued to an extreme while too hastily denied any objectivity." Latour asserts further that "[f]actual narratives do not differ from fictional ones as objectivity differs from imagination. They are made of the same material, the same figures" (251). While literary critics may agree with Latour's critique of the dismissal of the aesthetic as an airy world of the imagination, his according of ontological dignity and objectivity to fictional works may come as an unwelcome gift. According to Latour, the special status accorded to that which is understood to exceed the limits of the known world comes at too high a price. In a moment when traditional defenses of literary studies based on canonical value or the cultivation of civic virtue are in decline, critics may be inclined to agree with Latour's diagnosis: literature's association with the realm of the noninstrumental and with other forms of human singularity has not proven much of a bulwark against appeals to the bottom line in the contemporary university.

Still, if literary critics agree with Latour's account of the denigration of aesthetic criticism, they are unlikely to agree with his solution, which is to erase the difference between reality and fiction, between world and word. Latour's elevation of fictional beings, while intended as a tribute to literature, is at odds with the disciplinary ethos of literary studies. Latour ignores the highly-developed disciplinary protocols for the interpretation and analysis of literary texts, suggesting that they might be replaced with ANT as a universal (if flexible) method. To assimilate aesthetic objects to the realms of existence, Latour converts them into the general currency of "beings of fiction." In order to argue for the dignity of these beings, Latour refuses Saussure's theory of the sign. Latour proposes that "the sign is 'arbitrary' only for those who, having agreed to lose the experience of relations, try to reinject relations on the basis of the 'human mind' into a 'material world' that has been emptied in advance of all articulations" (256). Latour takes articulation out of the realm of language and puts it back in the world, proposing that "it is *the world itself that is articulated*" (256, italics in original). Rather than imagining a mute, meaningless world brought into expression through human agency, Latour suggests that humans' role is to capture processes of articulation that are ongoing. "If natural language takes itself in hand to take up the world, it is because the world has taken itself in hand and is still doing so, time after time, to persist in being. A linguist should never circumscribe the isolated domain of 'Language,' unless it is to interrupt this movement of articulation for a moment, to make the analysis easier" (257).

This flattening of the difference between sign and referent and the loss of the specificity of representation has drawn charges of positivism from Latour's critics. But if, from the perspective of humanists, Latour represents the dangers of positivism, we should remember that from the perspective of the sciences he is still associated with a rash relativism. In the bigger picture, Latour is an important ally, because of his decades-long challenge to the positivist ethos that sees no value in the humanities. With *AIME*, he has further challenged the stark division between hard empiricism and a beloved but irrelevant aesthetic sphere, linking this argument to his broader polemic against the bifurcation of nature, the separation between word and world. Latour's mediation between the sciences, the social sciences, and the humanities is thus helpful in mounting a defense of the humanities. But it is also useful in suggesting how literary studies might be expanded by an encounter with empirical methods.

In his elaboration of a "second empiricism" in "Why Has Critique Run Out of Steam?," Latour calls for critics to return to "the realist attitude"; at the same time, he urges them to cultivate less violent relations to their objects. He writes,

I want to show that while the Enlightenment profited largely from the disposition of a very powerful descriptive tool, that of matters of fact, which were excellent for debunking quite a lot of beliefs, powers, and illusions, it found itself totally disarmed once matters of fact, in turn, were eaten up by the same debunking impetus. After that, the lights of the Enlightenment were slowly turned off, and some sort of darkness appears to have fallen on campuses. My question is thus: Can we devise another powerful descriptive tool that deals this time with matters of concern and whose import will no longer be to debunk but to protect and to care, as Donna Haraway would put it? (232)

Acknowledging the utility of matters of fact, Latour nonetheless suggests the importance of moving beyond this framework. The drama of the passage, which traces a path from the clearing away of prejudice to a darkness encroaching on college campuses, is amplified by the invocation of affect and ethics. The point of matters of concern is not only to analyze or explain, but also to protect and to care. The fact this impulse and this phrase are attributed to Haraway is not incidental, since the distinction between fact and concern obey a gendered logic. Care is a name for the feminine as a generative matrix in Latour's work; it constitutes a background of unquestioned value in the form of generosity, self-sacrifice, and responsiveness to the object. Because care is invoked but not elaborated, Latour's allusions to it can look like an act of outsourcing.

Latour's blending of matters of concern as a methodology and as an ethics is belied by his durable interest in questions of epistemology and knowledge production. Furthermore, the affective register of Latour's scholarship does not reflect an attitude of tender regard or anxiety for the well-being of the object. Latour's relation to his objects is more generally characterized by curiosity, or cheerful interest. This mode is explicitly valued in "Why Has Critique Run Out of Steam?" In a discussion of the distinction between what he calls the fact and fairy position, Latour argues that, while critics are happy to see some ideologies debunked, they refuse such skepticism in the case of their own commitments. People do not usually recognize this combination of skepticism and belief as a contradiction, because they hold these views in relation to different objects, distinguishing, for instance, between (false) ideology and (true) science. But alongside this strict binary of fetishism and iconoclasm, Latour introduces a third category, which he describes an attitude of affection, cultivated in relation to activities that are probably best described as hobbies. He writes,

> The whole rather poor trick that allows critique to go on . . . is that there is never any crossover between the two lists of objects in the fact and fairy position. This is why you can be at once and without even sensing any contradiction (1) an antifetishist for everything you don't believe in—for the most part religion, popular culture, art, politics, and so on; (2) an unrepentant positivist for all the sciences you believe in—sociology, economics, conspiracy theory, genetics, evolutionary psychology, semiotics, just pick your preferred field of study; and (3) perfectly healthy sturdy realist for what you really cherish—and of course it might be criticism itself, but also painting, bird-watching, Shakespeare, baboons, proteins, and so on. (241)

In decrying the rigid binarism of the fact-fairy distinction, Latour risks falling into binary thinking himself. By introducing a third category of "cherished" activities and objects, Latour provides a way out of this dead-end. Detached from the realm of science and ideology, "painting, bird-watching, Shakespeare, baboons, proteins, and so on" invite neither debunking nor worship. The amateur or hobbyist provides the model for a non-instrumental love of one's objects: devotion and affection replace the torturous epistemologies of belief and skepticism, allowing for the development of a "healthy sturdy realism."

This attitude of "healthy sturdy realism" appears as an exemplary disposition to the material and social world. Throughout his work, Latour favors an engaged, worldly, and cheerful pragmatism. Latour models this attitude in *AIME* through an extended account of a hike in the Vercors Plateau. Latour describes both the difficulty of navigation ("I was having trouble finding the starting point for the path leading to the Pas de l'Aiguille" [74]) as well as his pleasure in overcoming this difficulty, with the help of French geological survey map 3237 ("I unfolded the map and, by looking from plasticized paper to the valley, located a series of switchbacks that gave me my bearings despite the clouds, the confusion of my senses, and the unfamiliarity of the site" [74]). He ultimately uses the example of the hike to suggest how a too-strict demarcation between the map and the territory, or between semiotic systems and material reality, undermines the reality of both. Reflecting on his account of his hike, Latour writes,

> I did not try to make my reader resonate with the warmth of my *feeling* for Mont Aiguille, a feeling that "will never be captured by the frozen knowledge of geologists or mapmakers." Quite to the contrary: the establishment of chains of reference, the history of cartography, geology, trigonometry, all this was just as warm, just as respectable, as worthy of attention as my pale expressions of admiration, as my emotions as an amateur hiker and as the shiver I feel when the wind comes up and chills the sweat running down my chest. By splitting Mont Aiguille into primary and secondary qualities, making it bifurcate into two irreconcilable modes, what is neglected is not only subjectivity, "lived experience," the "human," it is especially Mont Aiguille itself, in its own way of persisting, and, *equally*, the various sciences that have striven to know it and that depend on its durability to be able to deploy their chains of reference. In this matter, it is not only humans who lack room, it is first of all Mont Aiguille itself, and second, the various sciences that allow us access to it! If the splitting had caused only the neglect of human feelings, would the loss be so great? The danger is that this loss threatens to deprive us of both the map and the territory, both science and the world. (120–121, italics in original)

In a familiar argument against phenomenological overvaluation of experience as well as reflexive resistance to the "coldness" of science and technology, Latour criticizes the assimilation of the human to the qualities of richness, animation, and warmth. Valuing the human as the opposite of the built world will lead to the reification of the human. Yet Latour does not dismiss the image of the human altogether; instead of the Romantic subject sublimed through an encounter with nature, Latour offers an image of the human as a tool-user, curious and persistent. In what sense can Latour's resourceful amateur be said to care for or about Mont Aiguille? Or for the geological survey that allows him to navigate it, while still looking around? It is much more apt to say that this figure *cherishes* the mountain and the techniques that make it accessible to him. This relation of cherishing is characterized by attention and appreciation rather than solicitude, vigilance, or empathy. The world in this frame is not in need of care; sturdy and approachable, it is a problem to be solved rather than an enigma to be contemplated.

Latour greets both the known and the unknown with this spirit of cheerful pragmatism. In *Laboratory Life* (1979), his early collaboration with Steve Woolgar, Latour ironically rewrites the scene of the first encounter, describing the sociologist as an alien being landing in Roger Guillemin's laboratory at the Salk Institute. Latour avoids both the fear and loathing typical of colonial encounters and the deference accorded to the production of scientific truth. Instead, he uses a style of exhaustive in situ recording of the micro-behaviors of the scientists in the lab. The book opens with an "excerpt from observer's notes," a moment-by-moment recording of activity from a resolutely naïve perspective:

> *5 mins.* John enters and goes into his office. He says something very quickly about having made a bad mistake. He had sent the review of a paper. The rest of the sentence is inaudible.
>
> *5 mins. 30 secs.* Barbara enters. She asks Spencer what kind of solvent to put on the column. Spencer answers from his office. Barbara leaves and goes to the bench.
>
> *5 mins. 35 secs.* Jane comes in and asks Spencer: "When you prepare for I.V. with morphine, is it in saline or in water?" Spencer, apparently writing at his desk, answers from his office. Jane leaves.
>
> *6 mins. 15 secs.* Wilson enters and looks into a number of offices, trying to gather people together for a staff meeting. He receives vague promises. "It's a question of four thousand bucks which has to be resolved in the next two minutes, at most." He leaves for the lobby.

> *6 mins. 20 secs.* Bill comes from the chemistry section and gives
> Spencer a thin vial: "Here are your two hundred micrograms,
> remember to put this code number on the book," and he points to
> the label. He leaves the room.
>
> Long silence. The library is empty. Some write in their offices,
> some work by windows in the brightly lit bench space. The staccato
> noises of typewriting can be heard from the lobby.[13]

This slab of unprocessed field notes clarifies what Latour has in mind when he describes turning everything into data, and keeping track of all the moves of both the observed and the observer. At this point in the account—at the very beginning—it is not clear what John and Barbara and Spencer and Jane and Bill are doing. Or rather, it is clear what they are doing, since their visible activity and speech is meticulously recorded by the observer. But it is not clear *why* they are doing it, and this is the virtue of the perspective of the newcomer. As Latour and Woolgar go on to argue, it is by not knowing the inner meaning of these activities or the "'thought processes' which 'underlie'" (42, n. 2) them that they are able to see things about them that the scientists cannot: that the production of scientific knowledge is the product of a specific setting (the laboratory), and that it is achieved through processes of literary inscription.

The "staccato noises of typewriting" and the scratching of the observer's pencil give *Laboratory Life* its soundtrack. As Latour and Woolgar note, the laboratory technicians are "compulsive and almost manic writers" (48). Although they compare these workers to novelists, there is no evidence of a hushed respect for human creativity or expression in this account of inscription as a practice. Nor is there any sublime mystery in the making or discovery of scientific knowledge. Instead there is activity: "It seems that whenever technicians are not actually handling complicated pieces of apparatus, they are filling in blank sheets with long lists of figures; when they are not writing on pieces of paper, they spend considerable time writing numbers on the sides of hundreds of tubes, or pencilling larger numbers on the fur of rats" (48). The reduction of knowledge productivity to a sequence of observable actions might be understood as a kind of debunking, a refusal to hold sacred the human intention behind the automatism of behavior. And yet, as Latour and Woolgar suggest, the creation of knowledge is impossible without these material supports and practices.[14] Furthermore, stripping science of its halo allows for "an appreciation of how science is done," a perspective that is obscured by a too thorough familiarity with the content of science. "It is therefore necessary," Latour and Woolgar argue, "to retrieve some of the craft character of scientific activity through in situ observations of scientific practice" (28–29).

The attitude toward scientific practice that is on display in *Laboratory Life* has little to do with care. It is rather an attitude of curiosity, enabled by knowing a little but not too much about what is behind the visible actions in the laboratory.[15] Scholars in the humanities have tended to value care over curiosity as more attuned and a less violent attitude to objects of study. Latour, despite his indictment of unthinking or reflexive affirmation of humanist values, nods to such an attitude in his invocations of concern as care in his work on the limits of critique. But Latour's work across his career is a testament to what can be apprehended through the refusal to care too deeply. The observer's perspective in *Laboratory Life* recalls the violent ignorance of colonial encounter as well as the detached perspective of mid-century behaviorism. But it is not identical to either of these, nor are its ethical implications. The anthropological perspective that Latour and Woolgar adopt is indebted more clearly to midcentury microsociology, with its emphasis on minute observation as a resource for accounting for human action and interaction. In their chapter, "The Microprocessing of Facts," Latour and Woolgar cite the work of Harold Garfinkel (the founder of ethnomethodology), Alfred Schutz (who blended sociology and phenomenology), and Harvey Sacks (pioneering researcher in conversation analysis), as they attempt to account for the inner logic of scientific thought through observational means. They argue that most accounts of the difference between common sense and scientific logic depend on tautology: scientific thought can be understood as "identical to its daily life counterpart but for the inclusion of the word scientific" (153). For this reason, they argue, the best place to look for the creation of scientific knowledge is not in "interviews, archival studies or literature searches," but rather in the "daily activities" and the "smallest gestures" on display in the laboratory (152). "Our position," they write, "is that if such differences exist, their existence must be demonstrated empirically" (153).

I have argued in this essay that Latour's concept of matters of concern is best understood in the framework of methodology rather than ethics. He is interested in the question of how to pay proper attention to one's object of study as a technical and pragmatic matter rather than as a means of cultivating ethical subjectivity. Although Latour's methodological injunctions overlap with ethical questions—for instance in his advocacy of attachment—he is nonetheless committed to a repeatable, shared, and public technique that makes care as a practice of love secondary (at least). Concern, for Latour, is essential to knowledge production, and is therefore bound up with the violence that is potential within all knowledge production. Latour acknowledges this violence at every turn, from explaining research as "steamrolling" to his discussions of "studying up" and the resistance he assumes on the part of his research subjects. However, despite his attention to these examples, Latour does not see knowledge production as inevitably violent. His pragmatism, emphasis on problem solving, and insistence on a thoroughgoing empiricism resist a totalizing or tragic view of scholarship as inevitably destructive. Furthermore, Latour offers a menu of possibilities for navigating that potential for violence, suggesting that some forms of attention, interest, and curiosity are not damaging to the object. In his interest in the craft character of science, and in the model of the hobby as a semidetached form of affection, Latour describes the non-instrumental pleasure that we can take in research. Latour's practices of attention and description offer a model of methodological ethics, without recourse to the "bifurcated" realm of care that appears increasingly in his late work. This is a good model for thinking methodologically, and for thinking about the ethics of knowledge production. But it is just as clearly inadequate for thinking about ethics more broadly, whether in the context of interpersonal relations or political economy.

For literary critics, who have tended to hypostasize the violence of knowledge production, and to see too much promise in the realm of interpersonal ethics, Latour's emphasis on method is salutary. For those outside literary studies, the neglect of method in the field may be surprising. Literary critics have a superabundance of theory—concepts and convictions—but few accounts of how we put these concepts into practice. Most critics point to close reading as the core method of the field, without necessarily being willing to commit to a definition, let alone a step-by-step protocol or directives for training students. A steady stream of scholarship traces the historical roots of close reading, addressing its place within the history of the discipline as well as some of its major and minor variants and offshoots. Nonetheless, in many discussions close reading is taken for granted, mystified, or understood as a kind of instinctive talent—an inborn sensitivity to language, a tolerance for contradiction, a capacity for acts of extreme attention—that can be cultivated but is also simply given (or not). The problem is not that we don't have methods, but that we tend not to make them explicit. Explicitness is a form of demystification, one that threatens the charisma of the text and the critic. The anti-democratic nature of this attitude is clear: both because it idealizes genius, and also doesn't pay attention to the class and other social determinations of literary and linguistic competence.

We find a powerful argument for method in two late essays by the American anthropologist Margaret Mead, "From Intuition to Analysis" (1969) and "Toward a Human Science" (1976). In these pieces, Mead grapples with the transition from a pre-WWII integrative and intuitive practice of pattern recognition among broadly trained humanists to a technocratic, evidence-based, and specialized set of practices after the War. This transition forms part of Mead's own biography, since she was trained by Franz Boas and Ruth Benedict, and did her most prominent work on the question of cultural patterns. But she also engaged in post-War collaborations in cybernetics and communications research and was excited about the potential for developing more exact methods in visual anthropology. In "Toward a Human Science," she considers the value of new recording devices for producing the "form of knowing that we call science—that is, with the knowledge that can be arrived at and communicated in such a way that it can be shared with other human beings, is subject to their independent verification, and is open to further explanation by investigation in accordance with agreed-upon rules."[16] It is only with the advent of film and tape, Mead writes, that observation of human behavior can become a "human science": "in the past," she writes, "each great integrator of knowledge had to rely chiefly on his own capacity as a whole human being to observe the behavior and speculate about the past of members of his own species in ways that were—and are— unique to the human mind and dependent on the development of human culture. In more complex cultures, sharing the same traditions and education opened the way to an understanding of the insights of a philosopher, a historian, or an ethical leader who reported his observations in a shared language or demonstrated his ideas through artifacts or great works of art familiar to everyone involved. But just as in communication among physical scientists, more than a shared natural language is essential" (906).

A related conversation is underway in literary studies in debates about reading methods. But it is striking across these conversations that, even when the topic is method, there is a tendency to change the subject: from technique to ethics. This is not true across the board, and there has been a turn to method, perhaps most prominently in digital and quantitative projects and in bibliographic and textual studies. Nonetheless, Latour's call for scholars to treat their objects not as matters of fact but as matters of concern has been taken up within literary studies not as a methodological imperative but as an ethical one. It is not necessarily a problem that literary critics are trying to cultivate a more generous attitude toward their objects of study. Repair and love are hard to argue with. The problem is rather that, in the framework of the literary humanities, repair and love are disciplinary doxa. If, as a scholar of literature, you say you want to be generous toward the object, give yourself over to it, cultivate openness, allow yourself to be surprised by it, no one will challenge you—in fact, no one will have much to say at all, since you are simply repeating the core values of the field. This is what literary critics are supposed to spend their time doing, and one risks nothing in refusing objectivity or drifting away from the rigors of verification—no one wants that from you anyway.

This moment in literary studies recalls Latour's claims for the "high price" that fictional beings have paid for their honors: literary criticism is exceptional and beloved, but fungible. The university can do without it. Both as a matter of disciplinary survival and as a challenge to the ethical framework of our research practice, literary critics may need to practice caring *less* and learning to speak the language of method more fluently. Anna Tsing's *The Mushroom at the End of the World* (2015), provides an example of ethically and politically vital work that centers method. In her extended meditation on the matsutake mushroom, a form of life that springs up in "blasted landscapes," Tsing calls for a return to the "arts of noticing."[17] Cultivating fine-grained attention is both a method for producing knowledge and a way of getting by in an era of environmental destruction. "To live with precarity," Tsing writes, "requires more than railing at those who put us here" (3). It also requires, in her words, "stretch[ing] our imaginations to grasp" the contours of a new world (3). Tsing has many things in common with Latour: an emphasis on human and non-human entanglements; an alertness to the necessity of critical knowledge as a form of making; a refusal of the fetish of modernity; and a will to breach the division between the sciences and the humanities. Crucial in this last regard is her understanding of noticing *as a science*. She writes,

> To listen and tell a rush of stories is a *method*. And why not make the strong claim and call it a science, an addition to knowledge? Its research object is contaminated diversity; its unit of analysis is the indeterminate encounter. To learn anything we must revitalize the arts of noticing and include ethnography and natural history. But we have a problem with scale. A rush of stories cannot be neatly summed up. Its scales do not nest neatly; they draw attention to interrupting geographies and tempos. These interruptions elicit more stories. This is the rush of stories' power as a science. Yet it is just these interruptions that step out of the bounds of most modern science, which demands the possibility for infinite expansion without changing the research framework. Arts of noticing are considered archaic because they are unable to "scale up" in this way. (37–38)

For Tsing, the ability to scale up is a valuable feature of many knowledge projects, but it is not well suited to the indeterminacy of encounter. In such a framework, Tsing writes, "the nonscalable becomes an impediment." However, in the current moment, she writes, "it is time to turn attention to the nonscalable, not only as objects for description but also as incitements to theory" (38).

In Tsing's work I think we see many of the strengths of the interpretive humanities: attention to detail, ability to deal with complexity and contradiction, skepticism about scaling up. Literary scholars have been in the business of asking people to care more, and care better. This is the kind of thing that is easy to claim but hard to test, prove, or teach. We might focus instead on teaching people the arts of noticing, or concrete techniques for how to pay attention to texts and the world. This is not an abandonment of ethics or politics, but rather an insistence that ethics and politics must be instantiated through practice.

Notes

1. María Puig de la Bellacasa, *Matters of Care: Speculative Ethics in More than Human Worlds* (Minneapolis, MN: Univ. of Minnesota Press, 2017), 35.

2. Bellacasa uses these phrases in describing the utility of the concept of matters of concern: "The notion became popular as a renaming that could help to emphasize engaged ethico-political responsiveness in technoscience in an integrated way, that is, within the very life of things rather than through normative added values" (31).

3. Bruno Latour, "Why Has Critique Run out of Steam? From Matters of Fact to Matters of Concern," *Critical Inquiry* 30 (Winter 2004), 225.

4. On the "acid tools of critical discourse," see Donna Haraway, "Situated Knowledges: The Science Question in Feminism and the Privilege of Partial Perspective," *Feminist Studies* 14, no. 3 (Autumn 1988), 577. I discuss this aspect of Haraway's work at greater length in Heather Love, "The Temptations: Donna Haraway, Feminist Objectivity, and the Problem of Critique" in *Critique and Postcritique*, ed. Elizabeth S. Anker and Rita Felski (Durham, NC: Duke Univ. Press, 2017).

5. I am referring to a short version of this essay that Sedgwick published as the introduction to her a collection she edited, *Novel Gazing*. But a short version appeared in 1995 in the journal *Studies in the Novel*, and a later, better known, revised version in her 2003 collection *Touching Feeling: Affect, Pedagogy, Performativity*. Eve Kosofsky Sedgwick, "Paranoid Reading and Reparative Reading; or, You're So Paranoid You Probably Think This Introduction is about You," *Novel Gazing: Queer Readings in Fiction* (Durham, NC: Duke Univ. Press, 1997). Robyn Wiegman offers a helpful discussion of the publication history of this essay in Wiegman, "The Times We're In: Queer Feminist Criticism and the Reparative 'Turn,'" *Feminist Theory* 5, no. 1 (2014), 8–12.

6. Bruno Latour, *Aramis, or the Love of Technology*, trans. Catherine Porter (Cambridge, MA: Harvard Univ. Press, 1996), viii.

7. See Ian Hunter, *Rethinking the School: Subjectivity, Bureaucracy, Criticism* (St Leonards, Australia: Allen & Unwin, 1994).

8. Bruno Latour, *An Inquiry into Modes of Existence: An Anthropology of the Moderns*, trans. Catherine Porter (Cambridge, MA: Harvard Univ. Press, 2013), 247.

9. Bruno Latour, *Reassembling the Social: An Introduction to Actor-Network-Theory* (New York: Oxford Univ. Press, 2005), 126.

10. Although the concept of experiment has gained traction in recent years as a method for literary studies. See in particular Franco Moretti, *Graphs, Maps, Trees: Abstract Models for Literary History* (New York: Verso, 2005) and the work of the Stanford Literary Lab, https://litlab.stanford.edu/.

11. Bruno Latour, "Powers of the Facsimile: A Turing Test on Science and Literature," in *Intersections: Essays on Richard Powers*, ed. Stephen J. Burn and Peter Dempsey (Champaign, IL: Dalkey Archive Press, 2008), 2.

12. Although Latour does not cite Howard Becker (and there are no footnotes at all in *AIME*), Becker's *Art Worlds* exemplifies many of the qualities Latour praises in this passage. Howard S. Becker, *Art Worlds* (Berkeley, CA: Univ. of California Press, 1982).

13. Bruno Latour and Steve Woolgar, *Laboratory Life: The Construction of Scientific Facts* (Princeton, NJ: Princeton Univ. Press, 1986 [1979]), 15.

14. Cf. Gilbert Ryle, "The Thinking of Thoughts: What is 'Le Penseur' Doing?" on writing and thin description: "None the less it may still be true that the only thing that, under its thinnest description, Euclid is here and now doing is muttering to himself a few geometrical words and phrases, or scrawling on paper or in the sand a few rough and fragmentary lines. This is far, very far from being all that he is doing; but it may very well be the only thing that he is doing. A statesman signing his surname to a peace-treaty is doing much more than inscribe the seven letters of his surname, but he is not doing many or any more things. He is bringing a war to a close by inscribing the seven letters of his surname." Gilbert Ryle, "The Thinking of Thoughts: What is 'Le Penseur' Doing?" *Collected Papers*, vol. 2 (New York: Barnes and Noble, 1971), 496.

15. In a discussion of the role of the "ideal observer," Latour and Woolgar write, "observers steer a middle path between the two extreme roles of total newcomer (an unattainable ideal) and that of complete participant (who in going native is unable usefully to communicate to his community of fellow observers)" (44).

16. Margaret Mead, "Toward a Human Science" *Science* 191, no. 4230 (Mar. 5, 1976), 905–906.

17. Anna Lowenhaupt Tsing, *The Mushroom at the End of the World: On the Possibility of Life in Capitalist Ruins* (Princeton, NJ: Princeton Univ. Press, 2015), 3, 17.

CHAPTER FIVE

Redistributing Critique

ANDERS BLOK AND CASPER BRUUN JENSEN

> Critique has not been critical enough
> in spite of all its sore-scratching.
>
> Critical proximity, not critical
> distance, is what we should aim for.

Redistributing Critique with Bruno Latour?

IN RECENT YEARS, Bruno Latour has become famous across the intellectual firmament in no small measure due to his critique of critique. Even the hurried reader is likely to know that, in Latour's view, critique has recently "run out of steam."[1] Just a few pages into his magnum opus *We Have Never Been Modern*, Latour (even if confessedly "a bit unfairly") dismisses the positions of Pierre Bourdieu and Jacques Derrida, two icons of post-war French critical thinking.[2] In *Reassembling the Social*, similarly, critical sociology is memorably cast aside as a conspiratorial and "vampirical" threat to any real reckoning with shifting empirical realities.[3] It is thus uncontroversial to declare that Latour's intellectual stance runs directly counter to certain versions of critical theory, regardless of whether this is seen as a key attraction[4] or as exhibiting what is problematic, or naïve, about it.[5]

And yet, tendencies toward what we might call an "anti-critical reduction"—in which Latour's work is read as *primarily* asserting a critique of critique—have their own problems. Specifically, they risk losing track of the often quite critical-reconstructive intellectual and political projects to which his work has also been dedicated over the years.[6] These are projects in which Latour often lays claim to better-suited or in any case different forms of critical inquiry. They are meant to help "induce criticality" into all the major and contentious issues of our times, from postcolonialism to ecological disruption.[7] In the process, critique will be cut down to size but this does not mean it will disappear. Instead, it will be *redistributed* within a non-modern knowledge space.

In this chapter, we characterize this non-modern topology as the terrain of the "a-critical," and we describe Latour's intellectual stance as one of a constant, though variable, attempt to specify a new a-critical attitude. In analogy with how a- or non-modernity designates neither anti- nor post-modernity,[8] this terrain and the a-critical disposition it beckons is neither simply anti- nor un-critical. However, it is also not quite a matter of becoming post-critical. It might perhaps be said that the a-critical is located before or next to the critical, intersecting with it but not engulfed by it. The term thus points to a seriously reshuffled intellectual arena whose main parameters have ceased to be modern and whose new relations among intellectual genres, critique included, must be recalibrated.[9]

In the following, we outline the relations between Latour's a-critical terrain and the redistribution of critique in four steps. First, we examine his early encounter with Michel Serres's philosophy and distinctive form of textual interpretation. Second, we consider his defense of critical proximity as a mode of social science. Third, we turn to his political-theoretical engagement with ecological issues. Finally, we offer some thoughts on his somewhat belated (post-)postcolonial reckoning with (Euro-American) Moderns at the present time of planetary destruction. Rather than a singular anti-critical stance, we show, Latour's a-critical attitude has taken quite mixed forms and generated divergent effects. Rather than taking Latour hostage in general disputes over the status of contemporary critical theory, we suggest that the a-critical terrain he has mapped throughout his career offers a rich set of resources for the humanities as they continue to redistribute their own critical attachments.

An A-Critical Prelude: On the Interpretive Strategies of Michel Serres

To situate Latour's stance, it pays to go further back in time than his much-cited proclamations about critique's loss of steam. Of particular significance is the late 1980s and early 1990s, the time of *We Have Never Been Modern* and, also importantly, the time of Latour's most sustained engagement with the philosophy of Michel Serres.[10] Indeed, a diagnosis of the "waning capacities"[11] of modern critical theory was already integral to his analysis of the "modern constitution," as characterized by a continuous if ultimately unsuccessful effort to distinguish nature from culture, science from politics, and truth from "mere" belief. These failed separations, in turn, bear directly on critical theory.

Casting a wide net, Latour offered a typology of critiques—naturalization, socialization, and deconstruction—that he associated ("a bit unfairly," as noted) respectively with E. O. Wilson, Pierre Bourdieu, and Jacques Derrida.[12] Naturalization aims to reveal nonscientific forms of understanding as superstitions to be replaced with causal material explanations, while modern critique—along a route that leads to Bourdieu and diverse social constructivisms—"soon began to move in the other direction, turning to the newly founded social sciences in order to destroy the excesses of naturalization."[13] Finally, deconstruction and postmodernism are ridiculed as "disappointed enlightenment plus a disappointed critique coming from the social sciences."[14] Indeed, Latour depicts the latter as involuntarily scientistic since they never question the common-sense understanding of science as rational and objective, but merely reverses the conventionally positive evaluation of these traits.[15]

Critics of Latour occasionally delight in pointing out that his diagnosis of critique's waning capacities is itself critical. What looks like a performative self-contradiction in plain view, however, is addressed directly by Latour in *We Have Never Been Modern*. "To be sure," he writes, "by affirming that the [modern] Constitution, if it is to be effective, has to be unaware of what it allows, I am practicing an unveiling,"[16] and as everyone knows unveiling is part of the modern repertoire of critique. Latour's unveiling, however, "no longer bears upon the same objects as the modern critique and is no longer triggered by the same mainsprings." It therefore proceeds "without resorting to the modern type of debunking." This a-critical procedure, in which the modernist commitment to purification (of science from politics, etc.) is connected by Latour with the other half of hybridization (of nature with culture, etc.), has Michel

Serres as its most important model and source of inspiration. For Latour, one of Serres's greatest attractions was his complete disinterest "in the notion of a critique, be it transcendental or social."[17]

Commentators routinely note the difficulty of classifying Serres's writings. Equally at ease discussing mathematics, history and literature, his *oeuvre* has covered—and redrawn—an enormous terrain. In *The Birth of Physics*, he found in Lucretius's cosmological poem *De Rerum Natura* a precursor of chaos theory.[18] In *The Parasite*, he uncovered a sophisticated communication theory in the fables of Aesop and La Fontaine.[19] In neither case is the argument that these authors were out to solve the problems of contemporary science. Instead, Serres depicts a landscape of heterogeneous knowledge forms and mixed temporalities, where invention is liable to happen in unexpected places.

For instance, Serres suggested, it is precisely when Jules Verne gives himself up to his stories, and immerses himself "in areas where science officially has no place," that it is possible for him to anticipate "from afar the gesture, the thought, the system of the scientist."[20] Instead of imagining a landscape of mutual suspicion and hostility, this is an image in which knowledges are relayed between natural and social sciences, philosophy, literature and art. As such, Serres offers a stark contrast to the widespread "two cultures" trope, in which the natural sciences (and critiques based on naturalization) are pitted against humanistic interpretations (and critiques based on socialization or deconstruction) as entirely distinct and basically antagonistic forms of knowledge. Only a few years later, precisely these antagonisms would underpin the so-called Science Wars,[21] much to Latour's dismay.[22]

For firm believers in critique, Latour observes, Serres's outlook cannot but look "naïve and gullible beyond description."[23] In Latour's view, however, this "gullibility" is entirely commendable, because it signals Serres's refusal of the "distinction between beliefs on the one hand and knowledge on the other, or between ideologies and science, or between democracy and terror,"[24] distinctions central to any critical-theoretical endeavor. To further elucidate Serres's a-critical reading practice, Latour turns to literary criticism—arguably, as we suggest, the core genre of critique out of which his own alternative is gradually developed.

The premise of literary criticism, Latour states, is that textual commentary and interpretation is guided by questions defined by the critic. The critic establishes an analytical "meta-language"—centering for example on ideology, the play of différance, or the unconscious—that is used to account for the "infra-language" of the text itself. Crucially, the meta-language consists of a much smaller set of terms than the infra-language. Thus, terms like "ideology" or "the unconscious" can be used not only to explain (away) many events within a given text, they can also be used to make sense of an enormous and exceedingly diverse corpus. What is meant by "explanatory power" is precisely this capacity of replacing the many words of the infra-text with the few of the meta-language.

From a different angle, however, such explanations can also be seen as ways of dominating the text, which never gets to speak back, and therefore does not acquire the capacity to surprise the critic or exceed her expectations. Indeed, for Serres, critics act like they are spying on the text, as if they are "looking over the shoulder of someone with something to hide."[25] And since it is always possible for someone else to look over one's own shoulder, this opens a "vista of ongoing cunningness, like a succession of policemen and felons." Thus, the landscape of knowledge ends up resembling a "police state," characterized by an ethos of mutual vigilance.[26]

It will then come as no surprise that one of the main difficulties of reading Serres is that he proceeds without any meta-language. Rather than providing theoretical explanations, he attempts to allow texts to speak about other matters in their own idioms. This enables, for example, a reading of Lucretius's *De Rerum Natura* not as a quaint poem but rather as a physics of tomorrow, capable of dealing with "clouds, flows, fluxes, meteors, fluctuations, chaos, the world and its emergence."[27] Since this interpretative procedure is premised on rendering ambiguous who or what is commenting on what, the threat of domination dissolves. Instead, Serres's readings effect a "*crossover*" between registers or languages of science, literature, myth and religion, "whereby characters of one language are crossed with attributes of a different origin."[28]

As Latour later argued in a discussion of the novelist Richard Powers, the power of a-critical readings is to enable the activation of novels, fables, or myths as "resources for giving our own [empirical] descriptions the kind of grasp on reality that we can understand conceptually through the work of William James and Alfred North Whitehead."[29] In other words, Latour suggests, literary skills are needed to enact his own ideal of an "empirical philosophy," that will attend closely to variable worldly ontologies. Similar to Serres's refusal to view Lucretius as pre-scientific, Latour defines the importance of Powers's novels in terms of his refusal "to situate fiction in an easy position *in addition or in contrast* to science."[30]

As this suggests, Latour's extension of and transformation of Serres's refusal of critique, most forcefully presented in *We Have Never Been Modern*, is part of a comprehensive overhaul and redistribution of the modern landscape of knowledge practices in general. Indeed, it is because this non-modern ecology does not conform to a modern division of labor—with science/nature on one side and culture/society/politics on the other—that the role of the critic must necessarily change. Tracking the movements between supposedly incommensurable infra-languages, the new Serres-Latourian "anthropologist of science" has to be far more agile than the modern critic. She must furthermore cultivate a different kind of "enlightenment" which does not deny the import of science but refuses its role as objective arbiter, instead experimenting with language to add variety to the world and to nurture unexpected, promising hybrids.

Thus, the a-critical task becomes one of juxtaposition, relation, and the discovery of new connections. As Latour would later write, now that the old securities of the modern constitution are gone, at issue is a search for new common worlds; a problem of "composition" rather than "critique."[31] Henceforth, critics—literary and otherwise—need to pursue new paths and examine differently composed objects.

Critical Proximity as a Mode of Social Science

Against this backdrop, it comes as little surprise that neither Serres nor Latour fit into any default slot on the checkerboard of modern sociopolitical positions. Some might take offense at their lack of left-wing credentials, since one hears almost nothing about ideology, hegemony or class struggle, yet they are hard to dismiss as "reactionary."[32] And although their a-critical position might be mistaken for a-political quietism, it in fact instantiates a mode of inquiry based on the claim that the modern domains of science, society, culture and politics are all badly analyzed (and indeed badly formed) hybrids.[33]

It is not insignificant that Serres has described his work as a response to the shock of Hiroshima, which made clear that modern science harbors not only progressive but also destructive tendencies. Indeed, he depicted the post-Hiroshima situation in terms of the emergence of a *thanatocracy* consisting of a "black triad" of scientists, politicians and industrialists.[34] Yet, rather than extending a blanket critique of science, politics and capital, both Serres and Latour aim for finer discriminations. In place of a seamless web in which a homogeneous Science is spun into homogeneous Politics, Serres compared the landscape of knowledge with the Northwest Passage,[35] "a jagged shore, sprinkled with ice, and variable."[36] Here, the possibilities for making non-modern knowledge—including of politics—is less a matter of keeping a "juncture under control than an adventure to be had."[37]

Latour's most detailed exposition of the procedures required for such adventures is found in *Reassembling the Social*, an account of actor-network-theory (ANT) developed against the backdrop of sociology staged as a troubled discipline. In the present context, this book is further noteworthy because it ends up restoring to the adjective "critical" a much more affirmative sense than the one left behind by Latour in *We Have Never Been Modern*. Against the ideal of "critical distance," Latour now depicts his own version of social science as in pursuit of a new ideal of "critical proximity."[38]

The contrast between critical distance and critical proximity, in turn, parallels a number of other distinctions, which Latour makes in order to distinguish his Tarde-inspired "sociology of associations" from the established Durkheimian "sociology of the social" (of which critical sociology marks an extreme polarity). Notably, the distinction between infra- and meta-language originally borrowed from Serres's practice of literary interpretation reemerges as a contrast between "good" and "bad" forms of social analysis.[39] As a minimal infra-language, Latour suggests, the role of ANT "is simply to help them [the analysts] become attentive to the actors' own fully developed metalanguage."[40] Thus, it is meant to facilitate but neither to preordain nor to prescribe "a reflexive account of what they are saying." To finally become scientific, Latour argues, sociology must let go of its all-purpose "social forces," such as society, capitalism, and empire.[41] Becoming scientific is here seen as *premised* on achieving critical proximity to its varied objects of study.

Latour offers a number of reasons for his skepticism toward "social explanation" as usually practiced. In general, they resemble the Serres-inspired critique of standard literary criticism: by replacing the actual relations out of which the social is composed, such explanations "have of late become too cheap, too automatic; they have outlived their expiration dates—and critical explanations even more so."[42] The problem, however, is not simply epistemological but also moral, Latour suggests, as it ties into ANT's well-known skepticism towards treating the distinction between "micro" (small, local) and "macro" (big, global) as fixed and inherent. Notably, Latour here invokes the "sociology of critique" formulated by Luc Boltanski and Laurent Thévenot,[43] who are said to have shown that, "if there is one thing you cannot do in the actor's stead it is to decide where they stand on a scale going from small to big, because at every turn of their many attempts at justifying their behavior they may suddenly mobilize the whole of humanity, France, capitalism, and reason while, a minute later, they might settle for a local compromise."[44]

It is precisely in order to respect and ultimately understand the moral-critical contingencies of social life that the ANT analyst must refrain from deploying any critical-theoretical meta-language and instead stick closely to description.

Nevertheless, within the realm of infra-analysis, Latour still maintains that *some* descriptive approaches are better than others. And the best ones are those that "induce criticality" by paying close attention to non-obvious details, shifts, cracks and differences in the sociotechnical landscape that may "make a difference" to politics.[45] This is indeed the meaning Latour gives to critical proximity.[46] On this point, he thus shares an important assumption with various critical-theoretical positions: in the end, he agrees, there is no such thing as a "value-free" social science, safely removed from the contingencies of politics. Yet, infatuated by the idea of a metalanguage that ascribes to them a position of dominant explanatory power, from which flows a tendency to seek also to arbitrate disputes, critical sociologists misunderstand what it means for a social science to have political relevance.

By contrast, Latour argues that achieving political relevance is not about eliminating but about *enriching* the repertoire of agencies and mediators that must find their place in the collective. To capture this difference, Latour connects critical proximity with the idea that sociology does not study fixed and stable "matters of fact" but rather unfolding and controversial "matters of concern," a notion that has since been complemented by "matters of care."[47] Accordingly, a good, proximately critical description is one that articulates its object in richly textured and ultimately open-ended ways, thereby also making it receptive to new differences. Along these lines, Latour argues that Alan Turing's descriptions of the computer as a variegated sociotechnical matter-of-concern are more critical than analyses that center on its reifying or dehumanizing effects.[48] Similarly, Donna Haraway's redescription of dogs as significant companion species may ultimately be more politically consequential than the nth critique of neoliberalism.[49]

As these examples make clear, the notion of critical proximity as a mode of social science ultimately also serves as a bridge between social-scientific inquiry and politics as such. Indeed, it articulates a pragmatic democratic ideal according to which the social sciences are participants in, but never replacements for, what Latour elsewhere dubs "due political process."[50] As we will now see, this argument also functions as a prelude to Latour's *own* subsequent and quite explicitly normative-political engagements, most of which have been set on the terrain of ecological politics.

From Gaia (back) to Critical Zones: Ecological Politics

In May 1998, Latour gave a talk outlining a "(philosophical) platform for a left (European) party" at the invitation of the German social-democratic party (SPD).[51] In this talk, two main parameters of his re-distribution of the political landscape stand out. On the one hand, the New Left party must come to terms with the implications of a prolifer-ating range of techno-scientific controversies (over nuclear energy, gene-tic manipulation, and so on). On the other, it must learn from environ-mentalists to orient to the question of who or what is important or not among all these new entanglements of people and things. Crucially, La-tour insists, these ecological concerns have nothing to do with nature.[52] Instead, they pertain to what, following Isabelle Stengers, he calls a "cos-mopolitics," the object of which is to determine how collectives of pos-sibly antagonistic people and things can be led to peaceful co-existence via the deployment of a form of democratic due process that is ade-quately attuned to the importance of nonhuman agency.[53] Where the object of politics was once to modernize the world, Latour states, it must now be how to ecologize it.[54]

If literary criticism via Serres provided Latour with an alternative to philosophical Critique (capital C), and if critical proximity remains op-posed to critical theory (mainly of the Bourdieuan variety), then Latour's political writings are openly and affirmatively "critical." Before gaining its modernist connotations of indignation, Latour observes, "being crit-ical" had a much more object-oriented sense, pertaining to a particular state of affairs.[55] For the Greeks, as Boltanski and Thévenot[56] explain, *kritikê* was the art of passing decisive judgment, to differentiate and di-vide (*krínein*) during critically important but indeterminate situations. And, indeed, this is central to Latour's much discussed matters of concern.

Despite the fact that detractors have read Latour's exclamation, "why does it burn my tongue to say that global warming is a fact whether you like it or not?"[57] as evidence of (prior or complicit) climate denial, La-tour's argument thus never questions either the reality or severity of current-day ecological disruptions. Quite differently, the real question raised by his controversial quote is how to accept global warming with-out returning to a naïve idea of objective facts.[58] Subsequently, Latour has argued that the historian of climate change Paul Edwards has decid-edly accounted for the robustness of climate science, not in consequence of its practitioners' access to unmediated facts but rather due to the gradual extension of enormous infrastructures and the convergence of "more and more powerful models."[59] Contrasting with how climate deniers claim to access "reality" directly, Latour thus asserts that, "This tapestry [of climate science] is amazingly resilient because of the way it is woven—allowing data to be recalibrated by models and vice versa. It appears that the history of the anthropocene (climate sciences are by definition a set of historical disciplines) is the best documented event we have ever had."[60] To Latour, then, the reality of the new climatic regime is never in doubt.[61]

What his eco-political pursuit aims at, rather, is the far more troublesome possibility that the ruins of the modern constitution and its accompanying critical spirit leaves us unequipped to deal with a situation in which humans actually do *not* master our own creations, climate change and the Anthropocene included.[62] The previously mentioned types of naturalizing, socializing or deconstructive critique have made it impossible to achieve critical proximity with the earthly gatherings of people and things that have created the trouble with which we are confronted today. What is so deeply worrying about climate change, Latour asserts, is in fact that we remain too *indifferent*, insufficiently shaken up.[63] Correspondingly, what is needed are new ways of assembling and gathering many incongruent forces in order to create a sufficient disturbance to render our eco-political situation properly critical. This is the true message behind the notion that critique has run out of steam.

In recent years, Latour has increasingly exerted himself as a public intellectual trying to compose such new eco-political gatherings. Most visibly, he has joined Isabelle Stengers to reinvent an "intellectually serious" yet "extra-scientific" figure of Gaia.[64] While Gaia, in James Lovelock's version, was a scientific concept for a nonstandard Earth science of feedback loops, in the hands of Latour and Stengers it also designates a ticklish, easily irritable entity akin to the vengeful Greek goddess. The point, however, is not an unlikely unification of the scientific and the mythical. Rather, Gaia's constitutive hybridity is meant as a dramatization of our current ecological predicament as demanding an intellectual and public-political coming to terms with a massive "multiplication of agencies,"[65] an animation of the world which it is futile to criticize but important to learn to become sensitive to.

In this context, moreover, it is not coincidental to find Latour preoccupied with the public-political capacities of the humanities. Because they have always dealt with "a bewildering number of *figurations*, only some of which have looked like realistic portraits—or rather clichés—of human subjects,"[66] Latour argues that the humanities are better equipped to describe and articulate the many needed animations of the new climatic regime than the social sciences. Attention to figurations and textuality, then, is as important, as critical, as ever. What has changed, however, is the public-political aim and direction of such textual-intellectual capacities. Humanists, Latour suggests, may come to act as important chiefs of protocol for the many diplomatic encounters needed to prevent conflicting collectives to descend into "runaway violence" in the face of Gaia.

On top of this new diplomatic role, the skills of humanists are also needed to render Gaia practically legible and visible, in terms of what Latour has tellingly come to call critical zones.[67] Taking the term from the ecological sciences, Latour uses critical zones to describe the way in which the very ground(s) of life—the territory, the soil, the organisms, the biosphere—has entered a state of flux, from which it awaits rediscovery and re-composition. As such, the critical zone literally "engages all its inhabitants in a narrative of history, crisis, conflicts and transformations," very different from the time when it was still possible to talk "proudly of having one's feet firmly "on the soil."[68] Because these zones are about co-dependent and fragile worlds, their criticality is interobjective,[69] pertaining to planet-wide states of potentially disastrous affairs.[70] As ecological disruptions render more and more sociocultural situations critical in this sense, sociocultural scholarship must learn to participate in new cosmopolitics, a search for a new common ground and a new "common" world.[71]

As these formulations suggest, Latour's political reasoning renders the question of "the common" newly problematic. Doing so, they enter the terrain that has most richly been explored in postcolonial debates. In the final section, we explore their implications for the redistribution of critique.

Postcolonial Reckonings

As Latour has become increasingly articulate about the planetary politics, ecological and otherwise, of our times, postcolonial studies—with their attention to questions of historical inequities, exploitations and subjugations—become a crucial test for his a-critical attitude. And while Latour has been on the receiving end of much criticism for his failure to engage with postcolonial concerns, his recent work does in fact engage with such issues in its own way, and in apparent response to at least some of the critiques.

It can be argued—and Latour has come close to doing so—that anthropological and historical studies of modern science and its naturalism contributes to the provincialization of Europe famously called for by the historian Dipesh Chakrabarty.[72] Indeed, Chakrabarty's observation that concepts—including "the state," "public sphere," "the idea of the subject," "scientific rationality," and so forth—"all bear the burden of European thought," making them inseparable from the notion of "political modernity,"[73] aligns with Latour's effort to seriously recast the idea of modernity by reconceiving just such notions.[74] Moreover, Latour's recent work on Gaia and on modern modes of existence extends this move, since it seeks to provincialize notions such as the global and universal that lie at the heart of colonial aspirations.[75]

This, however, is not how Latour's critics appraise the situation. From one angle, the anthropologist Kim Fortun attacks *Modes of Existence* for remaining aloof from the real problems of late industrialism.[76] While offering abstract discussions of how psychic metamorphosis intersects with technology, or habit with fiction, Fortun observes, it contains practically no reference to pressing concerns such as disasters, toxic waste, or the many other destructive effects of global capitalism to which postcolonial studies alert us.

Meanwhile, in a postcolonial extension of the standard critiques directed at Latour's supposed power blindness, he is taken to task for denying the coevalness of the other.[77] This is exemplified by the indigenous feminist anthropologist Zoe Todd who details her disappointment in listening to Latour speak of climate change as a common cosmopolitical concern without once commenting on indigenous ways of conceiving the issue.[78] For Todd, Latour thus offers an illustration of how the "trendy" discourses circulating in "Euro-Western thought" systematically devalues other forms of knowledge.[79] While that claim is debatable at best, her pointed question is relevant: If food for thought, according to Latour, can—and should—be drawn from anywhere, why is his own repertoire almost invariably European? What kind of diplomacy is this?

In an interview with John Tresch, Latour offers one answer: "Because the Moderns have been so busy expanding, they never met themselves. They never had the chance to figure out what they were up to."[80] This basically reiterates the central idea from *We Have Never Been Modern*: that the moderns were able to create and proliferate hybrids at an accelerating pace *because* of their inability to see that this is indeed what they did. Although the claim that the moderns never had the chance to figure out what they were up to must obviously be taken with a pinch of salt (in general, all of them?), it does go some way to situate Latour's recent conceptual orientation towards the many modes of existence of the Moderns. Locating alternatives *within* modernity, Latour suggests, will create "more space to breathe . . . but also to enter into connection with the others, who are prisoners of modernization's limits."[81]

As this suggests, Latour would readily acknowledge that the Inuit Sila or other entities crucial to indigenous cosmologies are missing from his own current inquiries. They are missing precisely because the modes of existence with which he is dealing are specifically modern ones. Presumably, however, his defense, not least the idea that indigenous people are but prisoners of modernization's limits, would not satisfy Todd and other critics. Indeed, Martin Savransky has argued that Latour's cosmopolitics is seriously limited by its "facile dismissal of criticism" from postcolonial and feminist theorists.[82] Latour's aspiration to construct a common world, argues Savransky, unavoidably leads him to subsume a multiplicity of voices "into his own allegoric project."[83] Since admission into the common world is defined in terms of the democratic procedures of Latour's own "parliament of things," the common world eventually turns out to be a "normative political project" functioning on the principle of "no opposition, only integration."[84]

Such critiques exhibit a curious reversal, since Latour's earlier work was most commonly criticized for being overly antagonistic, if not positively "Machiavellian."[85] Indeed, some of his cosmopolitical writings, such as *War of the Worlds*,[86] are so acutely attentive to conflict and controversy that they center on a state of ontological warfare. Even so, more friendly critics, too, have argued that Latour's pragmatist political-theoretical bearings may require certain extensions and reconfigurations.[87]

Specifically, postcolonial encounters may create *particular* challenges to Latour's composition of the common world—and thus to his own maneuvers within the critical zones of current-day eco-politics. Arguably the best example is found in the work of the anthropologist Marisol de la Cadena, which explores the implications of the presence of earth-beings for modern politics in Peru.[88] What for the moderns of Peru are mountains, like Ausangate, de la Cadena explains, are *Tirakuna*, earth-beings, for the *Runakuna* inhabitants of the Cuzco area. Because mountains and earth-beings are radically different kinds of entities, and since the latter is generally seen as nothing but an "infantile"[89] belief by modern Peruvians, this is a scene rich in potential for what de la Cadena dubs "indigenous cosmopolitics."

The entry of earth-beings as political entities thus signals a plurality that simultaneously goes beyond the concerns with coloniality, gender and race that motivate Fortun and Todd's critiques of Latour *and* Latour's common world constructed by building blocks extracted from the dregs of modernity and reorganized in the parliament of things. While de la Cadena is also after ontological pluralization, she is pursuing it from a position of dampened faith in any form of progressive composition of commonality, since indigenous ontologies tend to sooner or later run up against the rigid limits of modern politics, at which point they will once again be relegated to the realm of superstition.[90]

Interestingly, in a review of the anthropologist Eduardo Kohn's *How Forests Think,* Latour is quite articulate about the need to remain on the lookout for other voices, pointedly remarking that the author's Peircean ontology has after all "not become for everyone else the definition of their common world."[91] Noting that Kohn's biosemiotic monism precludes the possibility of the Runa engaging in diplomatic encounters with others—forest engineers, the agronomists, or other "whites"[92]—Latour thus offers practically the same objection that others direct at his own work. However, while Todd and others argue that Latour fails to take into account non-modern people, Latour reverses the argument by insisting that Kohn fails to take into account modern ones.

Latour, in other words, stridently maintains his basic focus on the moderns.

And, indeed, he insists that rethinking specifically European modes of existence is of particular importance. In an article directed at an audience of international relations scholars, he thus refers to Peter Sloterdijk's argument that the defining feature of modern Europe has been its "mechanism to transfer the Empire" across generations.[93] Even if Europe has by now been "thoroughly 'provincialized'" by Chakrabarty and others, it is in nobody's best interest, Latour suggests, that Europe retreats from its previous hubristic aspirations only to hide in a corner and "nurse its wounds."[94] Instead, it has now become the specifically European burden "to think of an alternative to the principle of sovereignty that it has imposed everywhere."[95]

As noted, this requires one more push at provincialization, aimed this time at dethroning the very idea of "the Globe" through which imperial dominion was imagined in the first place. The problem today, Latour argues, is that the Globe, as a container for a geo-politics dominated by nation states, is continually conflated with the Earth, thereby obscuring the centrality of ecological disruptions. By highlighting the multiple ways in which "collectives *collect* the entities with which they enter into relations,"[96] the notion of "multi-naturalism" will make it possible to differentiate the two and to relativize "the global." Multi-naturalism, in Latour's version, thus facilitates a display of ontological variability *across* the world's peoples or collectives.

Here it is not coincidental that the notion of multi-naturalism was originally developed by the anthropologists Eduardo Viveiros de Castro to characterize the "endogenous" ontological distinctiveness of

Amerindian peoples.[97] Whereas Western naturalist ontology distinguishes between nature as a given "out there" and culture as what makes it possible to have different perspectives on nature, among Amerindians in the Amazon the situation is radically different. According to this ontology, everyone was originally human, but animals and spirits gradually diverged in bodily forms and dispositions, even while maintaining knowledge of themselves as human. Thus, humans, pigs and the spirits of the dead all see themselves *as* human, but because their bodies are different and capable of being affected in different ways, the worlds they see and inhabit are different from each other. What humans see as blood, jaguars see as manioc beer, for example, and when tapirs enter a large ceremonial house, humans see them as being in a pool of mud. For Amerindian ontologies then, culture is one and given (it is always human), while "natures" are many and variable.

It is not difficult to see why the multi-naturalist model appeals to Latour. Yet, when he defines it as a general cosmopolitical principle with which to wage war on the Globe, a tension analogous to the one between the modern Peruvian state and Tirakuna earth-beings reappears. What de la Cadena and Viveiros de Castro describe in terms of ontological incommensurability, Latour redescribes within a framework of the progressive composition of a common world—however many conflicts and diplomatic encounters such a composition may entail.

Isabelle Stengers, Latour's long-time discussion partner, has addressed this tension in an illuminating reflection on the relation between her own understanding of cosmopolitics and de la Cadena's adaptation.[98] While de la Cadena uses the term to designate a situation of ontological conflict between modernity and non-modernity, Stengers notes, her own original problem (along with Latour's) had been to *disentangle* modern sciences from their internal ontology-destroying tendencies. At issue, she suggests, are thus different conceptions of the political situation. While Stengers's speculative version of politics affirms the possibility of diplomacy with some modern practitioners, de la Cadena's ontological politics entails a general struggle against a destructive Western machine with which no peace is possible.[99] At issue, Stengers thus indicates, is another form of the postcolonial question about what counts as "progress," and what as "regression," in globally entangled encounters on a critically endangered planet.

On this point, Latour's (quasi-)exchange with Déborah Danowski and Eduardo Viveiros de Castro on the "criticality" or otherwise of indigenous experiences is telling.[100] Writing from their position as observers of Amerindian multi-naturalists, whose worlds have been progressively destroyed rather than composed for centuries, Danowski and Viveiros de Castro suggest that the smaller populations and weaker technologies of indigenous peoples and other sociopolitical minorities could become "a crucial advantage and resource in a post-catastrophic time" on an increasingly wrecked, crisis-stricken planet.[101] For his part, Latour signals skepticism about whether the wisdom of traditional peoples can be scaled up to the level of today's "giant technical metropolises."[102] But, Danowski and Viveiros de Castro counter, Latour fails to entertain the idea that it might be the Moderns who will have to "scale down" their ways of living.[103] Along these lines, a redistributed critique will have to concern itself with a redistribution of worlds, since there are presently "too few people with too much world and too many with too little."[104]

While these exchanges point to real differences in positionality (Europe, Latin America) and ethical-political orientation (around global inequities), the shared intellectual coordinates within which they take place are also noteworthy. What both sides take for granted is the demands nowadays placed on any progressive intellectual project by the twin conjuncture of postcolonial legacies and planetary ecological disruption. Importantly, they also take for granted what we have dubbed the space of the a-critical: today attention to glaring inequities and subjugations happen on the same plane of inquiry as attention to emerging alternatives and resistances, and without any recourse to a final arbiter or metalanguage. What is on display, in this sense, are varieties of a-critical attitudes, differentially attuned to the criticality of the diverse situations in which they find themselves. In this sense, the exchange serves as an interesting model for how the humanities might seek to sort out its own critical attachments.

Toward A-Critical Humanities?

In this chapter, we have retraced the steps of Bruno Latour's (a-)critical and reconstructive intellectual projects. Contrary to currently circulating interpretations, we have shown that they are not adequately understood in terms of a general "anti-critical reduction." Latour has encouraged a nonhierarchical version of literary criticism; aimed to attune the social sciences to critical proximities with shifting realities; staged the criticality of current eco-political predicaments; and engaged in diplomatic encounters across postcolonial terrains. None of these efforts suggest satisfaction with, or complacency towards, the state of the world today, neither intellectually nor practically. They are, however, indicative of a serious attempt to redistribute the critical energies of multiple kinds of inquiry, and to reorient the critical attachments and capacities of the contemporary humanities and social sciences. Placing our account amidst other emerging conversations around the import of Latour's work (as in this volume), we hope to contribute to the conversation about what it means to demodernize the humanities.[105]

For this project to have any purchase, we argue, it is crucial to distinguish the term a-critical from the un- or anti-critical, as well as from the post-critical. For Latour, after all, the non-modern terrain of the a-critical always came before and, at the very least, kept running in subjugated parallel to the critical spirit(s) of "modernity" that has recently, in his view, "run out of steam" to much fanfare. His "Copernican counter-revolution"[106] in thinking, which seeks to dismantle the modernist critical-theoretical apparatus, instead aims to unearth a wider-reaching sense of intellectual-political crossovers, indignations and future aspirations. Such crossovers, in turn, license an a-critical attitude that neither generally condemns, nor, to be sure, condones, the passing of situated judgments on important matters of collective concern. Instead it insists on the importance of "launching the arrow"[107] of the critically inquiring spirit in new directions, towards new aims and objectives.

Rather than aiming for the removal of all critical impetus, then, demodernizing the humanities is about finding ways of achieving sufficient collective criticality around the many issues, including viable ways of inheriting the past, on which the future of the planet hinges. It is, in short, about redistributing critique. From this perspective, planetary ecological crises and postcolonial legacies both stake out broad a-critical areas of knowledge and action, in which the varied skills of the humanities—environmental and otherwise—are only likely to gain in importance. In short, remapping the knowledge space along non-modern Latourian lines makes it possible to see many of the skills of the humanities as, if anything, *more* critical (as in *decisive*) than ever before.

It is, of course, also hard to imagine that Latour's own intellectual projects would have ever succeeded without such skills. His significant debt to Serres the philosopher should be described here as only the visible tip of an iceberg. Further down, we encounter as variable figures as the semiotician Algirdas Greimas, the theologian Rudolf Bultmann, and a whole host of critical-enough philosophers, Friedrich Nietzsche included.[108] Yet, it is equally hard to imagine Latour's moves without inspirations gleaned from Louis Pasteur, Alan Turing and, more recently, James Lovelock and a host of critical zone-ecologists. At stake is thus also the redistribution of a whole set of assumed tensions and hierarchies among the (so-called) "two cultures" of the sciences and the humanities. Latour's entire project, we might say, seeks to bring these internally variegated legacies into closer, more co-equal, and more demanding conversation for the future. Conversations that, as Serres would have it, run in all directions at once, beyond any scientism.

In our view, therefore, Latour's work deserves better than to be taken hostage in a generalized dispute over the pros and cons of critical theory. To paraphrase his own take on the Science Wars:[109] ultimately, we are not at war—we just have to collectively work out how to increase our chances of flourishing on a dangerously disrupted earth.

Notes

Epigraphs. Bruno Latour, "Why Has Critique Run out of Steam? From Matters of Fact to Matters of Concern," *Critical Inquiry* 30, no. 2 (2004): 232; Bruno Latour, *Reassembling the Social: An Introduction to Actor-Network Theory* (Oxford: Oxford Univ. Press, 2005), 253.

1. Latour, "Why has Critique Run out of Steam?"

2. Bruno Latour, *We Have Never Been Modern* (New York: Harvester-Wheatsheaf, 1993), 5.

3. Latour, *Reassembling,* 50.

4. Rita Felski, *The Limits of Critique* (Chicago, IL & London: Univ. of Chicago Press, 2015).

5. Benjamin Noys, *The Persistence of the Negative: A Critique of Contemporary Continental Theory* (Edinburgh: Edinburgh Univ. Press, 2010), 80–106.

6. By "anti-critical reduction," we are seeking to capture a broad tendency with many degrees and nuances. An example is the introduction to Jeffrey R. Di Leo, ed., *Criticism After Critique* (2014: 2ff), in which Latour's critique of critique features prominently, yet turns out to play practically no role in the subsequent chapters.

7. Bruno Latour, "Critical Distance or Critical Proximity?" unpublished paper (2003), 8, available at http://www.bruno-latour.fr/sites/default/files/P-113-HARAWAY.pdf.

8. Latour, *We Have Never Been Modern.*

9. On the notion of a nonmodern knowledge space and its novel relations and translations, see further Anders Blok, "The Anthropologist, the Moralist, and the Diplomat: Bruno Latour's Conceptual Figures and the Contemporary *Ethos* of Knowledges" (forthcoming in *Common Knowledge*) and Casper Bruun Jensen, "Disciplinary Translations: Remarks on Latour in Literary Studies and Anthropology" (forthcoming in *Common Knowledge*).

10. Bruno Latour, "The Enlightenment without the Critique: A Word on Michel Serres' Philosophy," in *Contemporary French Philosophy,* ed. A. Phillips Griffiths (Cambridge: Cambridge Univ. Press, 1987), 83–99.

11. Latour, *We Have Never Been Modern,* 35.

12. Latour, *We Have Never Been Modern,* 5.

13. Latour, *We Have Never Been Modern,* 35.

14. Hugh T. Crawford, "An Interview with Bruno Latour" *Configurations* 1, no. 1 (1993): 258.

15. Crawford, "An Interview with Bruno Latour," 254.

16. Latour, *We Have Never Been Modern.*

17. Bruno Latour, *The Pasteurization of France* (Cambridge, MA: Harvard Univ. Press, 1988), 258, n29.

18. Michel Serres, *The Birth of Physics* (New York: Clinamen Press).

19. Michel Serres, *The Parasite* (Baltimore, MD, & London: The Johns Hopkins Univ. Press, 1982).

20. Michel Serres, "Literature and the Exact Sciences," *SubStance* 18, no. 2 (1989): 7.

21. See Barbara Herrnstein Smith, *Belief and Resistance: Dynamics of Contemporary Intellectual Controversy* (Cambridge: Harvard Univ. Press, 1997).

22. See Bruno Latour, *Pandora's Hope: Essays on the Reality of Science Studies* (Cambridge, MA: Harvard Univ. Press, 1999). Although the non-modern

Serres-Latour approach is distinct, its general contours are not unique. The influential historian and literary critic Mary Poovey, for example, described the modern fact as an "epistemological unit" that enabled the elucidation of "connections between knowledge projects as different as rhetoric, natural philosophy, moral philosophy, and early versions of the modern social sciences." Mary Poovey, *A History of the Modern Fact: Problems of Knowledge in the Sciences of Wealth and Society* (Chicago, IL, & London: Univ. of Chicago Press, 1998): xiv–xv. A rich literature spanning science and technology studies, anthropology, the history of science and technology, and related areas pursues similar interests.

23. Latour, "Enlightenment Without the Critique," 83.

24. Latour, "Enlightenment Without the Critique," 85.

25. Michel Serres and Bruno Latour, *Conversations on Science, Culture, and Time* (Ann Arbor, MI: Univ. of Michigan Press, 1995), 133.

26. Serres's depiction interestingly resembles the "hermeneutics of suspicion" identified and criticized in Paul Ricoeur, *Freud and Philosophy: An Essay on Interpretation* (New Haven, CN: Yale Univ. Press, 1977).

27. Latour, "Enlightenment Without the Critique," 87.

28. Latour, "Enlightenment Without the Critique," 90–91 (italics in original).

29. Bruno Latour, "Powers of the Facsimile: A Turing Test on Science and Literature," in *Intersection: Essays on Richard Powers*, ed. A Stephen J. Burns and Peter Demsey (Champaign, IL: Dalkey Archive Press, 2008), 264.

30. Latour, "Powers of the Facsimile," 273.

31. Bruno Latour, "An Attempt at a 'Compositionist Manifesto,'" *New Literary History* 41, 471–490.

32. See Graham Harman, *Bruno Latour: Reassembling the Political* (London: Pluto Press, 2014).

33. See Latour, *Reassembling the Social*.

34. Latour, "Enlightenment Without the Critique," 92.

35. The Northwest Passage is a sea route that connects the Atlantic and Pacific Oceans through the Canadian Archipelago. Historically very dangerous to navigate due to icebergs, quite apropos of Latour's concerns with global climate change, as the ice melts the Northwest Passage is becoming progressively more attractive as a trade route.

36. Serres and Latour, *Conversations*, 70.

37. This point resonates with what has been described as "epistemologies of surprise" by Jane Guyer, "Epistemologies of Surprise in Anthropology: The Munro Lecture," *Hau: Journal of Ethnographic Theory* 3, no. 3 (2013): 283–307, and with the call for a new "ethos of adventure" for the sociocultural sciences found in Martin Savransky, *The Adventure of Relevance: An Ethics of Social Inquiry* (London: Palgrave MacMillan, 2016).

38. Latour, *Reassembling the Social*, 252ff.

39. Latour, *Reassembling the Social*, 221.

40. Latour, *Reassembling the Social*, 49.

41. Latour, *Reassembling the Social*, 137.

42. Latour, *Reassembling the Social*, 221.

43. See Luc Boltanski and Laurent Thévenot, *On Justification: Economies of Worth* (Princeton, NJ: Princeton Univ. Press, 2006).

44. Latour, *Reassembling the Social*, 184.

45. Latour, "Critical Distance or Critical Proximity?" 8.

46. Latour, *Reassembling the Social*, 253.

47. Maria Puig de la Bellacasa, "Matters of Care in Technoscience: Assembling Neglected Things," *Social Studies of Science* 41, no. 1 (2011): 85–106.

48. Latour, "Powers of the Facsimile."

49. See Donna Haraway, *When Species Meet* (Minneapolis, MN & London: Univ. of Minnesota Press, 2008).

50. In Bruno Latour, *Politics of Nature: How to Bring the Sciences into Democracy* (Cambridge, MA: Harvard Univ. Press, 2004). Latour adopts the term from John Dewey, *The Public and Its Problems* (New York: Holt Publishers, 1927).

51. Bruno Latour, "Ein ding ist ein thing: A (Philosophical) Platform for a Left (European) Party," *Soundings* no. 12: 12–25.

52. Latour, *Politics of Nature.*

53. Latour borrows the notion of cosmopolitics, along with a range of substantive ideas about its implications, from his long-time discussion partner Isabelle Stengers, e.g., Isabelle Stengers, *Cosmopolitics I* (Minneapolis: Univ. of Minnesota Press, 2010).

54. Bruno Latour, "To Modernize or to Ecologize? That's the Question," in *Remaking Reality: Nature at the Millenium*, ed. N. Castree and B. Willems-Braun (London and New York: Routledge, 1998), 221–42.

55. Latour, "Critical Distance or Critical Proximity?" 8.

56. Luc Boltanski and Laurent Thévenot, "The Reality of Moral Expectations: A Sociology of Situated Judgement," *Philosophical Explorations* no. 3 (2000): 208–31.

57. Latour, "Why has Critique Run Out of Steam?" 227.

58. This quote alone would probably put Latour among the most-cited intellectuals of our time, even if many citations come from outraged critics. For an extreme case, see Lee McIntyre, *Respecting Truth: Willful Ignorance in the Internet Age* (New York and London: Routledge, 2015), 109, which portrays Latour's argument as "the feeble bargaining of a thoughtless bully." This and related characterizations speak volumes about the ideological passions—indeed, the "criticality"—which Latour's reworking of science-society relations remains capable of generating even after decades of science studies.

59. Bruno Latour, "Waiting for Gaia," lecture at the French Institute, London (November 2011), 6, available at http://www.bruno-latour.fr/sites /default/files/124-GAIA-LONDON-SPEAP_0.pdf. Latour's reference is to Paul Edwards, *A Vast Machine: Computer Models, Climate Data, and the Politics of Global Warming* (Cambridge, MA & London: MIT Press, 2010).

60. Latour, "Waiting for Gaia," 6.

61. Bruno Latour, *Facing Gaia: Eight Lectures on the New Climatic Regime* (Cambridge: Polity Press, 2017).

62. Bruno Latour, "A Plea for Earthly Sciences," in *New Social Connections: Sociology's Subjects and Objects*, ed. Judith Burnett, Syd Jeffers. and Graham Thomas (London: Palgrave Macmillan, 2010), 72–84.

63. Latour, "Waiting for Gaia."

64. See Isabelle Stengers, *In Catastrophic Times* (Open Humanities Press, 2015), and for the quote, Bruce Clark, "Rethinking Gaia: Stengers, Latour, Margulis," *Theory, Culture & Society* 34, no. 4 (2017), 4.

65. Latour, *Facing Gaia*, 70.

66. Bruno Latour, "Life among Conceptual Characters," *New Literary History* 47, no. 2 & 3 (2016), 474.

67. Bruno Latour, "Some Advantages of the Notion of 'Critical Zone' for Geopolitics," *Procedia: Earth and Planetary Science* (2014), as well as Latour, *Facing Gaia*.

68. Latour "Some Advantages of the Notion of 'Critical Zone,'" 2.

69. Bruno Latour, "On Interobjectivity," *Mind, Culture, and Activity* 3, no. 4 (1996): 228–45.

70. Latour, "Life among Conceptual Characters," 471.

71. Bruno Latour, "Turning around Politics," *Social Studies of Science* 37, no.5 (2007): 811–20.

72. Bruno Latour, "Onus Orbis Terrarum: About a Possible Shift in the Definition of Sovereignty," *Millennium: Journal of International Studies* 44, no. 3 (2016): 309. The reference is to Dipesh Chakrabarty, *Provincializing Europe: Postcolonial Thought and Historical Difference* (Princeton, NJ: Princeton Univ. Press, 2000).

73. Chakrabarty, *Provincializing Europe*, 4.

74. Put forth respectively in Latour, *We Have Never Been Modern* and Latour 2016, "Onus Orbis Terrarum."

75. See Bruno Latour, *An Inquiry into Modes of Existence: An Anthropology of the Moderns* (Cambridge, MS, & London: Harvard Univ. Press, 2013).

76. See Kim Fortun, "From Latour to Late Industrialism," *Hau: Journal of Ethnographic Theory* 4, no. 1 (2014): 309–29.

77. These critiques usually ignore Latour's long-standing *reconceptualization* of power. See Bruno Latour, "The Powers of Association," in *Power, Action and Belief*, ed. John Law (London: Routledge and Kegan Paul, 1986), 264–280. The critique of anthropology for deleting "co-evalness" was originally developed in Johannes Fabian, *Time and the Other: How Anthropology Makes Its Object* (New York: Columbia Univ. Press, 1983).

78. Zoe Todd, "An Indigenous Feminist's Take on the Ontological Turn: 'Ontology' Is Just Another Word for Colonialism," *Journal of Historical Sociology* 29, no. 1 (2016): 4–22.

79. Todd, "An Indigenous Feminist's Take," 7–8.

80. John Tresch, "Another Turn after Ant: An Interview with Bruno Latour," *Social Studies of Science* 43, no. 2 (2013): 302–13.

81. Tresch, "Another Turn after Ant," 312.

82. Martin Savransky, "Worlds in the Making: Social Sciences and the Ontopolitics of Knowledge," *Postcolonial Studies* 15, no. 3 (2012): 364.

83. Savransky, "Worlds in the Making," 364.

84. Savransky, Worlds in the Making," 364, with reference to Latour, *Politics of Nature*. For further discussion see Casper Bruun Jensen, "Experimenting with Political Ecology-Review Essay: Latour, Bruno, 2004, *Politics of Nature: How to Bring the Sciences into Democracy* (Cambridge, MA, & London: Harvard Univ. Press)" *Human Studies* 29, no. 1 (2006): 107–22, and Matthew Watson, "Derrida, Stengers, Latour and Subaltern Cosmopolitics," *Theory, Culture and Society* 31, no. 1 (2014): 75–98.

85. See Olga Amsterdamska, "'Surely, You Are Joking, Monsieur Latour!'" *Science, Technology and Human Values* 15, no. 4 (1990): 495–504, also Harman, *Bruno Latour*.

86. Bruno Latour, *War of the Worlds: What About Peace?* (Chicago, IL: Prickly Paradigm Press, 2002).

87. See Anders Blok, "War of the Whales: Post-sovereign Science and Agonistic Cosmopolitics in Japanese-Global Whaling Assemblages," *Science, Technology, & Human Values*, 36, no. 1 (2010): 55–81; Noortje Marres, "The Issues Deserve More Credit," *Social Studies of Science*, 37, no.5 (2007): 759–89; but also Latour, "Turning around Politics."

88. Marisol de la Cadena, "Indigeneous Cosmopolitics in the Andes: Conceptual Reflections Beyond 'Politics,'" *Cultural Anthropology* 25, no. 4 (2010): 334–70; Marisol de la Cadena, *Earthbeings: Ecologies of Practice across Andean Worlds* (Durham, NC, & London: Duke Univ. Press, 2015).

89. De la Cadena, "Indigenous Cosmopolitics," 336.

90. It should be noted that ANT-inspired scholars have in fact sought to open up ANT to non-Western kinds of otherness. See e.g., Casper Bruun Jensen and Anders Blok, "Techno-animism in Japan: Shinto Cosmograms, Actor-Network Theory and the Enabling Powers of Non-Human Agencies," *Theory, Culture and Society* 30, no. 2 (2013): 84–115; John Law and Wen-yuan Lin, "Provincializing STS: Positionality, Symmetry and Method," *East Asian Science, Technology and Society* 11, no. 2 (2017): 211–27.

91. Bruno Latour, "On Selves, Forms and Forces. Comment on Kohn, Eduardo. 2013. *How Forests Think: Toward an Anthropology Beyond the Human*. Berkeley: Univ. of California Press," *Hau: Journal of Ethnographic Theory* 4, no. 2 (2014): 264.

92. Latour, "On Selves, Forms and Forces," 264.

93. Latour, "Onus Orbis Terrarum," 306, with reference to Peter Sloterdijk, *Si l'Europe s'éveille* (Paris: Mille et une nuits, 2003).

94. Latour, "Onus Orbis Terrarum," 306.

95. Latour, "Onus Orbis Terrarum," 306.

96. Latour, "Onus Orbis Terrarum," 309.

97. See Eduardo Viveiros de Castro, "Cosmological Deixis and Amerindian Perspectivism," *The Journal of the Royal Anthropological Institute* 4, no. 3 (1998): 469–88.

98. Isabelle Stengers, "The Challenge of Ontological Politics," in *A World of Many Worlds,* eds., Marisol de la Cadena and Mario Blaser (Durham, NC, & London: Duke Univ. Press, 2018), 83–112.

99. Stengers, "The Challenge of Ontological Politics," 86.

100. Déborah Danowski and Eduardo Viveiros de Castro. *The Ends of the World* (Cambridge & Malden, MA: Polity Press, 2017) engages at length with Latour, *Facing Gaia* and Bruno Latour, "Antropólogo Francês Bruno Latour Fala sobre Natureza e Política." Interview with Fernando Eichenberg. O Globo, December 28, 2013. Available at http://blogs.oglobo.globo.com/prosa/post /antropologo-frances-bruno-latour-fala-sobre-natureza-politica-519316.html.

101. Danowski and Viveiros de Castro, *The Ends of the World*, 95.

102. Bruno Latour, "Facing Gaia: The Gifford Lectures," unpublished manuscript, 2013, 128. Interestingly, this quote seems to have disappeared from the subsequently published version, i.e., Latour, *Facing Gaia.*

103. Danowski and Viveiros de Castro, *The Ends of the World*, 96.

104. Danowski and Viveiros de Castro, *The Ends of the World*, 97.

105. By Graham Harman, "Demodernizing the Humanities with Latour," *New Literary History* 47, no. 2 & 3 (2016): 249–74.

106. Latour, *We Have Never Been Modern*, 89.

107. To adopt a phrase from Savransky, *The Adventure of Relevance*, 5.

108. See Latour, "Life Among Conceptual Characters."

109. Latour, *Pandora's Hope*, 23.

CHAPTER SIX

Decomposing the Humanities

STEVEN CONNOR

ONE OF THE THINGS that anyone involved in the murky business of humanities scholarship comes to know, without knowing how, or perhaps without even knowing that they know it, is the difference between criticism and critique. It was necessary to internalize this understanding because it might be said that the decisive and defining shift that took place with the reorganization of the humanities from the 1980s onward around the various forms of theory was the shift from criticism to critique. Terry Eagleton once gently mocked a certain kind of literary criticism as the practice of inscribing "could do better" in the margins of literary texts, but that mode of criticism already seemed mistily antiquarian when I entered my own life of intellectual crime under Terry's witty tutelage at Wadham College in 1973. The thing known as criticism, or at least a certain ideal conception of it, was governed by the Arnoldian injunction to try to see the object as in itself it really is. Criticism required one to inhabit the text, to learn to see things in its own terms. There was a time, let us remind ourselves with amazement, when "criticism" might also have been known as "critical appreciation." The alumni who studied English in Cambridge in the 1950s and 1960s recognize clearly enough in the pained letters they occasionally write to me that literary appreciation, now thought of as credulous and soft-headed weakness, has given way to analysis in the service of deprecation.

Critique, by contrast, was characterized by a kind of deliberated recoil from the voluptuous temptations represented by texts, the cultivation of an alert, no-flies-on-me vigilance about the acts of reading and interpretation. Critique was underpinned by theory in a way that criticism never could be, because critique always both required and, more importantly in terms of thymotic satisfactions, reliably delivered the sense that the text had been outwitted. Critique allowed one to come at one's object of analysis from some higher ground, or rather perhaps from some cunning subterranean passage, which enabled one to tunnel behind or underneath its presumptions, articulate its silences, to see it, in short, not as in itself it really was, but as it was unable to see itself.

The growth of critique is not entirely due to the desire for dominion or for aristocratic immunity from guilt—or at least from gullibility. It is surely also a response to the sense that the humanities need to account for themselves, whether to justify their continued funding or secure the self-esteem of their exponents. A humanities dedicated to the work of critique can seem, or at least feel itself to be, an altogether more earnest and self-denyingly austere affair than a humanities dedicated to footling frolics of mere appreciation. One of the signs of this purposiveness was and is the absence of a noun form for the one who practices critique. Where it was clear that somebody engaged in the act of criticism could be called a "critic," it was not at all clear how one should refer to the person engaged in the work of critique. A critiquer? Critiquist? "Critical theorist" is the closest we have come. This invisibility or unsayability is odd, given that critique tended to be regarded as a much more systematic affair of disciplinary self-formation, requiring not just high levels of training but also unresting, out-of-hours zeal. I have no ready explanation to offer for this odd unnameability, unless it is that critique prefers to keep its deep psychic satisfactions unacknowledged.

The growth of an aggressive hermeneutics of suspicion made for a danger, that the flame of denunciation would burn through the entire canon of cultural objects. But this was defended against by the structurally necessary discovery that texts, or the best ones, the ones that would survive torching by critique, were the ones that could be seen to be themselves engaged in the very same work of critique, of themselves or some other object. So the show-trials of the critical humanities turned into mass reprievals of those texts which actively cooperated with the work of their interrogators and coughed up the names and addresses of their co-conspirators.

Bruno Latour's work has been for some considerable time propelled by a dissatisfaction with this work of critique. If there is one principle associated with the kind of work that Latour wants to encourage, though his addressees tend to be those in the social sciences rather than the humanities, it is the abandonment of critique, in favor of more affirmative postures and actions. Latour can be generous in his appreciation of his intellectual heroes—Gabriel Tarde, William James, and A. N. Whitehead are the names that recur in his recent work—but there is none as heroic, none as appreciated, and none whose work anticipates and enables as much of Latour's own, as Michel Serres. Latour's appreciation of Serres depends upon the fact that he sees in him a philosopher who abjures critique. "A 'critique' philosopher sees his task as that of establishing a distance between beliefs on the one hand and knowledge on the other, or between ideologies and science, or between democracy and terror—just to take three avatars of the 'Critique.' To be taken in, that is the main worry of a critique philosopher The Critique work is that of a reduction of the world into two packs, a little one that is sure and certain, the immense rest which is simply believed and in dire need of being criticized, founded, re-educated, straightened up."[1]

Latour praises Serres for not joining in this kind of bisecting, adversarial game. Critique is an agonistic, even a martial affair, and holding back from critique is part of Serres's oneiric sense of the vocation of philosophy, which he understands less as the love of knowledge than the knowledge of love. In the only conversation I have ever had with him, knowing of his interest in sport, I attempted some lamely matey banter about international rugby, with a teasing remark about the superior performance (at the time) of the English rugby team compared with the French. He replied with a shrug, accompanied by a "what-can-you-do?" half-smile, that gently but definitively declined to play the vulgarly adversarial game to which I was trying to recruit him. I had been reproved and felt I had something to expiate, a pettiness. For Serres, intellectual life cannot be conducted in the mode of attack and defense. There is in fact a great deal more ferocity in Serres than one might expect, but what he is most ferociously is an intellectual noncombatant. One might go even further and say that Serres's work gently and for the most part un-aggressively, like the wise parent distracting the toddler from its imperious tantrum, encourages its reader to become less interested in the self-indulgent agonies associated with epistemology—what can we know for sure? How can we avoid being made fools of?

Latour returns repeatedly to Serres's work for an example of how to do without the voluptuously austere pleasures of critique. In a later, oft-cited essay, Latour writes mockingly of the way in which the austere and aristocratic sense of distinction offered by the exercise of critique has become universally available. "Isn't this fabulous? Isn't it really worth going to graduate school to study critique? 'Enter here, you poor folks. After arduous years of reading turgid prose, you will be always right, you will never be taken in any more; no one, no matter how powerful, will be able to accuse you of naiveté, that supreme sin, any longer?'"[2] And yet passages like this are an unmistakable indication that Latour is as much driven by the libido of critique as anyone else. Time and again his work lays out the egregious errors of our own condition or our understanding of it that must be rinsed away by clear-eyed reflection. Latour's signature text in this respect is *We Have Never Been Modern*, which seems to take vast and mischievous pleasure in dismantling the core assumption of social theory, economic thought, and anthropological inquiry that, whatever else we may be, we can be sure at least that we are modern, in the sense that we have somehow been propelled into a condition in which we must stand outside and opposed to merely natural existence. To be sure, Latour is more inclined to encourage a sort of retoxification—the acknowledgement of complex, natural/cultural entanglements and confederations—than the detoxification of purifying differentiations of culture and nature. But the impulse to expunge and escape from erroneous thinking is as strong in him as in anyone, not least in the stylish comedy of his polemic, comedy never being far away from cruelty.

One of the difficulties with the project of showing repeatedly and ever more definitively that we have never been modern is that one must start to wonder who precisely the people are who are supposed to inhabit this noncategory (or, rather, perhaps, not to inhabit it). The hyphenated not-modern people, those who wrongly assume they are modern, come closer and closer to the unhyphenated folk who are just not modern, or have never been thought to be so in the first place. In *On the Modern Cult of the Factish Gods*, Latour surprisingly adopts the term "Blacks" to refer to such unmodern people.[3] And yet, this failure to make it as a hypothecated modern may be the very thing that marks you most definitively as a modern (or White).

Latour has characterized the project announced in *An Inquiry into Modes of Existence* as an attempt to move away from the negative mode of his own earlier work, as sternly intimated in the title of his *We Have Never Been Modern*, in order to "at last be able to give a positive, rather than merely a negative, version of those who 'have never been modern.'"[4] One wonders if there is not a memory in this characterization of his difficulty in finding a term to describe Serres's philosophy: "I am struggling for a word that would best describe Michel Serres' philosophy. 'Positive' would come to mind if Comte had not given this word a dubious posterity."[5] *An Inquiry into Modes of Existence* sets out to ask, if we have never modern, what is is that we have been? It is for this reason that the book is subtitled *An Anthropology of the Moderns*. The Moderns must be taken to be those who take themselves (erroneously) to be modern. It is perhaps a little like the distinction between white noise and pink noise. Both of them are, and sound like, kinds of noise: but where white noise is a purely random distribution of frequencies, pink noise is a distribution of frequencies in which the power-frequency relationship aligns with human hearing. Pink noise is a homelier, fleshier, less noisy sort of noise. The way of being not-modern that Moderns, indeed *"the* Moderns," exhibit, sounds reassuringly modern.

One might say that, for Latour, what are ever more definitively called, not just modern persons, but "The Moderns," are persons for whom being modern is somehow an issue, in something of the same way as an existential philosopher defines the human as that form of being for whom being is in question. So the Moderns are not not-modern in any common-or-garden way of being not modern, the way characteristic of people (and nonpeople) to whom it has never occurred to think of themselves as modern. Rather, they are not-modern in a fractiously reflexive way that involves them assuming that they are in fact modern, when they are not. So, for Latour, only moderns whose modernity is qualified in this way in fact qualify as any kind of Modern at all. It is a kind of set-theoretical hokey-cokey, in which, by setting foot in the category of the modern, you step outside it, only to find yourself at that very moment back inside it. This is all made even more exquisite by the fact that what it means to take yourself to be modern is precisely a set of negations in the first place, since to be modern is to be in the condition evoked in Yeats's "Sailing to Byzantium": "Once out of nature I shall never take / My bodily form from any natural thing." If to be modern is to be not-natural, to count as one of Latour's Moderns one must be not not-natural.

For Latour, the most important feature of a move beyond critique, or, if one wanted to avoid a preposition with which critique itself is so besotted, away from critique, in some other direction, is that it thereby becomes more able to accommodate and transact with objects. Critique, by contrast, carves objects away and carves the wielders of critique away from objects. Whatever might replace critique, by contrast, conveys us "not *away* but *toward* the gathering, the Thing."[6] Subtraction and abstraction thereby give way to accumulation and attraction: "Can we devise another powerful descriptive tool that deals this time with matters of concern and whose import then will no longer be to debunk but to protect and to care, as Donna Haraway would put it? Is it really possible to transform the critical urge in the ethos of someone who *adds* reality to matters of fact and not *subtract* reality?"[7] Objects are etymologically what are thrown up against subjects, resisting or at least deflecting their equanimity. Relishing the opportunity it gives him to rebut the accusation that his work has tended to dissolve objects and the facts that represent them into social constructions, Latour argues that the remission or renunciation of critique is in fact in the interest of a renewed attention to objects, and so seems to call for the removal of the queasy quotation marks that seem to shimmer around words like "fact," "object," and "world." He has asked for a social science that might be described, not as worldly, where that word might imply a weary, jaded cynicism, but "earthly," meaning willing to concern itself with the nature of our embodied life on this planet: "While we might have had social sciences for modernizing and emancipating *humans*, we have not the faintest idea of what sort of social science is needed for *Earthlings* buried in the task of explicitating their newly discovered attachments."[8]

We may usefully wonder why object-orientation might be taken to be an invigoration for the humanities as for the social sciences with which Latour mostly concerns himself. One pressing reason would indeed be that the relation to objects precisely supplies the distinction that keeps the idea of "the humanities" in place and in one piece. From the outside, it appears that those in the humanities have objects of study in much the same way as other fields. Historians study the past, literary critics study things called literary texts (it doesn't matter that for some time we have had no idea what these are, because we can study the issue of our not-knowing), musicologists study music, and so, apparently, on. This looks isomorphic with the ways in which geographers study lakes and traffic systems, linguists study the workings of language, astronomers study celestial objects, biologists study living organisms (whatever that is meant to mean), and physicists study the nature of the matter whose various ways of being arranged go to make up the objects of all the other subjects.

In fact, though, the humanities are not so much absorbed in their apparent objects as absorbed in the nature of their absorption in them. Indeed, increasingly, the object of the humanities has been the condition and possibilities of the humanities themselves. My home subject, English literature, has been converted into a factory for the detection and denunciation of various kinds of social sin, and the affirmation of various kinds of social good. Students increasingly arrive in universities already knowing that the most important thing about literature is how far it conduces to the work of human emancipation from various kinds of unquestionable wrong. This means that the study of literature becomes self-explicating and self-promoting. We study literature, not because of an interest in what kind of thing literary texts might be or do, not so much for what literature may show or tell us about our condition, and certainly not for the sake of pleasure, but in order to demonstrate the value of developing the powers that studying literature is thought to give. The little objects represented by the individual texts that come under scrutiny are all really surrogates for the Big Object that is literature itself, or rather, literary texts as instructed and inflected by literary-critical analysis. Every reading is really an allegory of this kind of reading: it is an arena for the exhibition and performance of what literature, subject to the right kind of literary-critical attention, can do. This seems to apply across the humanities. Every now and again critical theory in the humanities is given a shot in the arm by the discovery of another kind of wrong (preferably an irremediable one) to denounce, such as, in recent times, the depredation of the earth. But the real question at issue in the humanities is always "what are the powers and responsibilities of the humanities?" Almost always, it seems, the answer to that question is that the humanities are sovereign but neglected or marginalized, the unacknowledged legislators. The humanities are anti-elitist, but are founded upon a kind of democracy of *ressentiment* that allows everybody the fantasy of aristocratic distinction that comes from the exercise of critique.

The humanities have succeeded very well in defining themselves as a form of attention that is itself the answer to every problem that may be encountered. Vastly overvaluing their name, the humanities see themselves immodestly and incredibly as the custodians of the value of the human itself, which, so the theory goes, has always and everywhere to be rescued and redeemed from the alleged inhumanity of every object of critique, from capitalism to climate change. Throbbing behind this is a raw kind of vitalism: in defending the value of the human against other more technical or mechanical kinds of proceeding, the humanities identify themselves as the high priests of life making a stand against death.

Maybe I am really describing here, not the humanities in general, but English in particular. It is certainly the case that English has often acted on the assumption that it is at once the supreme form and the most representative form of the humanities as such. If I am being parochial here, my defense is that this is another defining feature of the humanities, namely the willingness of particular subjects to depend on this kind of synecdoche. In my experience, scientists, who may consent to be drawn politely and intelligently into discussion of the nature of scientific inquiry over a pint or a paella, do not spend their time wondering how or whether what they are doing does or does not constitute science, or actively embodies the spirit, value, or destiny of Science as such. But the only way to work in the humanities is to be continuously attuned to the question of what the humanities do and are. Everything done in the humanities bears on "the humanities." The circuit-diagram of the sciences, we might say, presents a complex and distributed picture. In the humanities, by contrast, every circuit seems to come straight off the fusebox, with the result that every overload or crossed wire seems to jeopardize the whole system. I think the disciplines known as "the humanities" could usefully learn to give up this obsessive self-reference, along with the pressure to autotrophic allegory, according to which every enterprise is justified as a proof or affirmation of what "art," "literature," "emotion," or any of the surrogates for the humanities themselves, can do.

Latour aims to make it possible for the humanities and social sciences to take more responsible account of earthly objects—of rivers and birds, of climate change, environmental damage, and species destruction. The most important thing here is to unlearn what it is to be modern, or rather to recognize that we can never have been the kind of moderns that we had taken ourselves to be, that is, creatures exiled in an empire of discourse and culture that means we must remain at a fascinating, but tragic distance from the natural existence that is no longer possible for us. Instead, Latour would have us learn to recognize that "Cultures—different or universal—do not exist, any more than Nature does. There are only natures-cultures."[9] Ultimately, the project of showing that modern humans can never have been nonnatural is intended to ensure that we rapidly understand that what we do has consequences for nature, and therefore for us. We may thus be freed from the illusion of our freedom from nature and recognize, in Serres's frequently repeated formula, that "we depend on what depends on us."

The project of opening up awareness of ecological issues runs in parallel with a rather different sense in which modernization might give way to ecology. This concerns not the object of the humanities and social sciences, but rather their own form, or mode of organization. In *An Inquiry into Modes of Existence*, Latour proposes a work of redescription that may allow us "to give more space to *other values* that are very commonly encountered but that did not necessarily find a very comfortable slot for themselves within the framework offered by modernity: for example, politics, or religion, or law, values that the defense of Science in all its majesty had trampled along its way but which can now be deployed more readily. If it is a question of ecologizing and no longer of modernizing, it may become possible to bring a larger number of values into cohabitation within a somewhat richer ecosystem."[10] If we can all agree that such a proposal has a soothing sound, we should allow ourselves to wonder why. Perhaps it is because it seems so Hippocratic, adhering to the physician's principle of *primum non nocere*, "first do no harm." Ecologizing means including, comprehending, accepting, tolerating rather than deciding. We might well say that deciding in the sense of cutting off (*de + caedere*), has become deciduous (*de + cadere*), and therefore cyclical.

And yet, of course, Latour's language has become, if anything, even more urgently martial, more militantly decisive than ever before. In *War of the Worlds*, we read that "the West has to admit to the existence of war in order to make peace."[11] But this making of peace has a greater dimension. We need to stop making war in order to wage it, in order to be able to take arms against the unwitting war that, according to James Lovelock as summarized by Latour, we are waging against the world: "He is not talking about one of those antiquated wars that so many humans wage against one another, but of another war, the one that humans, *as a whole*, wage, without any explicit declaration, against *Gaia*."[12] This war cannot be won, for "either we come out on top of Gaia, and we disappear with her; or we *lose* against Gaia, and she manages to shudder us out of existence. Now that's 'terror' for real."[13]

In this bellicose conception of our new relation to nature, Latour follows his conservative master, Serres, even to the point of borrowing his metaphor in *La Guerre mondiale*, of a war against the world: "The reader must forgive my audacity in changing the meaning of the expression: World War. Instead of giving to it the signification it has in the two conflicts which involved a majority of people, or nations, I speak here of a war that sets the whole world (*tout le monde*) against the World."[14] But this mobilization is different from other mobilizations, which were always temporary. Populations were enjoined with an *encore un effort,* to gather their energies and resources together to pit them against an enemy that threatened their very existence. Always, there was the prospect of victory. The instant jettisoning of Churchill, the war leader, by the postwar British electorate seemed to be the proof that British people had come to think of the war as being fought to bring into being a different world, without or beyond the state of conflict. But the effort that the world is currently being enjoined to make is different, for, like the communist revolutions, this mobilization must be permanent. Jacques Derrida has called language "a machine for undoing urgency."[15] But we will need to ensure that the problem of climate change remains something with which we can never permit ourselves to be bored. Perhaps indeed, maintaining the level of interested stress might be the most important thing the humanities can do in relation to the question of climate change.

The War on Terror is insistently twinned with, by being opposed to, the World War, the war to end the War-Against-the-World. Latour lost no time in condemning the fuss made in response to the Paris attacks on November 13 as trivial compared with the real threat to civilization that would be debated at the World Climate Summit in Paris. For Latour, "this kind of thuggery is a law-and-order matter, not war, despite all the flag-waving and calls to arms."[16] Having distinguished terrorism and climate change, Latour immediately reassociated them, on the grounds that both of them involve nihilistic suicide: "Just like those who kill themselves in the act of killing, people in positions of responsibility who fail to take on the issue of global climate change with the greatest seriousness is [sic] shouting in unison with the terrorists: Long live death!"[17]

The rhyme between "earthly science," or a humanities attuned to questions relating to ecology, and this ecology of interconnected, rather than hierarchically divided, forms of life seems compelling and natural. But there is also something strange about it. This strangeness consists in the suggestion that an intellectual culture that has an ecological form will be more capable of generating and sustaining ecological content and having ecological effects. Modernization put us at a fatal distance from nature, reducing it to the demeaned and disenchanted condition of object. Surely then, so the concealed logic goes, anything that disallows that distance will make it impossible for us to continue with our work of environmental despoliation? An enriched ecology of intellectual practices, of the kind that Latour projects in *An Inquiry into Modes of Existence*, is taken to be ecological in its effects. Latour's work since the early 1990s has been conducted on two fronts that he believes ultimately converge, namely social epistemology and ecology. This is a magical operation that partakes of a defining fantasy of self-definition. We have somehow to transcend transcendence, since the willingness or desire to attain the high ground, or even to slip the surly bonds of earth altogether, is precisely what has left us up to our necks in the mire. Latour thinks, or writes as if he thinks, that we need a drastic change of philosophy. Though we might remind ourselves again that Latour's primary affiliation and addressee is not the humanities but the social sciences, this is a familiar role for what calls itself the humanities, one that science and technology, for all its juggernaut-like power, tends meekly to acknowledge whenever it agrees to enlist an ethics expert on its advisory board. But a change of heart or mind need make no difference at all. A difference to our chances of survival will be whatever results in a dramatic decrease in carbon emissions. That's it. Don't follow the objects or the actants, follow the numbers, for they are what will kill or cure us.

This move, of mistaking epistemology for effect, is one of the most common of the dream-machines of the humanities. But the problem is not one of how we understand what we are; it is a problem of what we decide to do, or do without deciding. It is a technical and not an epistemological question. It is not a question of how we come to feel about our being-in-the-world; it is a question of what kind of being in the world we manage to bring about or retain. Even "The Question Concerning Technology" is most impelling and urgent as a technical and not a philosophical question. Climate change is a technical problem. Changing how we think about our place in the world might do some good, but only if it helps with the job of engineering.

It is not that the human should be entirely evacuated from the humanities. Indeed, it is just the opposite. The problem of human involvement in nature will need to become ever more prominent and unignorable. But taking account of the human—of human entanglements and effects—is different from referring every question to "the human." The question of whether and how humans are to survive and prosper, if they/ we are, can be usefully decoupled from the question of what "the human" might be, and what "the humanities" might take as their mission.

The way to make yourself important and necessary is to define a problem in such a way that only your involvement or intercession will solve it. Thus, science and engineering can be made to seem like the source of a problem that only the balm of letting-be can mitigate. One can see something similar in the ways in which the problem of the fantasy-object known as "capitalism" is construed in the humanities. Much political thinking is motivated by the effort to anthropomorphize capitalism—and one can tell this is going on whenever capitalism is referred to as "capital." This discursive move allows one to imagine capitalism as an ideology or even an intentional and self-willing subject rather than a set of structures and conditions. That is, it allows one to think in terms of what capitalism wants and what it does in the furtherance of its wants. The psycho-epistemological payoff for this projection of intentionality is immense. For now, one is faced not with the problem of a set of complex conditions that need to be understood and reconstructed without worsening the problem, but with the problem of a will, indomitable or insidious as it may be. And all that is then required to resist or, who knows, even to defeat capitalism, is to want something different from it, opposing your will to its. But if capital were not thought to want things any more than the Ebola virus or a tropical cyclone, one might have to set to the task of understanding and reconstructing mechanisms rather than reforming persons or forcing them into compliance. No wonder the humanities are so convinced that persons are much more complicated than machines, even though everything they say about them in the storybooks and nursery rhymes they love so much proclaims the opposite. One reforms the problem in fantasy in order to make it susceptible to fantasy solutions.

It looks as though our survival may depend upon an act of engineering greater and more extensive than any ever before undertaken, one that encompasses all aspects of social, political, economic, and psychological life. Whether we seek to slow or reverse climate change (wind turbines, solar panels, carbon capture), or simply to adapt to it (flood defenses), our response is going to, one way or another, have to be engineered. If things go well, the thing called the humanities may negotiate some kind of role in the work of stressory maintenance and affect management that may enable us, at the very least, to stay focused on a problem that is going to go away only if we are even more completely wrong about things than we have ever been. The greenhouse effect will be answered only by some heightening of the kind of "greenhouse-effect" that Peter Sloterdijk has identified as the work of culture, the creation of artificially maintained spheres of security and well-being to protect against "the cosmic frost infiltrating the human sphere" and the "shellessness in space" that ensued upon the Enlightenment banishment of divine providence from nature.[18] We will all of us need to understand that we are involved in engineering instead of emancipation.

The measure of failure and success will not be: have we at last understood the truth of our embeddedness in the world? It will be: have we helped decrease or increase carbon emissions? We are going for some time yet to have to live by numbers, watching the emission levels going up more or less quickly as one watches the taxi meter. What Latour offers is a kind of allegorical letting be: a project in which the humanities become the shepherds of social being, carefully conserving rather than brutally massacring "modes of existence." This is a pseudo-ecological exercise, one that is isomorphic with ecological thinking and action, without in fact being it.

Latour is right to want to sustain a diversity of styles of thought and forms of life. But it would be wrong to think of the humanities as naturally equipped to provide this diversity (not that anyone could accuse Latour of thinking this). It is often supposed that the pursuit of scientific or technical subjects of concern has tended to produce a grimly technicist monoculture, which it is the job of the humanities to diversify. But monomania is a general problem for human beings, not just for scientifically minded ones, and humanists can be just as one-eyed and obsessive as scientists. We need engineering and mathematics to rescue us from religious or poetic obsessiveness just as often as we need our medical students to be "humanized" by taking courses in the European novel.

The humanities routinely offer a dramatization and a glamorization of minority, an exiled marginality magnified into the *condition humaine*. "The humanities" often seem to mean just the same thing as "the Celts," which began its long and ludicrous career as a word meaning "not us," or "those over there." But as the proliferation of Burns suppers and St. Patrick's Day Parades all over the world attests, who does not want to be part of "the Celtic fringe"? Similarly, the humanities make a cult of self-Celtification. Everyone wants to be a Celt, where a Celt means someone on the fringe, someone driven out from the centers of power. We aspire to wear, as a badge of pride, the epigraph that Matthew Arnold took from Macpherson's Ossian for his lectures on Celtic literature: "They always went to battle and they always fell." (The much-reproduced Wikipedia page remarks of the word "keltoi," with an immaculate piece of *petitio principii*, that "several authors have supposed it to be Celtic in origin." But the question of who these Celts were and what the inside definition of being "Celtic" might be, is exactly what is in question when one investigates the use of the word.)

The humanities will become significant only when they really accept their marginality, rather than bloating it into a *folie de grandeur*. There may indeed be a public relations role for the humanities, in softening people up, or toughening them up, for the kind of world it looks as though we will need to bring about or put up with. One thing is certain, that the prodigious growth in the humanities (we are not supposed to notice this, since the humanities can only exist in a condition of defensive and resentful outrage at being cut) was sustained by a hydrocarbon-fueled high-growth economy that may have gone forever, or, if it has not gone, may henceforth have to be held at bay. The humanities have not shown much appetite for the kind of austerity that might have to go along with the low-growth economy forming part of a sustainable world. The humanities will become useful at the point at which they learn to be useful occasionally and in part, rather than existing contemptuously and uselessly apart. They may learn to take part in the composition of a new phase of human life, a negotiated rather than a martial Anthropocene, if they learn to give up their immodest excitability in favor of a kind of modest composure.

When asked in an interview what he thought the future of the humanities might be, Serres paused for a moment, then replied, simply, "Death."[19] But there might be a more hopeful way of seeing this imminent demise, a way presaged by the sly Brechtian adage "where there's death, there's hope." To be sure, there may be a cost. The humanities may get to join the epistemological party only at the cost of the principle that has sustained them. In other words, the vauntingly impotent humanities may earn a slice of power if they give up a large measure of their presumptuous dominion over the realm of "the human." A new name, or a discrediting of the old one, would really help a lot. Can there ever have been a more absurd claim than that the vast and proliferating range of things undertaken by humans—all forms of industrial production, economic activity, scientific research, technical development, and mathematical speculation—are somehow more incidental to being human than listening to music or reading stories? What kind of insanity is it to imagine that mathematics is not part of the humanities? Given the conspicuous absence of mathematical capacity or curiosity among badgers and bacteria, it would seem that nothing could be more essentially human than mathematics, along with everything that it makes sense of and makes possible. Yet those who affiliate themselves with "the humanities" persist in equating numbers with death and the inhuman.

Latour has found more important things with which to occupy himself than the nature or future of the humanities. This is precisely the reason why his work might be salutary for them. The most important thing that the humanities need to have said to them is "this is not about you." Latour's work allows us to wrench the question "what future is there for the humanities?" into the question "what future is there but the humanities?"—meaning by this that the many different ways in which the engineering of the human, and the regulation of the relations between the human and natural worlds, must proliferate. If the humanities can give up their perfervid fantasy that they are the custodians of the human, they may have something useful to contribute.

Notes

1. Bruno Latour, "The Enlightenment without the Critique: A Word on Michel Serres' Philosophy," in *Contemporary French Philosophy*, ed. A. Phillips Griffiths (Cambridge: Cambridge Univ. Press, 1987), 85.

2. Latour, "Why Has Critique Run Out of Steam? From Matters of Fact to Matters of Concern," *Critical Inquiry* 30, no. 2 (2004): 239.

3. Latour, *On the Modern Cult of the Factish Gods*, trans. Catherine Porter and Heather Maclean (Durham, NC: Duke Univ. Press, 2010), 29.

4. Latour, *An Inquiry into Modes of Existence: An Anthropology of the Moderns* (Cambridge, MA: Harvard Univ. Press, 2013), xxvi.

5. Latour, "Enlightenment Without Critique," 91.

6. Latour "Why Has Critique Run Out of Steam?" 246.

7. Latour "Why Has Critique Run Out of Steam?" 232.

8. Latour, "A Plea for Earthly Sciences," in *New Social Connections: Sociology's Subjects and Objects*, ed. Judith Burnett, Syd Jeffers, and Graham Thomas (Houndmills: Palgrave Macmillan, 2010), 75.

9. Latour, *We Have Never Been Modern*, trans. Porter (Cambridge, MA: Harvard Univ. Press, 1993), 104.

10. Latour, *Inquiry into Modes of Existence*, 11.

11. Latour, *War of the Worlds: What About Peace?*, ed. John Tresch, trans. Charlotte Bigg (Chicago: Prickly Paradigm, 2002), 29.

12. Latour, "A Plea for Earthly Sciences," 1.

13. Latour, "A Plea for Earthly Sciences," 1.

14. Michel Serres, *La Guerre mondiale* (Paris: Le Pommier, 2008), 137–38.

15. Jacques Derrida, "*Ja*, or the *faux-bond*," in *Points . . . Interviews 1974–1994*, ed. Elisabeth Weber, trans. Peggy Kamuf et al. (Stanford, CA: Stanford Univ. Press, 1995), 30–77.

16. Latour, "The Other State of Emergency," trans Jane Kurtz, Reporterre, November 23, 2015, available at http://www.bruno-latour.fr/sites/default/files /downloads/REPORT-ERRE-11-15-GB_0.pdf.

17. Latour, "The Other State of Emergency."

18. Peter Sloterdijk, *Spheres*, vol. 1, *Bubbles: Microspherology*, trans. Wieland Hoban (Los Angeles: Semiotext[e], 2011), 24.

19. "Michel Serres at Stanford" (lecture, Stanford University, Stanford, CA, 2011), available at https://www.youtube.com/watch?v=zb5-l45dbow.

CHAPTER SEVEN

Humanities in the Anthropocene

The Crisis of an Enduring Kantian Fable

DIPESH CHAKRABARTY

For Bruno Latour

Exordium

2015 WAS THE FIRST YEAR when the average surface temperature of the world rose by one degree Celsius above the pre-industrial average, thus taking us closer to the threshold of a two-degree rise, a Rubicon we are told we must not cross if we are to avoid what United Nations Framework Convention on Climate Change (UNFCC) of 1992 described as "dangerous anthropogenic interference with the climate system."[1] 2016, as one meteorologist put it, has so far been "off the charts" as far as global warming is concerned.[2] The historian Julia Adeney Thomas remarked in 2014 that the idea of being "endangered" could not be a purely scientific idea, for the planet has been through many other episodes of climate change—and five Great Extinctions of species—before.[3] "Dangerous" here is indeed a word that scientists, politicians, and policy makers use as concerned citizens of the world, glossing "danger" as a threat to human institutions. In Thomas's words: "Historians coming to grips with the Anthropocene cannot rely on our scientific colleagues to define 'the endangered human' for us. . . . It is impossible to treat 'endangerment' as a simple scientific fact. Instead, endangerment is a question of both scale and value. Only the humanities and social sciences, transformed though they will be through their engagement with science, can fully articulate what we may lose."[4] Indeed, one of the first general books to be written on the problem of anthropogenic climate change around the time of the publication of the Fourth Assessment

Report of the Intergovernmental Panel on Climate Change (IPCC), Tim Flannery's *The Weather Makers*, pointed out that the entity to which climate change posed a real threat was human civilization as we have come to understand and celebrate it.[5] "Civilization," of course, is a value-laden and therefore contested word that humanities scholars in recent decades have done much to demystify.[6] I bring up the point here simply to show how central the concerns of the humanities and the human sciences have been to defining one of the gravest problems humans face in the twenty-first century. The point is underlined when moral philosophers such as Peter Singer describe climate change as the "greatest ethical challenge" ever faced by humanity.[7] True, we could not define "human-induced planetary climate change" except with the help of big science; and, true, the problem of the "two cultures" of the sciences and the humanities remains.[8] But the questions of justice that follow from climate-change science require us to possess an ability that only the humanities can foster: the ability to see something from another person's point of view. The ability, in other words, "to imagine sympathetically the predicament of another person."[9]

This moral demand on humans today acquires an additional twist from the thought that, seen in a long-term perspective, unabated global warming may very well accelerate the already growing rates of human-induced extinction of nonhuman species, with unhappy consequences for humans themselves. Voices have been raised, including that of Pope Francis, recommending that human justice be extended not just to animals that cross a certain threshold of sentience (as animal liberationists once argued) but to the entire world of natural reproductive life—what Aristotle called the *zoe*. This proposition that, in effect, subjects the domain of biological life to the work of the moral life of humans marks, I argue, a critical turning point for the humanities today, as it departs radically from a tradition—inaugurated by, among others, Immanuel Kant—that made a strict separation between our "moral" and "animal" (i.e., biological) lives, assuming that the latter would always be taken care of by the natural order of things. This separation, after all, is what has buttressed for more than a century the much-critiqued gap between the humanities and the physical or biological sciences. Strands of environmentalist thought have questioned and on occasions attempted to close this chasm, but the gap persists and has not been easy to overcome. However, to ask, as we do today, how humans might use the resources of their moral capacity to regulate their life as a biosocial species among other species, is to bring within the ambit of human moral life something that has always lain outside of its scope: the history of natural life on the planet. The assumption—made since at least the Enlightenment—that our animal life could take care of itself while we struggled, consciously, in search of a collective moral life is now under serious strain, with serious implications for the humanities that have traditionally served as the domain for the discussion of moral issues in separation from biological life. I argue this point by looking, first, at some relevant writings of Kant in the context of discussions on climate change and possible human stewardship of life on the planet. I then engage, in conclusion, with the work of Bruno Latour to show where his thoughts indicate a way forward.

Two Narratives of Climate Change

We generally find two approaches to the problem of climate change.[10] One dominant approach is to look on the phenomenon simply as a one-dimensional challenge: how do humans achieve a reduction in their emissions of greenhouse gases (GHGs) in the coming few decades? Climate change is seen in this approach as a question of how best to source the energy needed for the human pursuit of some universally accepted ends of economic development, so that billions of humans are pulled out of poverty. The main solution proposed here is for humanity to make a transition to renewable energy as quickly as technology and market signals permit. The accompanying issues of justice concern relations between poor and rich nations and between present and future generations: what would be a fair distribution of the "right to emit GHGs"—since GHGs are seen as scarce resources—between nations in the process of this transition to renewables? The question of how much sacrifice the living should make as they curb emissions, to ensure that unborn humans inherit a better quality of life than that of the present generation, remains a more intractable one, its political force reduced by the fact that the unborn are not present to argue to press their case. "The nonexistent has no lobby," as Hans Jonas once remarked, "and the unborn are powerless."[11]

Within this broad description of the first approach, however, are nested many disagreements, ranging from capitalist to noncapitalist utopia of sustainable futures. Most imagine the problem to be mainly one of replacing fossil fuel-based energy sources with renewables. Some others—on the left—would agree that a turn to renewables is in order, but argue that because it is capitalism's constant urge to "accumulate" that has precipitated the climate crisis, the crisis itself provides yet another opportunity to renew and reinvigorate Marx's critique of capital. And then there are those who think of actually scaling back the economy, degrowing it, and thus reducing the ecological footprint of humans while designing a world characterized by equality and social justice for all. Still others think—in a scenario called "the convergence scenario"— of reaching a state of economic equilibrium globally whereby all humans sustain more or less the same standard of living. The role of the humanities is confined here mainly to climate justice issues, with both political economists and philosophers (both in Rawlsian and utilitarian traditions) contributing to relevant discussions.[12] For all its shortcomings, however, the reduction of the climate crisis to the problem of renewable energy has the advantage that we can develop frameworks of both policy and politics around it.

If, however, one also sees climate change not simply on its own but as part of a family of interlocking problems—exponential population increase, food insecurity, water scarcity, expansion of resource industries, and increase of economic inequalities contributing to human-animal conflicts, habitat loss for other species, greenhouse gas emissions, and so on—all planetary in scope and all speaking to the fact of an overall ecological overshoot on the part of humanity that affects the distribution of natural life on the planet, global warming seems more like a shared predicament for all humans—not to speak of other species—than a problem that can simply be resolved by switching to renewables. Add to this the knotty question of human "agency" that many scientists have underlined: the new geophysical agency of humans on a scale that has allowed them already to change the climate of the planet for the next 100,000 years, putting the next ice age off by anything between fifty and five hundred thousand years.[13] Within this perspective that looks into both deep pasts and deep futures, a very particular challenge opens up for imagination in the humanities. After all, the problem of climate change arises out of our need to consume more energy than before, and the excess greenhouse gases in the atmosphere could easily be looked upon as the resultant "waste" that cannot be properly recycled. Since this human "waste" affects other lifeforms—by acidifying the oceans or raising the average surface temperature of the planet—the crisis requires us to "imagine sympathetically the predicament" of not just humans but of nonhumans as well.

It is, of course, not global warming alone that caused this shift in our moral orientation. The phenomenon of an ecological overshoot on the part of humans—due, perhaps, to the development of a big brain that has helped humans over tens of thousands of years to create attachments and affiliations to imagined communities far beyond the face-to-face scale of kin group or band—is now seen by many to have taken place over a very long historical period reaching back to times that Daniel Smail describes as our "deep history."[14] The Israeli historian Yuval Noah Harari explains the issue well in his book *Sapiens: A Brief History of Humankind.* "One of the most common uses of early stone tools," writes Harari, "was to crack open bones in order to get to the marrow. Some researchers believe this was our original niche." Why? Because, Harari explains, "genus *Homo*'s position in the food chain was, until quite recently, solidly in the middle."[15] Humans could eat dead animals only after lions, hyenas, and foxes had had their shares and cleaned the bones of all the flesh sticking to them. It is only "in the last 100,000 years," says Harari, "that man jumped to the top of the food chain."[16] This has not been an evolutionary change. As Harari explains: "Other animals at the top of the pyramid, such as lions and sharks, evolved into that position very gradually, over millions of years. This enables the ecosystem to develop checks and balances. As the lions became deadlier, so gazelles evolved to run faster, hyenas to cooperate better, and rhinoceroses to be more bad-tempered. In contrast, humankind ascended to the top so quickly that the ecosystem was not given time to adjust."[17] Harari mentions an additional significant fact. As a result of their quick ascent to the status of the top carnivore, humans themselves, writes Harari, "failed to adjust." He adds: "Most top predators of the planet are majestic creatures. Millions of years of domination have filled them with self-confidence. Sapiens by contrast is more like a banana republic dictator."[18]

Human ecological footprint, we can say, increased further with the invention of agriculture (more than 10,000 years ago) and then again after the oceans found their present level about 6,000 years ago and we developed ancient cities, empires, and urban orders while moving to every part of the planet. It was ratcheted up yet again over the last 500 years with European expansion and colonization of faraway lands inhabited by other peoples, and the subsequent rise of industrial civilization. But a further ratcheting up by several significant notches happened after the end of the Second World War when human numbers and consumption rose exponentially, thanks to the widespread use of fossil fuels not only in the transport sector but also in agriculture and medicine, allowing, eventually, even the poor of the world to live longer—though not healthy—lives.[19]

Scholars have carried forward the notion of "overshoot"—"instances in which populations of organisms so changed their own environments that they undermined their own lives"—that William R. Catton, Jr. put forward in a book of that name in 1980.[20] The literature on animal liberation/rights that extends the human moral community to include (some) animals recognizes issues of both cruelty to animals and the overshooting of human demands for consumption.[21] Scholars working on human-induced species extinction in the context of anthropogenic climate change have long acknowledged the "overreach" that humans have achieved, often to their own detriment, in the various ecosystems they inhabit.[22] In addition, well-known arguments about "the Great Acceleration" and "planetary boundaries" that some earth scientists and other scholars have put forward are statements, precisely, about ecological overshoot on the part of humans. As the authors of the "Great Acceleration" thesis put it: "the term 'Great Acceleration' aims to capture the holistic, comprehensive and interlinked nature of the post-1950 changes simultaneously sweeping across the socio-economic and biophysical spheres of the Earth System, encompassing far more than climate change."[23] Their data document exponential rise in human population, real GDP, urban population, primary energy use, fertilizer consumption, paper production, water use, transportation, and so on—all happening after the 1950s. And there is a corresponding exponential rise in "earth system trends" to do with the emission of carbon dioxide, methane, and nitrous oxide; ocean acidification; loss of stratospheric ozone; marine fish culture; shrimp aquaculture; tropical forest loss; terrestrial biosphere degradation; etc.[24] Similarly, the idea of nine "planetary boundaries" that humans should avoid crossing, put forward in 2009 by Johan Rockström and his colleagues at the Stockholm Resilience Centre, was also an exercise in measuring human ecological overreach.[25] Some earth system scientists reported recently that "the present anthropogenic carbon release rate [around 10 Petagram C per year; 1 $Pg = 10^{15}$ grams] is unprecedented during the [entire] Cenozoic (past 66 Myr)" and that "the present/future rate of climate change and ocean acidification is too fast for many species to adapt" and will likely result in "widespread future extinctions in marine and terrestrial environments." We are, effectively, in "an era of no-analogue state, which represents a fundamental challenge to constraining future climate projections."[26]

Not only have marine creatures and many other terrestrial species not had the evolutionary time needed to adjust to our increasing capacity to hunt or squeeze them out of existence, but our greenhouse gas emissions now threaten the biodiversity of the great seas, and thus endanger the very same food web that feeds us. Jan Zalasiewicz and his colleagues on the subcommittee of the International Commission on Stratigraphy charged with documenting the Anthropocene point out that it is the human footprint left in the rocks of this planet as fossils and other forms of evidence— such as terraforming of the ocean bed—that will constitute the long-term record of the Anthropocene, perhaps more so than the excess carbon dioxide in the atmosphere. If human-driven extinction of other species results— say, in the next few centuries—in a Great Extinction event, then even the epoch-level name of the Anthropocene may be too low in the hierarchy of geological periods.[27] The music historian and theorist Gary Tomlinson, writing recently in the context of climate change, has summed up the problem nicely from an earth system point of view:

> Across millions of years of biocultural evolution . . . certain systems remained *outside* the feedback cycles of hominin niche construction. Astronomical dynamics, tectonic shifts, volcanism, climate cycles, and other such forces were in essence untouched by human culture and behavior (or if touched, touched in a vanishingly small degree). In the language of systems theory, all these forces were in effect *feedforward* elements: external controls that "set" the feedback cycles from without, affecting the elements within them but remaining unaffected by the feedback themselves. The Anthropocene . . . registers a rearrangement in which *systems that had always acted as feedforward elements from outside human niche construction have been converted into feedback elements within it.*[28]

Viewed thus, as Zalasiewicz says in the concluding paragraph of a recent essay, "the Anthropocene—whether formal or informal—clearly has value in giving us a perspective, against the largest canvas, of the scale and the nature of the human enterprise, and of how it intersects ('intertwines' now, may be a better word) with the other processes of the Earth system."[29] Anthropogenic climate change is therefore not a problem to be studied in isolation from the general complex of ecological problems that humans now face on various scales—from the local to the planetary—and that are creating new conflicts and exacerbating old ones between and inside nations. There is no single silver bullet that solves all the problems at once, nothing that works like the mantra of transition to renewables to avoid an average rise of 2°C in the surface temperature of the planet. What we face does indeed look like a wicked problem, a predicament. We may able to diagnose it but not "solve" it once and for all.[30]

Modernity and Kant's Geology of Morals[31]

If, as I have claimed, the challenge posed to our moral life by the scale of problems created by our practices of consumption makes a breach in the assumed separation of our "moral" and "animal" lives, and demands of us that we find "moral" solutions to problems created by "natural history" of the human species, then clearly the human sciences, and in particular the humanities, face a novel task today. For it was this very separation between the animal and moral life of the human species that underlay, for a large part of the twentieth century, the separation of the human from the physical and biological sciences.[32] The subject deserves more research. But older readers will remember how vociferously—and oftentimes acrimoniously—sentiments in favor of this separation were voiced when in 1975 Edward O. Wilson published his book *Sociobiology*, making some strong claims about connections between biology and culture and managing to infuriate, in the process, Marxists and social scientists of many other persuasions.[33]

The enduring importance of the assumed separation of the moral life of humans from their animal life in post-Enlightenment narratives of modernity is perhaps best studied with reference to a fable that Kant spelled out in a minor essay called "Speculative Beginning of Human History," published in 1786. The opposition between the animal life of the human species and its moral life was at the heart of this essay. The essay provides a fascinating reading of the Biblical story of Genesis and the question of human dominion over the earth.[34] The aim of Kant's exercise was to bring "into agreement with one another and with reason" what he saw as "the oft misunderstood and seemingly contradictory claims of the esteemed Jean-Jacques Rousseau":

> In his works, [wrote Kant], *On the Influence of the Sciences* and *On the Inequality Among Men*, he [Rousseau] displays with complete accuracy the inevitable conflict between culture and the human race as a *physical* species whose every individual member ought fully to fulfill its vocation. But in his *Emile*, in his *Social Contract*, and in other works he seeks to answer this more difficult question: how must culture progress so as to develop the capacities belonging to mankind's vocation as a *moral* species and thus end the conflict within himself as ["a member of both a"] moral species and a natural species?[35]

Kant regarded this conflict itself—engendered by the human species possessing, at the same time, both a "physical/natural/animal" (these words are used in the same sense in his essay) life and a moral life—as a decisive influence on human history. For "impulses to vice" arose from "natural capacities" that were given to man "in his natural state"; they necessarily conflicted with "culture as it proceed[ed]." "The final goal of the human species' moral vocation" could not be reached until "art so perfected itself" that it became, in Kant's words, a "second nature" (SB 54–55).[36]

Many in the massive literature on Kant have discussed the philosopher's answer to the Rousseau puzzle, some tracing certain critical elements in his answer back to ancient principles, including those postulated by Aquinas.[37] My purpose here, however, is not a historical excavation of the roots of Kant's thoughts but to reconstruct Kant's argument in order to explicate how precisely he sought to understand the relationship between the animal and the moral aspects of the human species. Kant began his essay by explaining why he could take the liberty of reading the story of Genesis speculatively, while clarifying that the speculative was not the same as "fictional."[38] Speculation could be "based on experience," but the experience in question was that of "nature," something that, for Kant, remained constant in its essential structure. So if human history were a history of freedom, then a statement about its "first beginnings" could be read speculatively (i.e. guided by reason) if we based ourselves on our experience of nature (by definition a constant), and only insofar as the beginnings in question were made by nothing other than nature itself. As Kant put it: "A history of freedom's first development, from its original capacities in the nature of man, is therefore *something different* from the history of freedom's progression, which can only be based on reports [and thus become the historian's province]" (SB 49, emphasis added).

Kant, of course, made certain assumptions about this original condition of humans so that "one's speculation [would] not . . . wander aimlessly." He took a certain figure of the human for granted—"one must make one's beginning something that human reason is utterly incapable of deriving from any previous natural causes"—and hence began "not with [human] nature in its completely raw state" but with "man as *fully formed adult* (for he must do without maternal care)." He also assumed "man" to actually be "a *pair*, so that [man] can propagate his kind," and the pair had to be "*only a single pair*, so that war does not arise, as it would if men lived close to one another and were yet strangers." This latter assumption, it seemed to Kant, ensured that "nature might not be accused of having erred regarding the most appropriate organization for bringing about" what Kant saw as "the supreme end of man's vocation, sociability" (for the desire to socialize would be maximized "by the unity of the family from which all men should descend"). Besides, he made some further assumptions to keep his speculative logic straight: "The first man could thus *stand* and *walk*; he could *talk* (Gen. 2:20), even *converse*, i.e. speak in coherent concepts (v.23), [and] consequently, *think*." This threshold of assumptions regarding human skills, he reasoned, would allow him "to consider only the development of morality in [man's] actions and passions." Having thus reconstructed this original pair of humans, Kant placed them squarely in the middle of what we might today see as the geological Holocene period, with considerable advances already made in "human civilization": "I put this pair in a place secured against attack by predators, one richly supplied by nature with all the sources of nourishment, thus, as it were, *in a garden*, and in a climate that is always mild" (SB 49–50, emphases original). Kant did not know this: but the "man" of his assumptions could have existed only after the last ice age was over!

Kant's "man" began his journey completely absorbed in the animal life of the species when instinct alone—"that *voice of God* that all animals obey"—"first guided the beginner." But by the time Kant has the human being in his sights, reason, a faculty somewhat beyond animal life and yet put in place by some design of nature, had already begun to "stir" and to "cook up" in humans—in partnership with a companion human faculty, imagination—"desires for things for which there is . . . no natural urge," with the result that "man became conscious of reason as an ability to go beyond those limits that bind all animals" (SB 50–51). A critically important discovery followed: "[man] discovered in himself an ability to choose his own way of life and thus not be bound like other animals to only a single one" (SB 51). The deepening of this "inner" propensity gave human beings the capacity to refuse desires that were merely animal—thus developing the ability to love. "*Refusal*," wrote Kant, "was the feat whereby man passed over from mere sensual to idealistic attractions, from mere animal desires eventually to love and, with the latter, from the feeling for the merely pleasant to the taste for beauty." This, together with the development of a sense of "decency," "gave the first hint of man's formation into a moral creature," a small beginning that for Kant was "nonetheless epochal" (SB 52). Reason also led humans to "the reflexive *expectation of the future*," and then to a height that raised "mankind altogether beyond any community with animals" enabling humans to conceive of themselves—"though only darkly"—as "the true *end of nature*." Humans could now see that the pelt of the sheep "was given by nature" not for the sheep but for them. Their dominion over the earth that Genesis speaks of had thus begun. But this also led to the idea of equality of all humans—"[men] must regard all men as equal recipients of nature's gifts"—and, more importantly, to the idea that "man became the *equal of all* [other] *rational beings*, no matter what their rank might be (Gen. 3:22), especially in regard to his claim *to be his own end*" (SB 52–53). This formulation is, of course, a close cognate of the famous Kantian dictum about treating every human being not instrumentally but as an end in himself or herself.[39]

Kant was acutely aware that this "portrayal of mankind's early history" revealed "that its exit from . . . paradise . . . was nothing but the transition from the raw state of merely animal creature to humanity, from the harness of the instincts to the guidance of reason—in a word, from the guardianship of nature to the state of freedom" (SB 53). This, as Kant explains, had to be the story of a fall, morally speaking. Before reason stirred in the human breast, "there was neither a command nor a prohibition and thus no transgression either." But reason could ally itself "with animality and all its power" and thus give rise to "vices of a cultivated reason" (to produce wars, for instance). "Thus, from the moral side," writes Kant, "the first step from this last state [the state of innocence] was a *fall*; from the physical side, a multitude of never-known evils of life [natural disasters, hardship], thus punishment, was the consequence of the fall" (SB 53–54). Much of human history as we know it followed from the fall: there was hardship, inequality—"that source of so many evils, but also of everything good"—wars, and humans getting "drawn into the glistering misery of the cities" (SB 56–57). But this also complicated the role of reason in the story of human freedom. Humans could use reason in a way that hastened the vocation of their species—a species designated, according the Genesis story of "man's" dominion, "to rule over the earth, and not as one designated to live in bovine contentment and slavish certitude" (SB 57). But reason did not straightforwardly guide humans toward recognition of their vocation (though Kant in other essays will explain why humans would nevertheless end up fulfilling their destiny). Kant would thus write: "The history of *nature*, therefore, begins with good, for it is God's work; the history of *freedom* begins with badness, for it is *man's* work" (SB 54).

The key to human beings' success was "*to be content with providence*," wrote Kant in concluding this essay (SB 57–58). But this precisely was what was never easy for humans to do. Providence worked through what humans considered adversity: wars (which in the end generated "respect for humanity from the leaders of nations"), brevity of life (which guaranteed that improvement accrued to the species and not to individuals), and the absence of a golden age of all leisure and no toil (SB 57–59). As Kant put it: "Contentment with providence and with the course of human things as a whole, which do not progress from good to bad, but gradually develop from worse to better; and in this progress nature herself has given everyone a part to play that is both his own and well within his powers" (SB 59).

The late Kant would anticipate, repeat, elaborate on, and develop these basic points in the third *Critique* (the section on teleological judgment) and in several essays including "Idea for a Universal History with a Cosmopolitan Intent" (1784) and "On the Proverb: That May be True in Theory, But Is of No Practical Use" (1793). Here is Kant, in the third *Critique*, for example, on the subject of the separation of the moral life of humans from their natural history:

> External nature is far from having made a particular favorite of man. For we see that in its destructive operations—plague, famine, flood, cold, attacks from animals great and small, and all such things—it has as little spared him as any other animal. Besides all this, the discord of inner *natural tendencies* betrays him into further

misfortunes through oppressions of lordly power, the barbarism of wars, and the like. Man, therefore, is ever but a link in the chain of physical ends. As the single being upon the earth that possesses understanding, and, consequently, a capacity for setting before himself ends of his deliberate choice, he is certainly titular lord of nature, and, supposing we regard nature as a teleological system, he is born to be its ultimate end. But this is always on the terms that he has the intelligence and the will to give to it and to himself such a reference to [final] ends as can be self-sufficing independently of nature. Such an end, however, must not be sought in nature.[40]

The important point here is the separation that Kant effected—in order to put forward his theory of human freedom—between the animal and the moral lives of the human. He assumed that human beings' animal life was given, constant, and was to be provided for by the planet (the "biosphere," in today's terms). Human history and thinking were concerned mostly with the constant struggle of humans to meet their moral destiny of a "perfect" and just sociability: "Nature has given man two different capacities for two different ends, namely, an end for man as animal species and another end for man as moral species" (SB 55n).

The Entangled Moral and Animal Lives of Humans

The pressure that "the animal life" of the human species—our material and demographic flourishing (in spite of the gross inequities of human societies)—now puts on the distribution of natural, reproductive life on earth, endangering human existence in turn, is something that becomes clearer by the day. It is not surprising, then, that thinkers and philosophers should call climate change the greatest ethical challenge facing humanity. And they raise a critical moral-theological question, revisiting, in secular forms, the Biblical proposition of "man's dominion over earth": what should humans do, now that our animal/natural life overwhelms the natural lives of nonhumans? Indeed, the question of capitalism reemerges in this morally charged context. Should we continue with capitalism but without fossil fuels? Should we be seeking alternatives to capitalism? Should humans retreat back into small communities? Should the wealthy consume less?

These moral questions testify to the endurance of one of Kant's propositions: that the moral life of humans assumes that man can "choose his way of life and not be bound like other animals to only a single one" (SB 57). But if what I have argued above is right, then it could also be said that the Kantian fable of human history that I recounted is now coming under strain in unprecedented ways. On the one hand, many thinkers still work with (implicitly Kantian) ideas about our moral life representing a zone of freedom; but we cannot any longer afford the assumption that Kant, along with many others, made—that the needs of our animal life will be attended to by the planet itself. We now want our moral life to take charge of our natural life, if not of the natural lives of all nonhumans as well. The Biblical question of human dominion has now assumed the shape of secular questions about human stewardship of and responsibility to the planet.[41]

For reasons of space, let me work here with only two prominent examples of such thinking: Pope Francis's recent and prominent encyclical to Catholic bishops, and a recent essay by Amartya Sen. The Pope's encyclical is probably the only available Western/European attempt so far to read humanity's current climate crisis in terms of a deep-set spiritual crisis of modern civilization, albeit within the terms of Catholic theology, though that does not lessen its value. (For an Indian scholar, it is reminiscent of a famous essay Rabindranath Tagore wrote in 1941, the year he died, entitled "The Crisis of Civilization.") The Pope has quite a radical critique of the excesses of consumerist capitalism and especially of what he sees as the "misguided," "tyrannical," "excessive," and "modern" anthropocentrism of the "throwaway" civilization that capitalism has spawned and promoted.[42] In this context, he revisits the question of human "dominion": "An inadequate presentation of Christian anthropology gave rise to a wrong understanding of the relationship between human beings and the world. Often, what was handed on was a Promethean vision of mastery over the world, which gave the impression that the protection of nature was something that only the faint-hearted cared about. Instead, our 'dominion' over the universe should be understood more properly in the sense of responsible stewardship."[43] "We are not God," writes Pope Francis elsewhere in the book, opposing strongly, by implication, the view that humans are now the God-species. "This responsibility for God's earth means that human beings, endowed with intelligence, must respect the laws of nature and the delicate equilibria existing between the creatures of this world . . ."[44]

Sen makes a similar argument but within a non-Christian framework, drawing on some tenets of Buddhist thought. Writing on the climate crisis and on human responsibility to other species, Sen argues for the need for a normative framework in the debate on climate change: one that he thinks—and I agree—should recognize the growing need for energy consumption by humans if the masses of Africa, Asia, and Latin America are to enjoy the fruits of human civilization and to acquire the capabilities needed for making truly democratic choices. But Sen also recognizes that human flourishing can come at some significant cost to other species and therefore advocates a form of human responsibility towards nonhumans. Here is how his argument goes:

> Consider our responsibilities toward the species that are threatened with destruction. We may attach importance to the preservation of these species not merely because the presence of these species in the world may sometimes enhance our own living standards. This is where Gautama Buddha's argument, presented in *Sutta Nipata*, becomes directly and immediately relevant. He argued that the mother has responsibility toward her child not merely because she had generated her, but also because she can do many things for the child that the child cannot itself do.
>
> In the environmental context it can be argued that since we are enormously more powerful than other species, [this can be a ground for our] taking fiduciary responsibility for other creatures on whose lives we can have a powerful influence.[45]

There is, of course, some irony in the fact that one of the species "threatened with [at least partial] destruction" is the human species itself. Humans need to be responsible to themselves, which, as the history of humanity shows, is easier said than done. But think of the problems that follow from this anthropocentric placing of humans *in loco parentis* with regard to "creatures on whose lives we can have a powerful influence." We never know of all the species on which our actions have a powerful influence; often we find out only with hindsight. Peter Sale, the Canadian ecologist, writes about "all those species that may be able to provide goods [for humans] but have yet to be discovered and exploited, and those that provide services of which we simply are unaware."[46]

This applies even more to the life-form that constitutes the "sheer bulk of the Earth's biomass": microbial life (bacteria and viruses). As Martin J. Blaser observes in his book *Missing Microbes*, microbes not only "outnumber all the mice, whales, humans, birds, insects, worms, and trees combined—indeed all the visible life-forms we are familiar with on Earth—they outweigh them as well."[47] Could we ever be in a position to value the existence of viruses and bacteria hostile to us, except insofar as they influence—negatively or positively—our lives? Here again the question is complicated by the fact that ecology and pathology often give us changing and contrary perspectives. Bacteria and viruses have played critical and often positive roles in human evolution, such as the ancient stomach bacteria *Helicobacter pylori*. Since the rise of antibiotics and the consequent changes in the biotic environments of our stomach, however, *H. pylori* has come to be seen a pathogen.[48] We cannot be responsible stewards for these life-forms even when we cognitively know about the critical role they have played—and will continue to play—in the natural history of life, including that of human life itself.[49]

This would mean that humans could only ever discharge the responsibility that Sen tasks them with imperfectly, since they would never fully know who exactly their wards were or for whom they could assume responsibility in a fiduciary sense. But here indeed is evidence of the strain under which the Kantian fable of human history currently labors. Kant did not demand of human morality that it brought within its own conspectus the natural history of life. Needless to say, however, his framework was based on a pre-Darwinian understanding of the history of natural reproductive life and constructed long before humans began to discover and understand the roles of microbes in biological history. We are at a point, however, where we are debating the question of extending the sphere of human morality and justice to include the domain of natural reproductive life.

It is, of course, undeniable that questions of justice between humans have been central to the tradition of the humanities. The intensification, globally, of capitalist forms of social organization has sharpened the political instincts of scholars in the human sciences. Furthermore, given the history of human values in the second half of the twentieth century, we are committed in principle to securing the life of every human and to ensuring their moral and economic flourishing, regardless of the overall size of the human population and its implications for the biosphere.[50] Besides, any practical proposal for reducing the size of human populations in effect becomes an anti-poor proposition and is therefore morally repugnant. At the same time, a single-minded focus on human welfare and intrahuman justice increasingly seems inadequate. This is the dilemma to which the humanities need to respond. The question is: since what the humanities and the human sciences provide are perspectives from which to debate the issues of our times, can they overcome their hallowed and deeply set human-centrism and learn to look at the human world also from nonhuman points of view?

Turning to Latour, Looking Ahead

Latour developed his art of thinking long before many of us woke up to the problem to which he was responding: the problem posed to modern thought by the unsustainable opposition between nature and science on the one hand and culture and society on the other. He has developed his thinking over a number of texts including the recent *An Inquiry into Modes of Existence*.[51] Since I have been discussing microbial life in this essay, however, let me turn to the classic book of his that speaks of microbes, *The Pasteurization of France*, to show how his thinking clears a path for developing an approach that challenges human modes of being and knowing, and where the human receives intimations of the nonanthropocentric precisely through the rustle of a language that no doubt remains, ultimately, all too human.[52] Additionally, it remains a nice coincidence for this essay that Latour's anticolonial humor in this book is aimed in part at the good old philosopher from Königsberg whose titanic presence in all discussions of modernity, for all the barbs we can throw at him, is impossible to escape.

Quite early on in his study of Louis Pasteur's work, Latour draws our attention to the agential presence of microbes not only within the constrained conditions of the laboratory, but in everyday human life. "A salesman sends a perfectly clear beer to a customer," writes Latour, but "it arrives corrupted." Why? Because "between the beer and the brewer there was something that sometimes acted and sometimes did not. A *tertium quid*: 'a yeast,' said the revealer of microbes" (*PF* 32, 33). The presence of microbes tells Latour that "we cannot form society with the social alone": we have to add in "the action of microbes" (*PF* 35). Thus, "you organize a demonstration of Eskimos in the museum. They go out to meet the public, but they *also* meet cholera and die. This is very annoying, because all you wanted to do was to show them and not to kill them." "Traveling," similarly, "with cow's milk is another animal that is not domesticated, the tuberculose bacillus, and it slips in with your wish to feed your child. Its aims are so different from yours that your child dies" (*PF* 33–34). Thus it is only after the milk has undergone the process of Pasteurization and the microbe has been "extirpated" that it will come to represent the purely "social," i.e. "economic and social relationships in the strict sense," which can only happen in some very limited and technologically produced conditions (*PF* 39).[53] Latour concludes the first part of this book by remarking that "as soon as we stop reducing the sciences to a few authorities that stand in place of them, what reappears is not only the crowds of human beings, . . . but also the 'nonhuman'" (*PF* 149–50). His project becomes that of "the emancipation of the nonhumans" from what he calls "the double domination of society and science" (*PF* 150).

Microbes speak of deep time in the history of life. "For about 3 billion years," writes Blaser, "bacteria were the sole living inhabitants on Earth. They occupied every tranche of land, air, and water, driving chemical reactions that created the biosphere, and set conditions for the evolution of multicellular life."[54] Emancipating such nonhumans from the "double domination of science and society" could not be a political task in any institutional sense of the political. Nor does it produce an immediate program of activism. It is a question, primarily and at the current state of development of the governing institutions of humans, of creating a nonanthropocentric perspective on the human world.

In the second part of the book, "Irreductions," Latour looks upon this project of "emancipation" of the nonhuman as something akin to an intellectual act of decolonization. "Things-in-themselves?" he asks, putting a rhetorical question to Kant with his characteristic wit, and retorts: "But they're fine, thank you very much. And how are you? You complain about things that have not been honored by your vision?" Latour's critique of the anthropocentrism of Kant's thinking uses the metaphor-concept of colonization to create agential space for the nonhuman. "Things in themselves lack nothing, just as Africa did not lack whites before their arrival," he writes. "However, it is possible to force those who did perfectly well without you to come to regret that you are not there. Once things are reduced to nothing, they beg you to be conscious of them and ask you to colonize them." And he proceeds to place Kant in a line of colonial heroes: "You are the Zorros, the Tarzans, the Kants, the guardians of the widowed, and the protectors of orphaned things" (*PF* 193). "What would happen," he asks further, "if we were to assume instead that things left to themselves are lacking nothing?"

This is also where the idea of deep time becomes a part of his critique: "For instance, what about this tree, that others call *Wellingtonia*? . . . If it is lacking anything, then it is most unlikely to be you. You who cut down woods are not the god of trees. . . . It is older than you. Soon you may have no more fuel for your saw. Then the tree with its carboniferous allies may be able to sap *your* strength." And Latour drives home the limitations of calculating on human time scales: "So far it [the tree] has neither lost nor won, for each defines the game and time span in which its gain or loss is to be measured" (*PF* 193).

And then comes the arrow of a question aimed at the heart of an ancient Biblical thought, one that declared humans to be specially destined to exercise dominion over the planet: "Who told you man was the shepherd of being? Many forces would like to be shepherd and to guide others as they flock to their folds to be sheared and dipped. . . . There are too many of us, and we are too indecisive to join together into a single consciousness strong enough to silence all the other actors. Since you silence the things that you speak of, why don't you let them talk by themselves about whatever is on their minds, like grown-ups? . . . Do you enjoy the double misery of Prometheus so much?" (*PF* 192–94). This I regard as the most important civilizational question of our times, the one that the Pope raised within the limits of his religion.

Latour's epochal question reminds us that deep pasts and futures are not amenable to human-centered political thought or action. This does not mean that our usual disputations about intrahuman in/justice, inequalities, oppressive relationships will not continue; they will. But now that the moral and biological lives of the species *Homo sapiens* cannot any longer be disentangled from each other, one has to learn to have recourse to forms of thought that go beyond—but that do not discard—the human-political. The connected stories of the evolution of this planet, of its climate, and of life on it cannot be told from any anthropocentric perspective. These other perspectives are necessarily anchored in stories of deep time, and they make us aware that humans come very late in the history of this planet, which was never engaged in readying itself for our arrival. We do not represent any point of culmination in the story of the planet. This is where Latour's—and some other scholars' attempts—to open up vistas of aesthetic, philosophical, and ethical thought help us to develop points of view that seek to place the current constellation of environmental crises in the larger context of the deeper history of natural reproductive life on this planet. This I see as a primary purpose of the "new" humanities of our times.[55]

Notes

A version of this essay was presented to the Centre for Policy Research, New Delhi, in March 2016. I am grateful to my hosts, Pratap Bhanu Mehta and Navroz Dubash, and to the audience for the comments they made. Thanks are also due to Rochona Majumdar, Rita Felski, and Stephen Muecke for comments on an earlier draft and to Gerard Siarny for help with research.
1. Tim Lenton, "2oC or not 2oC? That is the Climate Question," *Nature* 473, no. 7345 (2011): 7, available at http://www.nature.com/news/2011/110504/pdf

/473007a.pdf. For the exact wording of the phrase, see Article 2 of the *United Nations Framework Convention on Climate Change* (New York: United Nations, 1992), 4, available at https://unfccc.int/resource/docs/convkp/conveng.pdf.

2. Eric Holthaus, "When Will the World Really Be 2 Degrees Hotter Than It Used To Be?" *FiveThirtyEight*, March 23, 2016, http://fivethirtyeight.com/features /when-will-the-world-really-be-2-degrees-hotter-than-it-used-to-be/. I am grateful to my colleague, James Chandler, for drawing my attention to this article.

3. Jan Zalasiewicz and Mark Williams, *The Goldilocks Planet: The Four Billion Year Story of Earth's Climate* (Oxford: Oxford Univ. Press, 2012).

4. Julia Adeney Thomas, "History and Biology in the Anthropocene: Problems of Scale, Problems of Value," *American Historical Review* 119, no. 5 (2014): 1588.

5. Tim Flannery, *The Weather Makers: How Man is Changing the Climate and What it Means for Life on Earth* (2005; Melbourne: Text Publishing, 2008). See chap. 22, entitled "Civilisation: Out with A Whimper?"

6. See, for instance, the literature cited and discussed in Dipesh Chakrabarty, "From Civilization to Globalization: The 'West' as Shifting Signifier in Indian Modernity," *Inter-Asia Cultural Studies* 13, no. 1 (2012): 138–52.

7. Peter Singer, "Climate Change: Our Greatest Ethical Challenge" (lecture, Univ. of Chicago, Chicago, IL, October 23, 2015). See https://cie.uchicago.edu /event/climate-change-our-greatest-ethical-challenge-peter-singer.

8. For more on this, see my "The Future of the Human Sciences in the Age of Humans: A Note" in *European Journal of Social Theory* 20, no. 1 (2017): 39–43.

9. Martha C. Nussbaum, *Not for Profit: Why Democracy Needs the Humanities* (2010; Princeton, NJ: Princeton Univ. Press, 2012), 7.

10. In the following couple of paragraphs, I draw upon my "response" piece, "Whose Anthropocene? A Response" in "Whose Anthropocene? Revisiting Dipesh Chakrabarty's 'Four Theses,'" ed. Robert Emmett and Thomas Lekan, *RCC Perspectives: Transformations in Environment and Society* 2016, no. 2 (2016): 103–14.

11. Hans Jonas, *The Imperative of Responsibility: In Search of an Ethics for the Technological Age*, trans. Jonas with David Herr (1979; Chicago: Univ. of Chicago Press, 1984), 22.

12. See, for instance, Steve Vanderheiden, *Atmospheric Justice: A Political Theory of Climate Change* (New York: Oxford Univ. Press, 2008).

13. For further discussion, see my "The Climate of History: Four Theses," *Critical Inquiry* 35 (2009): 197–222; "Climate and Capital: On Conjoined Histories," *Critical Inquiry* 41, no. 1 (2014): 1–23; "Postcolonial Studies and the Challenge of Climate Change," *New Literary History* 43, no. 1 (2012): 1–18.

14. Daniel Lord Smail, *On Deep History and the Brain* (Berkeley and Los Angeles: Univ. of California Press, 2007).

15. Yuval Noah Harari, *Sapiens: A Brief History of Humankind* (New York: Harper, 2015), 9.

16. Harari, *Sapiens*, 11.

17. Harari, *Sapiens*, 11–12. A similar point is made by Jonas while comparing the speed of human-technological changes to that of changes brought about by natural evolution: "Natural evolution works with small things, never plays for the whole stake, and therefore can afford innumerable 'mistakes' in its single moves, from which its patient, slow process chooses the few, equally small 'hits.' . . . Modern technology, neither patient, neither patient nor slow, compresses . . . the many infinitesimal steps of natural evolution into a few colossal ones and foregoes by that procedure the vital advantages of nature's 'playing safe.'" Jonas, *Responsibility*, 31. See also the section on "Man's Disturbance of the Symbiotic Balance," 138.

18. Harari, *Sapiens*, 11–12.

19. The last big famine India saw, for example, was in 1943, though many in the country still die of hunger and malnutrition.

20. William R. Catton Jr., *Overshoot: The Ecological Basis of Revolutionary Change* (Chicago: Univ. of Illinois Press, 1980), 95–96; Doug Cocks, *Global Overshoot: Contemplating the World's Converging Problems* (New York: Springer, 2013).

21. See Lewis Regenstein, "Animal Rights, Endangered Species and Human Survival," in *In Defense of Animals*, ed. Singer (1985; New York: Blackwell, 1987), 118–32; Singer, "Down on the Factory Farm," in *Animal Rights and Human Obligations*, ed. Tom Regan and Singer (1976; Englewood Cliffs, NJ: Prentice Hall, 1989), 159–68.

22. See Jessica C. Stanton et al., "Warning Times for Species Extinction Due to Climate Change," *Global Change Biology* 21, no. 3 (2015): 1066–77; Rodolfo Dirzo et al., "Defaunation in the Anthropocene," *Science* 345, no. 6195 (2014): 401–06; Céline Bellard et al., "Impacts of Climate Change on the Future of Biodiversity," *Ecology Letters* 15, no. 4 (2012): 365–77; Gerardo Ceballos et al., "Accelerated Modern Human-Induced Species Losses: Entering the Sixth Mass Extinction," *Science Advances* 1, no. 5 (2015): 1–5.

23. Will Steffen et al., "The Trajectory of the Anthropocene: The Great Acceleration," *The Anthropocene Review* 2, no. 1 (2015): 1–18.

24. Steffen et al., "The Trajectory of the Anthropocene," 1–18.

25. J. Rockström et al., "Planetary Boundaries: Exploring the Safe Operating Space for Humanity," *Ecology and Society* 14, no. 2 (2009): 32, available at http://www.ecologyandso-ciety.org/vol14/iss2/art32/.

26. Richard E. Zeebe et al., "Anthropogenic Carbon Release Rate Unprecedented During the Past 66 Million Years," *Nature Geoscience*, March 21, 2016, http://www.nature.com/ngeo/journal/v9/n4/abs/ngeo2681.html.

27. "If global warming and a sixth extinction take place in the next couple of centuries, then an epoch will seem too low a category in the hierarchy [of the geological timetable]." Personal communication with Professor Zalasiewicz, September 30, 2015.

28. Gary Tomlinson, "Toward the Anthropocene: Deep-Historical Models of Continuity and Change," *South Atlantic Quarterly* 116, no. 1 (January 2017): 24–25.

29. Zalasiewicz, "The Geology behind the Anthropocene," *Cosmopolis* 2015, no. 1 (2015): 12. I am grateful to Professor Zalasiewicz for sharing this paper with me.

30. See the detailed and excellent discussion in Frank P. Incropera, *Climate Change: A Wicked Problem—Complexity and Uncertainty at the Intersection of Science, Economics, Politics, and Human Behavior* (New York: Cambridge Univ. Press, 2016).

31. The expression "geology of morals" was suggested to me by Bruno Latour on reading an earlier version of this essay.

32. The separation was formalized in the nineteenth century when the modern social sciences arose as identifiable disciplines. See Fabien Locher and Jean-Baptiste Fressoz, "Modernity's Frail Climate: A Climate History of Environmental Reflexivity," *Critical Inquiry* 38, no. 3 (2012): 579–98.

33. The saga of that intellectual war is recapitulated in Eric M. Gander, *On Our Minds: How Evolutionary Psychology is Reshaping the Nature-versus-Nurture Debate* (Baltimore: Johns Hopkins Univ. Press, 2003), chap. 3.

34. For a different, stimulating, and critical reading of this essay, see Bonnie Honig, *Political Theory and the Displacement of Politics* (Ithaca, NY: Cornell Univ. Press, 1993), 19–24.

35. Immanuel Kant, "Speculative Beginning of Human History," in Kant, *Perpetual Peace and Other Essays*, trans. Ted Humphrey (1983; Indianapolis, IN: Hackett, 1988), 54 (hereafter cited as SB). The words within parenthesis belong to this edition of the essay.

36. For the use of "animal," "natural," "physical" as synonyms, see 54n.

37. See the literature discussed in and the conclusions of Daniel P. Shields, "Aquinas and the Kantian Principle of Treating Persons as Ends in Themselves" (PhD thesis, Catholic Univ. of America, 2012), chap. 1 and 2.

38. I completely agree with Honig's remark: "The stories fables tell about the founding of a form of life invariably serve as powerful illustrations of the now more subtle and sedimented but no less active processes and practices that constitute and maintain our present, daily." Honig, *Political Theory*, 19.

39. See, again, Honig, *Political Theory*, 27–34, for a rich and complex reading of these Kantian injunctions.

40. Kant, *The Critique of Judgement*, trans. James Creed Meredith (1790; Oxford: Clarendon, 1973), 2:93–94.

41. For a critical discussion of some of the issues involved here, see Clive Hamilton, "The Delusion of the 'Good Anthropocene': Reply to Andrew Revkin" (June 17, 2014), available at http://clivehamilton.com/the-delusion-of -the-good-anthropocene-reply-to-andrew-revkin/.

42. Pope Francis, *Encyclical on Climate Change and Inequality: On Care for Our Common Home* (London: Melville House, 2015), 72–74.

43. Pope Francis, *Encyclical on Climate Change and Inequality*, 73.

44. Pope Francis, *Encyclical on Climate Change and Inequality*, 42, 43.

45. Amartya Sen, "Energy, Environment, and Freedom: Why we must think about more than climate change," *The New Republic* 245, no. 14 (August 25, 2014), 39.

46. Peter F. Sale, *Our Dying Planet: An Ecologist's View of the Crisis We Face* (Berkeley and Los Angeles: Univ. of California Press, 2011), 222. J. R. McNeill and Peter Engelke note in their book, *The Great Acceleration: An Environmental History of the Anthropocene since 1945* (Cambridge, MA: Harvard Univ. Press, 2014), 87, that of the quarter of a million species that went extinct in the twentieth century, most "disappeared before they could be described by scientists" and were creatures "unknown to biology."

47. Martin J. Blaser, *Missing Microbes: How the Overuse of Antibiotics is Fueling Our Modern Plagues* (New York: Picador, 2014), 13–14, 15, 16.

48. Blaser, *Missing Microbes*, chap. 9.

49. Luis P. Villarreal, "Can Viruses Make Us Human?" *Proceedings of the American Philosophical Society* 148, no. 3 (2004): 296–323; Linda M. Van Blerkom, "Role of Viruses in Human Evolution," *Yearbook of Physical Anthropology* 46 (2003): 14–46.

50. See the discussion in my "Climate and Capital."

51. See, in particular, the following three works by Latour: *We Have Never Been Modern*, trans. Catherine Porter (Cambridge, MA: Harvard Univ. Press, 1993); *Politics of Nature: How to Bring the Sciences into Democracy*, trans. Porter (Cambridge, MA: Harvard Univ. Press, 2004); and *An Inquiry into Modes of Existence: An Anthropology of the Moderns*, trans. Porter (Cambridge, MA: Harvard Univ. Press, 2013).

52. Latour, *The Pasteurization of France*, trans. Alan Sheridan and John Law (Cambridge, MA: Harvard Univ. Press, 1993), 193 (hereafter cited at *PF*). See also *Politics of Nature* and *Modes of Existence*.

53. See also *PF*, 43.

54. Blaser, *Missing Microbes*, 12–13.

55. Bill Brown's *Other Things* (Chicago: University of Chicago Press, 2015) is a notable example of recent work in this direction.

CHAPTER EIGHT

Fictional Attachments and
Literary Weavings in the Anthropocene

YVES CITTON

BY SUMMING UP thirty years of personal research, collective inquiries, and countless publications, Bruno Latour's AIME project, expanding on his 2013 book *An Inquiry into the Modes of Existence*, represents a remarkable attempt to reset modernity in order to reposition our forms of life and categories of understanding more in line with the disquieting demands of the Anthropocene. Its basic—even though antifoundationalist—question is: how are we to communicate between disciplines, between cultures, between human and nonhuman entities, so that what we hold dear in our modern ways of living can be preserved, nurtured, and fostered, while overcoming the epistemological and existential paradigm that has set modernity on a fast track to social and environmental collapse?

How can literary studies fit within such a program? Rita Felski, in her recent work, has already mapped out a few promising directions. Latour's sneaky irreverence toward "the Moderns" has led him to stress the shortcomings of the *critical* attitude that has come to be identified with "literary criticism" for a good number of decades.[1] From the avant-garde denunciation of bourgeois conformism to the demystification of mass-media stereotypes,[2] from the Sartrean critical intellectual to the Adornian critical theorist, from triumphant deconstruction to postcolonial denunciations, and all the way to emerging ecocriticism, texts have been consistently *read against*—against their author's intent, against their class of origin, against their social effects, against their grain. While, in spite of their apparently negative stance, most of these critical readings have been very positively productive, Latour helps us feel the latent presence of a self-defeating arrogance inherent in the position of superiority taken up by the critic-as-demystifyer. Critique is irretrievably modern insofar as it accuses (or suspects) the others naively to "believe" in something the demystifyer "knows" to be illusory. When asserting that *We Have Never Been Modern*, Latour suggests that we should never be naively critical—even though we can't fully avoid being so.

This critique of critique is not a mere negation of negation, bringing us back to our comfortable critical mode of reading, only one notch higher, more superior and self-confident than ever. It flattens the very structure on which critique and criticism rested in order to unveil *the* (hidden, repressed, underlying, profound) meanings of the texts. As Felski eloquently shows, "postcritical" literary practices debunk the implicit superiority of the critic, putting the interpreter's tinkering with the text on par with (if not below) the immanent agency of the text itself (via and beyond its author), a text treated as an ever-springing source of new affordance. The point is not exactly, in a Marxian twist, that texts can no longer be merely "interpreted," but need to be "acted upon" (used, performed, taken to task)—since this had already always been the case under the reign of critique. The point is rather to reorient the interpretive performance of the texts toward a more explicitly constructive use of its affordance. Instead of asking what our interpretation can *undo* (totalitarianism, capitalism, colonialism, sexism, mastery, fundamentalism, etc.), we are invited to ask what it can *make* (a platform of negotiation? a mapping of controversies? a handbook of strategies? a lexicon of sanity?). This same instruction will apply here: what can literary scholars and teachers make with/of Latour's work? And conversely: what can a Latourian approach make with/of literary skills?

My response will develop along the following lines: literary studies can find inspiration in Latour's AIME project insofar as it helps them map and locate where and how they can contribute to the debates that will shape our necessary turning away from the suicidal path of capitalist ecocide. In return, AIME can find in literary studies a field of research that will help it overcome its current attentional deficit toward the issues of media agency. Along the way, we will discover that the Humanities may indeed be the most dangerous and reactionary of all disciplines, unknowingly siding with our most intimate enemy in the anthropocenic war ("the Humans") . . .

Recoloring the Cows [PRE]

We can locate two promising sites in order to initiate a dialogue between Latour's work and literary studies in his conception of distributed agency and in his notion of attachments, as Felski noted in a recent article.[3] From the late 1980s until the early 2000s, Latour developed a highly successful cottage industry under the label of actor-network-theory (or ANT). While most of us envisaged action as performed and authored by a human subject (occasionally helped by various forms of instruments and assistants), his work taught us to attribute the efficiency of human actions to complex *networks* assembling heterogeneous bits of physical stuff, institutional leverage, symbolic tools, energy sources, intellectual credit, and financial montage. In parallel with Foucault's *dispositifs* and Deleuze's *agencements*, his *hybrids* have unhooked us from our romantic addiction to a heroic, unrealistic, and self-illusory model of personal agency, making us more aware of the distributed nature of human agency. Here again, one never really acts *against* one's environment, but always necessarily with it and through it. More often than not, our networks act through us, with our personal agency operating as a mere relay, no more and no less decisive than the other human and nonhuman elements activated in the network.

After thirtysome years of impressive mileage and increasing returns milked from the ANT farm, Latour attempted to relocate his operations on a slightly different, more discriminating, plane. The massive project devoted to *An Inquiry into the Modes of Existence*, launched years ago but rewritten and completed in the 2010s, was, among other things, an attempt to overcome some shortcomings and abuses in the popular success of ANT, where "everything equally becomes actor-network," with the consequence that, "as denounced by Hegel, all cows become grey. When I realized that, I thought something was wrong: it is very well to make an actor-network, it *unfolds* the associations. But it does not *qualify* them."[4] The goal of the AIME uplift was precisely to qualify, within a rich and subtle range of twelve categories (each identified by a three-letter acronym standing for the preposition it specifies—PRE), the various modes of existence that give its unique colors to each action-network.

Brutally summarized for those who may have missed this latest episode, the AIME recoloring of the ANT cows went somewhat like this (if you've read the book, fast forward to the next section; if you haven't, fasten your seat belt!).[5]

The ANTities composing our common universe are perceived as agents insofar as they enter into networks [NET] allowing them to reproduce their existence [REP], metamorphose their identity [MET], and develop habits [HAB] that tend to concatenate chains of heterogeneous operations into units of action. At this first, basic level, these ANTities demonstrate both a capacity to persevere in their own being (comparable to Spinoza's *conatus*) and a propensity to (self-)plasticity, which allows them to adapt to constantly changing environments.

In the course of this adaptation, they elicit the apparition of three types of "quasi-objects." Through the zigzagging invention of technical objects [TEC], they devise short-circuits allowing them to fold long series of operations into speedy and seemingly effortless tricks. With the help of fictions [FIC], they sustain worlds capable of living off of their own coherent (self-induced but not autonomous) systems of resonance. Thanks to more or less rigid procedures of reference [REF], they elaborate cognitive constructions allowing them to secure access to phenomena and causalities far removed in space and time. This second level provides the ANTities with various types of extensions of themselves, folding time, space, and agency along ever more complex lines and dynamics.

Such foldings generate three types of "quasi-subjects" that, in their turn, further the development of yet more unpredictable extensions of self-plasticity. Through politics [POL], the ANTities circularly convince each other about what ought to be the best common course of actions, speaking obliquely again and again about the same topics: always to be reconsidered under a slightly different light, never really agreeing, but producing along the way larger ANTities in which a collection of "I" tends to cohere in a collective assertion of "we." Through law [LAW], the ANTities devise (and conform to) certain means of enunciation meant to validate proper forms of translation through various domains of action, originally heterogeneous to each other. Through religion [REL], the ANTities feel called to be something else (or more?) than ephemeral networks; they gain in subjective consistency by being addressed as "persons," expected to respond for the purposes and implications of their actions, well beyond their brief individual existence on earth. At this third level, [POL], [LAW], and [REL] together invest the ANTities' agency with experiences of subjectivation, which provides them with a very relative, very dubious, but nevertheless very necessary sense of autonomy within the multiple levels of intra-actions constitutive of our multiverse.

In order to articulate more finely and more strongly quasi-objects with quasi-subjects, a fourth level of analysis focuses on the links that tie them together—with the explicit goal of providing an alternative to the operation devoted to the "economic science" at the turn of the twenty-first century. In spite of their necessary sense of relative individual autonomy, quasi-subjects cannot help but experience attachments [ATT] to countless other forms of beings: their emotions, passions, desires, needs, and interests constantly remind them how dependent they are on each other, as well as on a wide variety of other means of subsistence, comfort, and pleasure. [ATT] accounts for *an economy of (often unequal) interdependencies*. The management of such complex forms of attachments requires a great deal of organization [ORG]: quasi-subjects devise stacks upon stacks of ingenious scripts, in order to ensure that the appropriate elements of their environment will be at the right place at the right time to meet their desires, needs, and interests. All scripts, however, were not born equal. Some are broader, more intense, more commanding, more powerful than others: macroscripts absorb microscripts within vertical and entangled trees of inclusion, integration, and subordination. [ORG] accounts for *an economics of hierarchical management*. There seems to be an irreducible gap, delay, and *différance* between the ever more clever devices invented to manage the organization of our attachments and our intuitive perception of balance and fairness in the exchanges of goods, services, and favors. Morality [MOR] manifests itself through the nagging scruple that a transaction may have left one of the parties short-changed, while other parties gained more than their fair share. The face-value of procedural justice constantly needs to be readjusted to the fair value of a more substantive perception of justice, attentive to the singularity and relative weight of the contracting parties. [MOR] questions the dominant accounting procedures in the name of *a moral economy*.

AIME attempts to emancipate these twelve modes of existence (REP, MET, HAB; TEC, FIC, REF; POL, LAW, REL; ATT, ORG, MOR) from their current suppression under the collapsing weight of the economic ideology. This impressive project is guided by (at least) two highly ambitious goals. The first, *diplomatic* goal consists in producing a document that our Western modern culture could bring to the negotiation tables where the different inhabitants of planet Earth are already bound, willy-nilly, to discuss the way in which (as well as the values according to which) they are willing to share and, more urgently, to protect our common assets. The second, *anthropological* goal aims toward reversing, or bifurcating, the evolution that, under the domination of Western modernity, has led to the tyrannical and suicidal rule of one undifferentiated science—"economics." AIME analyzes the dismal science as an unstable, indiscriminate, and inconsistent mash-up of [ATT] and [ORG], pretending to have set itself free from [MOR], thanks to a supposedly value-neutral use of [REF], and imposing its totalitarian criterion of accounting onto areas of concern that, in reality, require other, very different and much more specific criteria of evaluation. Both in order to put on the negotiation table an explicit description of our Western values distilled through the modern period, and in order to open up the noose of the "economic rationality" currently strangling our sociopolitical evolutions, AIME relies on the use of the twelve prepositions [PRE] succinctly described in this summary: these three-letter operators are devised both to ensure the relative autonomy recognized in these various modes of existence (currently crushed under the tyrannical hegemony of the "economic science"[6]) and to help their pluralist articulation within our multileveled forms of collective agency.

Reformulating the Nets [FIC]

Where does literature fit within such a Big Picture? Its main location, of course, is to be found in FIC: the *Odyssey*, the *Divine Comedy*, *Jacques le fataliste*, *Sense and Sensibility*, *Molloy*, or *Ma Rainey's Black Bottom* all provide a certain form of presence to fictional ANTities, with which the readers and spectators develop certain forms of attachments. A human-invented plot, which never and nowhere "existed" in the first place (i.e., which necessarily escapes the procedures of [REF]), does indeed have a certain mode of "existence," since it does affect us, sometimes quite profoundly. We (really) "care about" fictional characters: we fear for them, we hope for them, we are happy when they end up happy. Our encounter with them often alters and shapes our worldview, our perceptions, our attention, our behaviors—sometimes much more significantly than do our encounters with "real" human beings.

[FIC], of course, is not restricted to the mere "content" of the fictions (their characters, their plot), but accounts for the mode of existence specific to what we, Moderns, identify as "works of art," where the medium, forms, and content constitute an inseparable unity. An abstract painting or a sonic organization, totally independent from any representational pretense, belongs to [FIC] as much as Emma Bovary does. Their felicity conditions are not to be found in the correspondence with an external reality, but in an immanent force of vibration (a property Bernard Stiegler calls "consistence," distinguishing it from subsistence, existence, and insistence).[7] The beings of [FIC] exist inasmuch as they are animated by the (mysterious and elusive) strength of resounding vibrations that allows them to resonate both internally, thanks to a certain degree of coherence between their different elements of composition, and externally, thanks to their attunement to concerns and issues that inhabit and structure our shared experience of reality. In order to benefit from the surplus value of sensitivity, sensuality, or intelligibility that they can bring to us, however, we need to "care for" them: they only exist insofar as we have the "disponibility" (the leisure, the time, the luxury) to invest our attention in them—for they live solely from our attention: their mode of existence vanishes when no human attention comes to fuel their terribly fragile and precarious life.

But [FIC] is not even restricted to what we commonly call "fiction," in a rigid and superficial opposition with "reality." Our modern definition of literature includes "nonfiction" works such as Montaigne's *Essays* or Rousseau's *Confessions*, as legitimately as Margaret Cavendish's allegorical flights of fancy in *The Blazing World*. In its broadest sense, [FIC] encompasses all of our expressive attempts to forge (*fingere*: invent, devise, design, craft) objects that *may* help our orientation within our puzzling experience—which they do as soon as they manage to resonate in and with it. [FIC] are therefore to be found everywhere, not only in literature and the arts, but in religion, politics, law, and even the sciences. They are ANTities that we throw into the world, in our expressive attempts to catch aspects of reality previously escaping from our grasp—like nets that sometimes are pulled back loaded with prey (when they are blessed with phasing into an external resonance), sometimes come back empty (when their tinkered assemblage is not graced with inner consistency, when they fail to match outer wavelengths, when nobody listens). Thus, there is [FIC] wherever humans (or even nonhumans[8]) attempt to formulate an expressive device that will gain another (firmer) mode of existence by returning home loaded with different properties: this device can be a judgement, when a well-argued case returns as [LAW]; a piece of legislation, when a demand manages to close the circle of [POL]; or a scientific discovery, when the conditions for the capture of the prey are made sufficiently explicit by [REF], so that the net systematically returns with its intended targets.

Hence emerges a first redescription of literature according to AIME, very much in tune with our professional common sense: well beyond the pages of novels, poems, or plays, well beyond the limits of the artistic sphere, there is literature as soon as an agent attempts to (re)formulate a new concatenation of letters, words, and sentences in order to meet a shared thirst to account for a yet-inexpressible nuance in our modes of existence. Although somewhat younger in terms of age, Latour clearly belongs to the glorious generation of Barthes, Deleuze, and Derrida, who promoted the practice of "writing" (*écriture*) as central both to social life and to theoretical-philosophical inquiry—as demonstrated by his witty personal style of writing, full of tongue-in-cheek puns and animated by a playful jubilation with the (inexhaustibly wise) poetic richness of our common languages.

Reweaving the ANTs [MET] [REL] [POL] [MOR]

If [FIC] extends well beyond the traditional limits assigned to fictions, similarly, the powers of literature extend well beyond the sole mode of [FIC]. The work accomplished by literary ANTities—actor-networks intricately enmeshing the writer and her readers, the text and its interpretations, the editor and the believer, the aesthete and the scholar, the learner and the teacher—fuels as well several other modes of existence.

When resonating at their highest power, literary experiences transform us by mobilizing affects that are originally out of our control, providing us with the means to tame or unleash them [MET]. They constitute us as persons, they call us to become more than we currently are, they lift us above our own expectations [REL]. They help us share a common ground through the back-and-forth movement of interpretation, through progressive and reciprocal attunements over the deeper meanings that are to be found in the choice of a certain word or in a certain twist of the plot, thus composing, maintaining, adapting the interpretive communities that are the underlying agents of social change [POL]. They excel in making us feel the unacceptability of certain forms of behavior that may be perfectly legal, but nevertheless socially repugnant, expressing scruples that haunt us in silence, undermining our common power to collaborate and coevolve in peace [MOR].

Literary experiences nurture these modes of existence through the way they *matter* to us. As Ranjan Ghosh recently wrote, "Literature is more important in its 'mattering' than in its 'matter,' in the unfolding of love than in the mere assertion of it."[9] The power of [FIC] is less to be located in its capacity to represent something that never actually existed before, than in its ability to make us feel that this inactual ANTity actually (although mysteriously) matters to us. The arts matter through their mattering, i.e., through their power to give actual existence to relations whose *relata* are still to emerge in our shared reality. [FIC] makes us feel attached [ATT] to ANTities that still escape our shared cognitive mapping. These fictional attachments pave the way for fuller recognition of the role played by such ANTities, within as well as without us. They instill or awaken the still-hollow forms of what will start to matter as soon as it is more strongly perceived.

This process of mattering is at the very core of the AIME project as a whole. Against the late modern tendency to attribute subsidiary existence to whatever cannot be translated into quantified financial value, AIME claims that there are at least twelve modes of existence that, for us (who have never been fully) Moderns, are irreducible to each other, even though they most often inextricably permeate each other. Each of these twelve modes matters: each of them suffices, on its own, to claim a certain form of existence, to provide some of the stuff our experience is made of. Each of them is important enough for us collectively to *care about* and *for* it. One could say that the AIME project is itself deeply literary, insofar as its dynamics rely on a recursive loop whereby we care about and for something because it matters to us, while at the same time it matters to us because we care about and for it.

Literature perfectly illustrates this self-feeding loop that provides the *ultima ratio* of our attention ecologies. Latour has often stressed how impossible it is to separate neatly matters-of-facts from matters-of-concern. We construct the facts in light of our concerns, so that the *facta* can simultaneously be described as "objective" (insofar as *they* bring their own responses to our questions independently of our desire to hear them say this rather than that) and as "relative" (insofar as they only respond to these questions that *we* have asked to them, in view of our selfish and limited current concerns). The first movement of the loop is intuitive enough: we pay attention to something when it matters to us. What is less intuitive, but equally true, is the reverse movement: something starts to matter to us when we pay attention to it. An archetypal literary writer expressed this principle more succinctly than anyone else: "For a thing to become interesting, it suffices to look at it for a long time."[10] In other words: concerns make facts matter, but attention to facts generates concerns.

It is this reverse movement of the attention loop that Jean-Marie Schaeffer has recently analyzed as the fundamental drive of the aesthetic experience, after having described it as the crucial spring of successful literary studies.[11] While it is highly questionable that only a specific set of texts deserves to be considered as literary, we can more easily acknowledge that there is a specific type of attention that constitutes a text (whatever it may be) as literary. Literature is less in the eye of the beholder than in his gaze, i.e., in the aesthetic attention he devotes to the text. Schaeffer provides a very rich description of the contrast between standard vs. aesthetic attention (serial vs. parallel treatment of information; ascending vs. descending; convergent vs. divergent; integrating vs. detailing; focused vs. distributed; task-oriented vs. freewheeling; economical vs. anti-economical; hierarchical vs. dehierarchical). His conclusion agrees with Flaubert's statement as to the importance of the time-factor: "To engage in an aesthetic experience means to adopt a particular attentional style, the divergent style," while "the disposition to adopt this divergent cognitive style is proportional to the individual's capacity to tolerate delayed categorization."[12] The facts that matter in an aesthetic experience only surface once the matters of immediate concern (along with their preexisting categorizations) have been temporarily put to rest, so that we can let unsuspected categories emerge from a freewheeling attention that discovers new facts and new concerns within the matter under scrutiny.

This suspensive time and space of delayed categorization requires us first to accept, and then hopefully to revel in, looking and listening without understanding: they need us to be receptive to what can be perceived and made sense of, beyond, above, and apart from what our preconceived categories lead us to identify. This privileged and luxurious time and space allows the reader, listener, spectator to develop a certain type of activity well described by Jacques Rancière in (non-Latourian) terms of "emancipation": the spectator's emancipation "begins when one understands that looking is also a form of action, which confirms or transforms the pre-existing distribution of positions. The spectator too acts, like the pupil or the scholar. He observes, he selects, he compares, he interprets. He links what he sees with many other things he has seen on other stages, in other types of places. He composes his own poem with the elements of the poem provided to him."[13]

What interests me in Rancière's formulation is his reference to the activity of linking, tying, or binding.[14] French poets have for the longest time played with the anagrammatic proximity between *lire* and *lier*: to read is to link, and to reread (*relire*) is to relink (*relier*). Literature— which vividly carries the image of the continuous thread linking the *litterae* traced by the pen on the paper—is a matter of relinking our facts into new forms of concerns, as well as it is a matter of tying new concerns onto previously unobserved facts. Yes, we are always already attached to each other, humans and nonhumans alike, whether we know (and like) it or not. But it takes the relinking activity of the literary tracings for some of these ubiquitous attachments to become matters of concern, to be mapped into actor-networks, to appear as modifiable *facta*, and hence to be reconcatenated into less unjust, more satisfying, less dangerous liaisons.

Latour has often repeated, over the last years, that our common world demands less to be defended, preserved, or protected, than to be *composed*.[15] It does not stand on its own: it is wonky, always on the verge of collapsing, it results from our clumsy efforts to compose it, to link and relink its countless heterogeneous threads, bits, and pieces. In this endless work of composition, linking and relinking, literature appears as a most powerful weaver. Epics, tragedies, comedies, poems, tales, and novels, along with, of course, films and TV series, constantly perform the daily weaving of our social fabric. Their mattering is a meshworking. The ANTities constituted by the countless networks of collaborations and coevolutions require an endless work of maintenance and update: they will be fortunate enough to matter only as long as they manage to weave and reweave the attachments that sustain their existence.

This activity of meshworking has been addressed with both depth and wit by anthropologist Tim Ingold, in a short article in form of a dialogue where he staged himself as a SPIDER while admittedly caricaturing Latour as an ANT. Before the AIME project came to fruition, he (sympathetically) criticized ANT for painting all cows in grey, i.e., for indiscriminately considering everything as a "network." More precisely,

Ingold-the-SPIDER pointed to Latour-the-ANT that the weaving of our daily existence should not be modeled after a network of heterogeneous objects, but after a fabric of threads constituting a *medium* or a *milieu*—ontologically different from a mere entity inhabiting this milieu:

> You imagine a world of entities—spider, web, stems, twigs and so on—which are assembled to comprise the necessary and sufficient conditions for an event to happen. And you claim that the agency that "causes" this event is distributed throughout the constituents of this assemblage. My point, however, is that *the web is not an entity*. That is to say, it is not a closed-in, self-contained object that is set over against other objects with which it may then be juxtaposed or conjoined. It is rather a bundle or tissue of strands, tightly drawn together here but trailing loose ends there, which tangle with other strands from other bundles. The world, for me, is not an assemblage of bits and pieces but a tangle of threads and pathways. Let us call it a *meshwork* . . . so as to distinguish it from your *network*.[16]

Aren't fictional attachments and literary weavings more in tune with the SPIDER than with the ANT? Does literature matter more as a network of entities or as a meshwork of threads? Does it link us together as juxtaposed and conjoined, or does it immerse us in the medium of a milieu? Latour's latest writings, focused on the figure of Gaia, may help us address such questions—and be more specific about what Latour can bring to literary studies and what literary studies can bring to Latour.

Rewiring the Humanities (as Alien Loops)

The series of Gifford lectures Latour delivered at the University of Edinburgh in February 2013 under the general title *Facing Gaia: A New Enquiry into Natural Religion* provide an interesting framework with which to recast the place of literary studies and the humanities in the age of the Anthropocene. The difficulty raised by Ingold is (implicitly) addressed through a (rather literary) problem of nomination, the choice of an adequate name for the-species-formerly-known-as-"humans." After pondering and rejecting a series of other options ("Gaians," "Terrestrials," "Earthlings"), Latour finally unveils the best candidate expected to face up to the irruption of Gaia: "I have chosen *Earthbound*—'bound' as if bound by a spell, as well as 'bound' in the sense of heading somewhere, thereby designating the joint attempt to reach the Earth while being unable to escape from it, a moving testimony to the frenetic immobility of those who live on Gaia. I know that it's terribly dangerous to state the matter this starkly, but we might have to say that at the epoch of the Anthropocene the Humans and the Earthbound should be at war."[17]

The clever choice to rename us "Earthbound," and to consider "the Humans" as our enemies, makes a whole lot of sense once we realize how intimately a certain form of (scientist and anthropocentric) modern humanism has been an accomplice to the careless and arrogant wrecking of our common environment. This choice cannot but question our habit to unite behind the flag of "the Humanities." Even though Latour devoted many semesters of his tenure at Sciences Po Paris teaching a course on *les humanités scientifiques*,[18] it seems difficult for us Earthbound to be simultaneously at war with "the Humans" and enrolled in "the Humanities." This may be where the binds and relinkings operated by literature could provide an alien alternative coming from within the hollow shell of the humanities.

What interests me the most in the baptismal sentence quoted above is the absence of reference to the most obvious connotation of the term "Earthbound": aside from the magic spell and from the direction of movement, the suffix *-bound* fully belongs to the vocabulary of attachments so strongly emphasized in the AIME project. We are bound to the Earth by the many forms of tying, linking, and binding that pull us together on the surface of this planet. Even more than the suffix, it is its linking with the *Earth* that may be the most striking feature of our new name. For the Earth can be understood as a planet, of course, but also, more interestingly, as an element and as a medium.

Or, translated into SPIDER vocabulary: the Earth can stand both for a network of *actants* bound to each other, "juxtaposed and conjoined," and for a muddy medium/milieu that immerses us in the elemental experience of a continuum. Isn't it a persistent mistake of our enemies, the Humans, to divide and conquer by individualizing and objectifying—and inevitably putting a price tag to—what should rather be perceived as a continuous agency? Our Earth is less a network than a bundle: we are less bound by distant links than tied (and crushed upon each other) by tight knots. As Arne Naess stressed many decades ago, environmentalism has got it all wrong from the very start when it claims to preserve an environment that surrounds its inhabitants. The individual *is* its environment: both exist as the set of relations that weave them together into one single piece of fabric.[19]

What should these Earthbound creatures do, with their feet in the mud and their relations tightly knit? The short answer is that they should write literature:

> What I propose to say is that, in this new cosmopolitical situation, those who wish to present themselves to other collectives have a) to specify what sort of people they are, b) to state what is the entity or divinity that they hold as their supreme guarantee and c) to identify the principles by which they distribute agencies throughout their cosmos. Of course, conflicts will ensue—but then also, later, some chance of being able to negotiate peace settlements. It is precisely these peace conditions that are *not even going to be looked for* as long as we believe that the world has *already* been unified once and for all—by Nature, by Society or by God, it doesn't matter which.[20]

This diplomatic endeavor—which constantly looms at the horizon of the AIME project as well as of the Gaia writings that followed it—deeply resonates with the function assigned to literature by a philosopher like Richard Rorty.[21] The necessity (and opportunity) for a plurality of cultures to coexist and coevolve on the surface of planet Earth calls every one of them to express (1) its conception of identities and becomings, (2) its figures of authority and validation, and (3) the modes of existence it is eager to assert and protect (a.k.a. its "final vocabulary"). Literature, as we have already seen, is intrinsically compositionist (a.k.a. "poetic"): we Earthbound need it to weave our lives and values together, thanks to the back-and-forth movements of narration, explanation, explicitation, interpretation, redescription, rewriting, relinking, precisely because we don't trust either Nature, nor Society, nor God to unify them for us.

Our countless literary stories (plays, tales, novels, epics, memoirs, autobiographies) are operators of unification, agents of worlding: our literary attention projects value on them not only according to the narrative lines they weave across our lives, but also according to their capacity poetically to express our perspectives of becomings, our rituals of validation, and our modes of existence. The Gifford lectures are quite explicit on this point: "As for the rites and rituals which are necessary to render this people conscious of its vocations, it is to the artists that we would have to turn."[22]

How should these artists take their turn? By developing a literary form of attention. The delayed categorization that is a precondition to aesthetic attention, according to Schaeffer, appears in the Gifford lectures as a capacity to name and account for our realities by letting them speak through us, by momentarily surrendering agency to the unexpected reactions coming from the Earth back to us. By calling ourselves "Earthbound," in the sense of "heading for the Earth," we remind ourselves that we must delay projecting our preexisting categorization upon our environment, in order to become more attentive to our milieu's weaker signals.

> But Earthbound are not land-surveyors, cartographers or geologists looking *from above* at the flat surface of their well-delineated maps. Their discipline is not geometry and optics but rather biology and natural history. The initiative of naming and surveying no longer comes *from them* to the land they have appropriated by a sovereign gesture of domination. As we have recognized in the third lecture, the lines that they have learned to trace, thanks to their instruments, have the shape of entangled and retroactive *loops*. Those loops don't start with them toward the map, but *from* the landscape *back to them*—and more often than not they come back with a vengeance! Each of those loops registers the unexpected reactions of some outside agency to human action.[23]

It is our literary attention that binds us to the medium of the Earth, considered as a source of unexpected recategorization. Our first imperative must be to control our habits of projecting our (hopelessly "human") mental maps upon our surrounding landscapes: instead of looking for what we know, as Humans have grown accustomed to, we must learn to listen to the noise, in order to let the soundscape reshape our minds—Earth-bound rather than task-oriented.

Such is the challenge of literature in the Anthropocene: the "entangled and retroactive loops" that weave our common lives must originate "from the landscape back to us." Even if, of course, as Rorty reminded us, "the world does not speak. Only we do,"[24] literary attention assumes that some form of Alien wisdom is speaking through the text, well above and beyond the mere intent of its all-too-Human author. The Earth—i.e., the meshwork of relations that sustain our common lives, whose constant interweaving simultaneously composes our milieu and our self—is the ultimate Alien to which artists and shamans have trained themselves to become attentive.[25] And it is this interweaving of echoing responses that gives parallel consistency to the subject of enunciation, to her interpreter, to the interpretive community to which they belong, and to the shared world within which they interact:

> Whatever is reacting to your actions, loop after loop, begins to take on a consistence, a solidity, a coherence, that, for sure, does not have the technical predictability of a cybernetic system, but which nonetheless weighs on you as a *force* to be taken into account. This is what happens when you keep adding the "response" of the ice sheet to the "response" of acidity of the oceans to the "response" of thermohaline circulation, to the "response" of biodiversity, and so on and so forth. Such an accumulation of *responses* requires a *responsible* agency to which you, yourself, have to become in turn *responsible*. Here again, the performances end up generating a competence: "behind" those cumulative responses, it is hard not to imagine that *there exists a power that does listen and answer*. To grant it a personhood is not to imply that it may speak and think or that it exists as one single substance, no more than you would do with a State, but that in the end it has to be recognized as a *politically assembled* sort of entity. What counts is that such a power has the ability to *steer* our action, and thus to provide it with limits,

loops and constraints, which is, as you know, the etymology of the word "cybernetic."[26]

What if Latour was reaching his most penetrating insights when he is apparently giving in to his most religious tendencies? It would be too easy to disqualify this long quote as carried away by a mystical drift. One would be right, of course, to object that the mere fact of "responding" does not suffice to make one "responsible"; that thermohaline circulation does not "respond" to anyone, since it cannot be considered as a subject of enunciation; that imagining a superior power endowed with "the ability to steer our action" and "granting it a personhood," even if a final twist requalifies this power as a "cybernetic" system, cannot but sound uncomfortably close to the countless dogma projecting final causes onto a divine "power that does listen and answer" to human needs and demands. Is Gaia merely a (female) avatar of the Good Lord?

What would happen, however, if we, Earthbound creatures who, as we know, have never been modern in the first place, were to overcome our visceral reluctance toward such mystical drifts? What if the relinking (*relier*) practiced by literature was profoundly analogous to the weaving of agency practiced by religion (*religare*)? What if religion itself was merely a form of literary criticism—since *religio* can also be derived from *relegere*: to read again, with care and devotion, the same canonized texts?[27] What if, within the context of the war against "the Humans," the humanities could only be saved by embracing the critical care and the literary exaltation of Alien loops?

Remediating the Spells [MED?]

We are finally getting a glimpse of where we, Earthbound, are heading to, in the epoch of the Anthropocene: our destiny is *to become-medium*. This is to be understood in the various meanings folded in the highly polysemic term of "medium."[28]

The Anthropocene calls us to become-medium, first and foremost, in the sense of becoming "milieu." Gaia is coming back with a vengeance, in terms of climate change, collapse in biodiversity, pesticidal and nuclear contaminations, because we have ceased to feel identified with the meshwork of relations composing our being-(in-)the-environment. This was indeed the main point stressed by SPIDER in its supplementing of the ANT theory, as staged by Ingold: the spider's web, like the "air and water are not entities that act. They are material media in which living things are immersed, and are experienced by way of their currents, forces, and pressure gradients. For things to interact they must be immersed in a kind of force field set up by the currents of the media that surround them. Cut out from these currents—that is, reduced to objects—they would be *dead*."[29]

This (deep) ecologic awareness resonates with recent currents in literary criticism. For a number of years, ecocriticism has taken up the challenge of reading literature precisely as "an accumulation of *responses* requiring a *responsible* agency to which you, yourself, have to become in turn *responsible*," staging a dense interplay from "the 'response' of the ice sheet to the 'response' of acidity of the oceans to the 'response' of thermohaline circulation, to the 'response' of biodiversity, and so on and so forth."[30] More generally, though, this ecological turn of literary criticism deserves to be understood within the broader perspective of the rich affordances inherent in the references made to "media."

For the Latourian take on the Anthropocene also calls us to become medium in the sense of becoming spiritual "mediums"—shamans and intermediaries between the various ghostly presences that haunt our down-to-Earth realities. After all, the first connotation stressed when baptizing us as "Earthbound" was indeed "as if bound by a spell." Within the AIME project, both [MET] and [REL] take on some of the properties the Moderns have dumped on the boogie man of mediumism.[31] A wonderful little book by Laurent de Sutter has recently suggested, via Gabriel Tarde (the nineteenth-century sociologist so close to Latour's heart), that [LAW] fully deserves to be added to the list, stating that "there is no law [*droit*] without magic, and there is no magic without law."[32] The theory of the "factishes," in its mix of scientific "facts" and magic "fetishes,"[33] already stressed the need to add a mediumic perspective to our supposedly disenchanted approach promoted by modernity. It was profoundly inspired by Tobie Nathan's ethnopsychiatry, which finds much therapeutic wisdom in the "magic" practices disqualified by Western science as "superstition."[34] We Earthbound are spell-bound through the meshwork of relations that influence us well beyond the few causal links we manage to become aware of: we are ubiquitously relinked by these rich religious spells that necessarily complement and strengthen our poorly explicated social ties.

This Latourian magic suggests it may be time for literary studies to account for their mediumic dimension: what are we doing, when we in-terpret literary texts beyond their explicit or historical content, if not becoming mediums, in-spired and in-spirited by dead authors' inscrip-tions whose endlessly unfolding meanings are carried through us by an unpredictable succession of Alien loops? What is the literary experience good for, if not for affording metamorphoses through which both the texts and our selves become something more than they were, something (deliciously or disturbingly) alien to what they used to be? The first and the second meaning of "medium" thus converge in a single necessity. As suggested in a popular song by Seal, "we are never gonna survive unless we get a little crazy": the need to become-medium attuned to the alien loops coming from our environmental milieu is nothing less than a matter of survival. In an age when (foreign and English) literature de-partments are being downsized, if not simply shut down, while narrow-minded GDP accounting, maddeningly driven by financial profit, steers us ever more rapidly against the anthropocenic wall of reckoning, mak-ing literary studies a matter of survival may come more naturally than it may have seemed fifty years ago.

But there is a third meaning for which the call to becoming-medium makes even more sense for Latourian-inspired literary studies— reconceived and *remediated*[35] within the larger field of "media" stud-ies. The notion of media is experiencing a highly stimulating redefini-tion itself, with authors such as John Durham Peters directing his latest inquiry "Toward a Philosophy of Elemental Media," while the New York trinity of Alexander R. Galloway, Eugene Thacker, and McKenzie Wark suggests we should go back to "the central question: *what is me-diation?*"[36] Fifteen years ago, Jeffrey Sconce had already published a masterful study of the constant weavings that tied together media tech-nologies and mediumic imaginaries in a "logic of transmutable flows"

between "1) the electricity that powers the technology, 2) the information that occupies the medium, [and] 3) the consciousness of the viewer/
listener."[37] Beyond the role of informational milieu played by mass-
media since the twentieth century, the most interesting definitions of media provided by younger theoreticians tend to be environmental: "Media
are an action of folding time, space and agencies; media are not the
substance, or the form through which mediated actions take place but
an environment of relations in which time, space and agency emerge."[38]

Literary studies have already provided a good number of the most interesting media theorists, from Marshall McLuhan and Friedrich Kittler
to N. Katherine Hayles and Michael Cuntz. The development of a "media archaeology" creatively hybridizing historical inquiries, science and
technology studies, aesthetic analyses, political philosophy, and artistic
practices allows for media studies both to craft fascinating new objects
of research-experimentation and for vital technopolitical issues to be revisited from the kind of reflexive standpoint to which literary studies
have so much to bring: "Media archaeology is introduced as a way to
investigate the new media cultures through insights from the past new
media, often with an emphasis on the forgotten, the quirky, the non-
obvious apparatuses, practices and inventions. In addition, as argued in
this book, it is also a way to analyze the regimes of memory and creative
practices in media culture—both theoretical and artistic. Media archaeology sees media cultures as sedimented and layered, a fold of time and
materiality where the past might be suddenly discovered anew, and the
new technologies grow obsolete increasingly fast."[39] Because of their
long and rich tradition of hermeneutics, because of their marginalized
situation within the current media landscape dominated by audiovisual
and digital products, because of their reflective and (yes!) critical temporality of rather slow motion, because of their deconstructivist and theoretical bend, literary studies occupy a privileged position, somewhat per-
pendicular to the development of mass- and new-media. Now that their
once-hegemonic position is being (brutally) reduced to quasi-minority
status—an opportunity in terms of intellectual inventiveness, as much as
a curse in terms of institutional status and economic livelihood—they
are well located to shed an orthogonal light on the rapidly evolving me-

diascape, helping to reveal its more occult dynamics—if only literary scholars are willing to recast their practices in the framework designed by Hayles under the label of *Comparative Media Studies*.[40] Such a field of inquiry provides a natural but curiously missing development to both literary studies and the Latourian project. The "media spells"[41] that structure our public debates and social networks are curiously left out of the modes of existence listed in the AIME project, even though they account for a unique and crucial modality of our social dynamics. Even more than what comes from the sensory experience of our proxemic material environment, what currently "matters" to us (individually and, even more so, collectively) is what the media draws our attention to. It is this process of technologically mediated mattering that we desperately need better to understand. Media archaeology provides an inspiring framework to figure out how, by "folding time, space and agencies," the old and new media simultaneously structure our perceptual milieu, cast magic spells once enacted by shamans, sorcerers, and mediums, and remediate the multifarious relations that constantly reweave our social meshwork. Even though skeptical about elevating the media [MED?] to the full status of a mode of existence, Latour does not condemn the project, leaving it to his continuators to build the case and do the job.[42]

Similarly, literary studies have not exploited their capacity to shed light on mediation as a spell, even though most of the analytical and experiential devices to pursue such an inquiry are readily available in our current methodological toolbox. In our anthropocenic age of ubiquitous media, a Latourian approach thus suggests that we consider the task of remediating our media spells as a most important and most promising perspective where literary studies and ecopolitical mediactivism may felicitously converge. If indeed "the Humans and the Earthbound should be at war," the "Humanities" need to be rebound to our earthly environment. And since the media now provide our most common intellectual and sensory milieu, becoming the very element in which we inhabit our world, the humanities need to be media-bound.

Literary studies can matter (again) if they manage to investigate and mobilize our fictional attachments in order to weave our inseparably mediatic and mediumic modes of existence into a reconstituted sustainable meshwork. If, as stated earlier, there is literature as soon as an agent attempts to (re)formulate a new concatenation of letters, words, and sentences in order to meet a shared thirst to account for a yet-inexpressible nuance in our modes of existence, our first task may be to intervene in the anthropocenic war with a baptismal gesture of diplomacy—remediating the humanities, irretrievably tainted by their siding with "the Humans," with a comparative and compositionist conception of the *medianities*, still to be invented.

Notes

1. See Rita Felski, *The Limits of Critique* (Chicago: Univ. of Chicago Press, 2015), esp. 162–85.

2. See for instance Philip Watts, *Roland Barthes' Cinema* (Oxford: Oxford Univ. Press, 2016) and Olivier Quintyn, *Valences de l'avant-garde. Essai sur l'avant-garde, l'art contemporain et l'institution* (Paris: Questions théoriques, 2015).

3. Felski, "Latour and Literary Studies," *PMLA* 130, no. 3 (2015): 740–41.

4. Bruno Latour, "Les Médias sont-ils un mode d'existence?" *INA Global* 2 (2014): 152. This 2013 interview with Michael Cuntz and Lorenz Engell was first published in German translation and in an expanded form as "Den Kühen ihre Farbe zurückgeben. Von der ANT und der Soziologie der Übersetzung zum Projekt der Existenzweisen," in *Zeitschrift für Medienund Kulturforschung (ZMK)* 4, no. 2 (2013): 83–100.

5. The rest of this section is shared with a more political article entitled "All You Need Is LOVE," published in *Alienocene*, Strata 3, 2018, available on https://alienocene.com/2018/10/23/all-you-need-is-love/.

6. This goal sets the AIME project in continuity with Michael Walzer's *Spheres of Justice: A Defense of Pluralism and Equality* (New York: Basic Books, 1983) and with Luc Boltanski and Laurent Thévenot, *On Justification: Economies of Worth* (Princeton, NJ: Princeton Univ. Press, 2006).

7. See Bernard Stiegler, *Technics and Time*, 3 vols. (Stanford, CA: Stanford Univ. Press, 1998–2010).

8. In his admirable latest book, Jean-Marie Schaeffer analyzes the way in which certain species of birds arrange and ornate their nests as a form of aesthetic behavior strikingly similar to our artistic practices—see *L'expérience esthétique* (Paris: Gallimard, 2015).

9. Ranjan Ghosh, "Literature: The 'Mattering' and the Matter," *SubStance* 131, no. 2 (2013): 42–45.

10. Gustave Flaubert, Letter to Alfred Le Poittevin, September 16, 1845, in *Correspondance* (Paris: Gallimard, Pléiade, 1973), 1:252. On these questions, see my *Ecology of Attention* (Cambridge: Polity, 2016).

11. Schaeffer, *L'expérience esthétique*, 48–112, and *Petite écologie des études littéraires: pourquoi et comment étudier la littérature* (Vincennes: Thierry Marchaisse, 2011), 112–14.

12. Schaeffer, *L'expérience esthétique*, 104–105.

13. Jacques Rancière, *Le spectateur émancipé* (Paris: La Fabrique, 2008), 19 (my translation).

14. This metaphor is further developed in one of his later publications on literature, *Le fil perdu* [The Lost Thread] (Paris: La Fabrique, 2014).

15. See Latour, "An Attempt at a 'Compositionist Manifesto,'" *New Literary History* 41, no. 3 (2010): 471–90.

16. Tim Ingold, "When ANT Meets SPIDER: Social Theory for Arthropods," in *Being Alive: Essays on Movement, Knowledge and Description* (London: Routledge, 2011), 91–92.

17. Latour, "War of the Worlds: Humans against Earthbound," *Facing Gaia: A New Enquiry into Natural Religion*, lecture 5 of the Gifford lecture series delivered between February 18 and February 28, 2013, at the invitation of the University of Edinburgh, available online at http://www.giffordlectures .org/lectures/facing-gaia-new-enquiry-natural-religion. The quotes are excerpted from a PDF document sent by Latour when he delivered the lectures.

18. See Latour, *Cogitamus: Six lettres sur les humanités scientifiques* (Paris: La Découverte, 2010).

19. Arne Naess, *Ecology, Community, and Lifestyle: Outline of an Ecosophy* (Cambridge: Cambridge Univ. Press, 1989).

20. Latour, *Facing Gaia*, lecture 4, "The Anthropocene and the Destruction of the Image of the Globe."

21. See for instance Richard Rorty, *Contingency, Irony, and Solidarity* (Cambridge: Cambridge Univ. Press, 1989).

22. Latour, *Facing Gaia*, lecture 6, "Inside the 'Planetary Boundaries': Gaia's Estate."

23. Latour, *Facing Gaia*, lecture 6.

24. Rorty, *Contingency, Irony and Solidarity*, 6.

25. See David Abram, *The Spell of the Sensuous: Perception and Language in a More-Than-Human World* (New York: Vintage Books, 1997).

26. Latour, *Facing Gaia*, lecture 6, "Inside the 'Planetary Boundaries': Gaia's Estate."

27. I thank Stephen Muecke for suggesting this alternative etymology, highly relevant indeed. A parallel between literary studies and theology may not bode well, however, for the future of the profession: by the end (or the middle?) of the twenty-first century, literature professors may be as few and marginalized within the university as theology professors have been for the most part of the twenti- eth century. This unappealing fate may lead us to consider ourselves as practitio- ner shamans no less than as erudite scholars.

28. On this question, see Thierry Bardini, "Entre archéologie et écologie: une perspective sur la théorie médiatique," *Multitudes* 62 (2016): 159–68.

29. Ingold, "When ANT Meets SPIDER," 92–93.

30. Latour, *Facing Gaia*, lecture 6.

31. Bertrand Meheust, *Somnambulisme et médiumnité*, 2 vols. (Paris: Empêcheurs de penser en rond, 1998).

32. Laurent de Sutter, *Magic: Une métaphysique du lien* (Paris: PUF, 2015), 82 (my translation).

33. Latour, "The Slight Surprise of Action: Facts, Fetishes, Factishes," in *Pandora's Hope: Essays on the Reality of Science Studies* (Cambridge, MA: Harvard Univ. Press, 1999), 266–92.

34. See for instance Tobie Nathan, *L'influence qui guérit* (Paris: Odile Jacob, 1994).

35. Jay David Bolter and Richard Grusin, *Remediation: Understanding New Media* (Cambridge, MA: MIT Press, 2000).

36. John Durham Peters, *The Marvelous Clouds: Toward a Philosophy of Elemental Media* (Chicago: Univ. of Chicago Press, 2015); Alexander R. Galloway, Eugene Thacker, and McKenzie Wark, *Excommunication: Three Inquiries in Media and Mediation* (Chicago: Univ. of Chicago Press, 2015).

37. Jeffrey Sconce, *Haunted Media: Electronic Presence from Telegraphy to Television* (Durham, NC: Duke Univ. Press, 2000), 8.

38. Jussi Parikka, "Media Ecologies and Imaginary Media: Transversal Expansions, Contractions, and Foldings," *The Fibreculture Journal* 17 (2011): 35.

39. Parikka, *What Is Media Archaeology?* (Cambridge: Polity, 2012), 2–3. See also Siegfried Zielinski, *Deep Time of the Media: Toward an Archaeology of Hearing and Seeing by Technical Means* (2002; Cambridge, MA: MIT Press, 2006) and Erkki Huhtamo and Parikka, *Media Archaeology: Approaches, Applications, Implications* (Berkeley and Los Angeles: Univ. of California Press, 2011). For a literary exercise in media archeology, in the form of a close reading of the works of Charles-François Tiphaigne de La Roche (1722–1774), see my *Zazirocratie: Très curieuse introduction à la biopolitique et à la critique de la croissance* (Paris: Éditions Amsterdam, 2011).

40. See N. Katherine Hayles, *How We Think: Digital Media and Contemporary Technogenesis* (Chicago: Univ. of Chicago Press, 2012) and Hayles and Jessica Pressman, introduction, "Making, Critique: A Media Framework," in *Comparative Textual Media: Transforming the Humanities in the Postprint Era* (Minneapolis: Univ. of Minnesota Press, 2013), vii–xxxiii.

41. On the notion of "media spell," see the dossier "Envoûtements médiatiques" in *Multitudes* 51, no. 4 (2012), available at https://www.cairn.info/revue-multitudes-2012-4.htm, as well as my *Ecology of Attention*.

42. See Latour, "Les Médias sont-ils un mode d'existence?"

CHAPTER NINE

Are the Humanities Modern?

SIMON DURING

THIS PAPER THINKS about the relationship between Bruno Latour's thought and the humanities as they exist today. It is motivated by the conviction that certain of Latour's key arguments lie at some distance from what the humanities actually do and have done. More specifically, it is motivated by the belief that the humanities are not "modern" in Latour's distinctive sense of that term and, furthermore, they are not "not modern" in his even more distinctive sense of *that* term either. These convictions cannot be discounted on the grounds that many of Latour's works themselves have recently helped shape the humanities at least in some of their more avant-garde forms. Nor can they be discounted on the grounds that Latour's writings resonate with broader intellectual streams and tendencies within the contemporary humanities. It is clear enough that Latour's thought chimes in with, for instance, French thinkers like Giles Deleuze, Michel Serres, Isabelle Stengers, and Michel de Certeau as well with recent political ecological thought in the *Naturphilosophie* lineage and also with more limited moments like object-orientated ontology, Manuel Castell's geographical sociology or Karen Barad's feminist philosophy of science.[1] But from my point of view, while his arguments and terminology may resonate with, and indeed help organize, some of the humanities' contemporary interests and procedures, nonetheless they belong to a conceptual universe at some remove from what the humanities mainly are and have been. In that light, their reception may be considered as evidence of a break in the humanities. Or as an extension. Or as signaling something new, a "post-humanities," we might call it.

I have, however, found my sense that Latour's model fails fully to connect to the humanities difficult to pursue. That is partly because his claims are both so wide-ranging and so hard to pin down.[2] But the main reason is that the humanities themselves remain so indefinite, their name from the beginning having been called on more for polemical than for analytical purposes. That is why I have pursued my argument by sidestepping the vexed question of what the humanities today actually are as well as by focusing just on a narrow tranche within Latour's oeuvre. This essay deals only with *Why We Have Never Been Modern,* the book of his that has reached furthest into the humanities and has most to say about them, and, then, principally with two of its claims, namely, that ever since the seventeenth century "we" have thought in the terms of the "modern constitution," and, second, more fleetingly, with its critique of critique.[3]

Although I will leave the humanities more or less unexamined here, certain features do provide my argument's background.

First their history is obscure. The humanities, named as such, only appear in the 1920s and 1930s and take hold, originally in the United States, only after the Second World War.[4] At first, the humanities had an intimate relationship to the liberal arts that preceded them, and a more distant but still real one to that humanism or *Litterae Humaniores* which dates back to Europe's early modern period.[5] But those genealogies and connections are faltering, and indeed, in the latter case, although they still operate, have all but disappeared from open view.

These losses of heritage are happening because, as their status and extent decline, the humanities are being thoroughly reshaped. For instance, they are decreasingly being thought about as a set of individual disciplines each with its own history and more as a "meta-discipline" all of its own. Students and teachers are increasingly just "in the humanities." At the same time, partly as a result, it is also becoming clearer how various the humanities are. It is impossible to find a single method or concept or value that define the current humanities as practiced in all kinds of departments, schools, programs and in all kinds of disciplines, interdisciplines, post-disciplines, subdisciplines, and fields. They do not just interpret. They are not merely concerned with value or meaning.

On one border, they merge into the social sciences. On another, they have swallowed theology. And they involve a great deal of practical, sometimes vocational, training whether, say, in communications, music therapy, or curatorship.

Nor are the humanities merely pedagogic and academic. They may still be defined by their place in the education system but they extend far beyond it. I am not thinking here just of the "public humanities" as it looks from within academia but rather at the way in which the subject and object division is being eroded in the humanities. So many of the humanities' primary *objects*, the things that are analyzed and celebrated (when it does analyze and celebrate)—music, novels, dance, movies, tv shows, paintings, installations, exhibitions, blogs, and so on—are also now saturated in the academic humanities' thought and values just because they are so often produced by arts graduates. Significant sections of our culture's various "worlds"—the worlds of the creative arts, literature, digital reviewing, show business and so on—can thus be understood as the pedagogical humanities' vehicles. In that way, they, too, are what the humanities now are.

This mode of thinking or, better, this analytic schema which thinks in terms of agglomerations and flows rather than hard borders and oppositions is foreign neither to Latour nor to the various humanities schools with which he resonates. But it also makes it harder to think of the humanities as limited by (or trapped within) "the modern." And, more importantly, it does not *weight* the various networks, groupings, methods, aim, and so on that constitute the humanities. Which is to say that in particular it does not acknowledge the importance of one key lineage for the humanities' understanding and presentation of themselves, namely that which concentrates on cultures as constructed and constructive signifying systems which are also agential powers on terms that do not necessarily fall under the sway of the principle of sufficient reason but rather proceed by way of "symbolic" or "figurative" logics, and which, in order to be thought about and assessed and extended, require practices of understanding, empathy, interpretation and evaluation.[6] It is a lineage which (to name some names) flows from Gottfried Wilhelm Leibniz and Giambattista Vico through Johann Gottfried

Herder to William Humboldt, Samuel Taylor Coleridge and Matthew Arnold and then to Ernest Cassirer, Erich Auerbach, Isaiah Berlin, Edward Said, and Martha Nussbaum. Among many others. Let us call this genealogy the "high humanities"—"high" because although it is being dispersed and marginalized it remains an essential point of reference for understanding, interpreting, evaluating, etc., the humanities themselves, i.e., for turning the humanities' protocols in on themselves. It also provides the necessary background of a particular *persona* whose aura has not yet quite vanished: that of the humanities scholar whose disciplined, nonscientific skills and sensitivities allow him or her privileged access to what our culture is. It is the high humanities, thought thus, that Latour lies furthest from.

Latour began his career at the intersection between anthropology and science studies, and his first monograph, *Laboratory Life: The Construction of Scientific Facts* (1979) (coauthored with Steve Woolgar), was based in observations on how a Californian neuroscience lab actually worked. On that basis he devised both his innovative actor-network-theory (ANT) and his repudiation of any hard human/nonhuman division. That led to him becoming convinced that science in general is driven not just by researchers and their hypotheses but by the *things* it uses and examines and the connections between things and people it creates. Associations between people are not thinkable outside of particular uses of things and technologies, and vice-versa. Thus a network has, as Latour, puts it, a "double constitution."[7] Science's power lies in its capacity to expose aspects of the world by extending these connections or networks into new zones and ceaselessly attaching to more objects and agents; and its findings, to which nonhuman things contribute, are thus simultaneously constructed and true or "real."

Latour's claim leads to an obvious question: how have we overlooked science's practical structure for so long? His answer is simultaneously theoretical and historical. He argues that, ever since the seventeenth century, we have been thinking under the spell of the modern constitution which purified nature by insisting on a strict separation of the nonhuman from the human. Our established picture of science relies on this disjunction between the nonhuman and human which then underpins further oppositions, especially those between things and intentional consciousness. This constitution has caused us not just to discount non-human agency generally but also to disavow the expansive associational modes of organization and connection between people, tools, things and technologies in which, according to Latour, scientific knowledge is produced.[8]

Latour's modern constitution is a complex device. It purifies the world by separating nature from the human only in such a way that that purification endlessly fails. For Latour, it does so primarily across three axes: society, nature and God. Under the modern constitution, society becomes not just autonomous but is gradually conceived of as the constructive force behind the human world in toto. For Latour, this sociological turn, which imagines even individuals as social constructs, is central to modernity. Under the modern constitution too, nature becomes ontologically inert, positioned just as the *object* of society, mind and experience and their constructive powers. Last, under the modern constitution, God retreats from the world: he becomes merely transcendent rather than immanently active among people and things.

For all that, Latour argues that we have never been modern because this modern constitution never actually worked, or rather (in a not quite deconstructive logic) it worked because it was also allowed not to. It can't work partly because it contains paradoxes and aporias: e.g., for it, society is both constructed and constructive; God is both distant and intimate. But, more importantly, it works because associations and hybridizations combining natural and social elements were always there in front of us as what Latour calls "quasi-objects."[9] New and complex "repertoires" have been required to reconcile quasi-objects to the modern constitution. These repertoires include the idea that meaning is fundamentally linguistic; that the world exists only "under description" or "under narration," as well as the (fundamentally Heideggerian) idea that beings forget that Being which is coextensive with existence and historicity. If I read Latour correctly, for him, this nostalgia for authentic *being* also underpins the holistic concept of culture, which he rejects: "Cultures—different or universal—do not exist, any more than Nature does," he roundly declares (104). But the point that pertains most to us here is that, were we to examine the humanities in a Latourian spirit, they would likely be positioned among those reconciliatory repertoires that enable the modern constitution and its nature/human split to survive. They would exist as a reaction to that modern constitution which science, in its emergent phase, has bequeathed to us.

Latour further argues, however, that the moderns that we have never been are now disappearing from view. Today, in a postindustrial society that is connected less spatially than digitally, and in a threatened society facing manmade ecological collapse, the pressures on the modern constitution have become overwhelming. In an often-cited phrase, Latour tells us that "the proliferation of hybrids has saturated the constitutional framework of the moderns." (51) So, a new ontological and historical picture of the world is taking shape, not least in Latour's own work of course. As the whole apparatus of purification, separation, hybridization erodes, along with its repertoires of reconciliation, we are, putatively, ready to think in terms of Latour's own actor-network-theory. That is to say: not in terms of ontological separations and oppositions within a cultural/social world itself organized into discrete zones but rather in terms of provisional but expansive associations and networks gathered to pursue limited projects and which can include objects or agents from any ontological or conceptual field whatsoever. As the modern constitution collapses so, too, presumably, do the humanities insofar as they are no longer required to deal with that constitution's structural problems. From this point of view, we are now in the era of the post-humanities.

Latour's work's relation to the humanities is difficult to pursue because his account of the modern only touches on them tangentially. Indeed, in *Why We Have Never Been Modern*, he forthrightly declares "modernity has nothing to do with the invention of humanism," and even if the humanities and humanism are by no means identical, his locating the modern at a distance from humanism remains a sign of how remote from his purposes they lie.[10] His notion that the humanities might exist as *reactive* formations becomes a particularly obdurate barrier here. For as it turns out, the humanities have had, and still have, their own positive programs, many inherited from eras that predate the "modern" by anyone's understanding of that term. As we will see, it is not the case, for instance, that the humanities are subsequent to, and dependent on, any modern break between nature and mind.

In this light it seems more useful to begin our enquiry into the humanities as a set of positivities not with the humanities as they exist today after centuries of elaboration, but at exactly the moment at which certain of their positive programs first appeared at the moment of Latour's modern constitution. Drawing on Steven Shapin and Simon Schaffer's *Leviathan and the Airpump* (1985), Latour situates the modern constitution's establishment in the mid-seventeenth-century quarrel between Robert Boyle and Thomas Hobbes over Boyle's demonstration that vacuums exist. So we can draw on this famous quarrel to center and focus our argument, namely, that important modes of what would become the humanities—the high humanities—in fact emerged in the work of those who participated in the development of mid-seventeenth-century new science and took full cognizance of the Boyle/Hobbes debate. Their purpose, however, was not to paper over incoherencies in the new science or natural philosophy but to join new knowledge practices to the humanist literary culture in place. In that effort, they developed positive programs that would flow into what became the high humanities.

This moment was also the moment of the first—late seventeenth century—"ancient versus modern" debate in which the so-called modern did not mean what it does for Latour but rather named the belief that contemporary knowledge and culture could match or surpass classical civilization's achievements, a belief that was underpinned by a not yet fully articulated concept of historical progress, and, thus, of a certain political progressivism in the contemporary sense. Originally, however, the intellectual programs that would come to form the humanities were neither ancient nor modern in those terms any more than they were either modern or not-modern in Latour's terms.

Before thinking about the twinning of the new science and an emergent humanities, a further clarification is useful. Latour finds the modern created in and through a specific *division*—that between the human and the nonhuman. Formally, this is a conventional move. The notion that modernity is constituted by a schism routinely comes into play whenever the humanities (especially those allied to orthodox Christianity) have tried to conceptualize the modern. From this point of view, the ultimate schism out of which modernity falls is the Reformation itself.[11] But we can also think, for instance, of the notion, most often developed by today's radical orthodox Christian theologians, that thirteenth-century nominalism invented a division between God and Being which sets the terms for all later "modern" understandings.[12] Or we might turn to R. H. Tawney's Anglican claim that the modern order came into existence in the seventeenth century when, under emergent capitalism, individualism broke off from communitarianism.[13] Or to Charles Péguy's notion that modernity involves the separation of the "the eternal" from "the temporal" (which, according to Péguy, happens in France and late—in 1881!)[14] Or of Friedrich Schiller's more secular idea that modernity is characterized by the division between the naïve and sentimental (i.e., [roughly] between the spontaneous and the self-conscious) and the latter's predominance.[15] Or a little closer to Latour, Anthony Giddens's argument from within theoretical sociology, that modernity is based in "the separation of time from space" and (as Thomas Pfau phrases it), "the consequent emergence of a "radical historicity."[16] Or most famously of all, we might cite T. S. Eliot's "dissociation of sensibility" by which feeling and thought were disjoined in an event which (at least in the argument's first iteration) Eliot also located in the mid-seventeenth century.[17] Belief in the dissociation of sensibility, of course, drove the emergence of what came to be one of the twentieth-century humanities' most vital disciplines: that mode of literary criticism based in close reading which in its original form was designed, like ANT, to overcome a disjunction which originated and structured modernity.

It seems to me that the very ordinariness of the view that modernity is founded on division is a reason to be skeptical of it. It is as if we *need* such a story to motivate understandings and critique of the modern. That is one reason why I want here to locate a moment in the development of the humanities that abuts Hobbes's argument with Boyle but does *not* return to this structure of emergence via division (or, in Latour's case, of a complex failure of a still maintained division). As I say,

I want rather to draw attention to positivities gathered from different heritages, rhetorics, interests, and methods to create not networks or assemblages or associations but rather methods, curricula, traditions, purposes, and values.

So, to the early modern moment.

Nobody was more involved in promoting the new science in such a way as to develop what would become the liberal arts and the humanities (both as pedagogical programs and as practices) than the English poet Abraham Cowley. Indeed, he was such an energetic and innovative agent in mediating between the new science and older humanism that, for our purposes here, we can restrict our analysis to his oeuvre and career.

Cowley was closely involved with the English proponents of the new philosophy, of which he became the most distinguished publicist. He met Hobbes when they were both living as Royalists in exile in Paris and wrote a famous ode in honor of the older man. More importantly, he often drew on Hobbes's epistemology in his own writing as well the theory of poetry that Hobbes spelled out in his Preface to William Davenant's epic *Gondibert* (1651) and elsewhere. His relation to Boyle was less personal but soon after the Restoration (in 1661) he published a pamphlet under the title *A proposition for the advancement of learning*, quickly reissued as *A proposition for the advancement of experimental philosophy*. It sketched out an ambitious plan to train talented youth into the Baconian/Boylean experimental method (a plan which, incidentally, downplays the "witnessing" so important to Shapin and Schaffer's account of Boyle). And when Cowley's young disciple (and biographer) Thomas Sprat was appointed "Historiographer" (i.e., official ideologue) for the new Royal Society in whose establishment Boyle was prominent, and went on to publish his famous *The History of the Royal Society* (1667), the book was prefaced by Cowley's celebratory "Ode To the Royal Society": "Bacon, like Moses, let us forth at last, / The barren wilderness he past, / Did on the very Border stand / Of the blest promis'd Land."[18] Late in his life Cowley made a pitch for an agricultural school (the first ever) which would train students into the experimental method with the particular intention of increasing national farm production. So just as Cowley publicly aligned with Hobbes, he supported the Boylean program too. In sum, it was he, more than anyone, who gave the new experimentalism and rationalism their humanist and *literary* legitimation in England.

Cowley was first and foremost a poet, and his acceptance of that office as it operated in the mid-seventeenth century was central to his career as well as to his relation to those concepts and practices which, centuries later, would help shape the humanities. He published his first collection, *Poetical Blossoms* (1633), when he was just fifteen. The book was a success: various editions of it appeared over the next few years, and it set him, as a London tradesman's son, on the path to patronage. As a precocious Latinist, he studied at Trinity College, Cambridge, where he wrote plays both in Latin and in English, one of which was performed for Charles I, the beginning of a long association with the Royal Family. Exiled as a Royalist in Paris, Cowley worked for Queen Henrietta Maria as a secretary and spy. There he became familiar with the *libertin*, skeptical and Epicurean philosophies current among Royalist circles in Paris (and he was later to become a close friend of Saint-Évremond who promulgated *libertin* Epicureanism in England).[19] From Paris, Cowley published *The Mistress* (1647), a suite of "metaphysical," spirtuo-erotic poems, heavily influenced by John Donne, which was widely read and admired. But it was the 1656 publication of a folio volume, bluntly entitled *The Poems*, which cemented his reputation as the period's leading poet.

The Poems began with a critical preface, explaining and defending the book. It republished *The Mistress* poems but then went on to show Cowley's mastery/invention of a number of other genres, based on a bold theory of translation that he sketched out and enacted. That theory rejected literal translation for "imitation": namely loose, free, creative versions either of an original text or, more loosely still, of a genre's spirit. *The Poems* first such imitations were the *Anacreontics*, a suite of light lyric poems on miscellaneous topics. But his most significant imitations were his so-called "Pindaric Odes" (some of which were translations of Pindar's original, some original poems by Cowley, some imitations of verses from the Old Testament). These Pindaric odes were "rough" in the sense that their rimes and rhythms could be irregular; they were digressive and disconnected from generic emotions and *topoi* (they were thus sometimes associated with "raving"); they could border on sublimity. They had a political undercurrent too: as Cowley was later to assert (in an ode, "On Liberty"), they should be thought of as expressions of individual—rugged—liberty.[20] As such they would form the framework not just for later odes by poets like Thomas Gray, William Wordsworth, and Alfred Tennyson but for a broader structure of feeling that would endure until at least the end of the nineteenth century.

The Poems also contained an (unfinished) epic in heroic couplets, *Davideis,* based on the life of the biblical David. For Cowley, this is a story which can ennoble poetry and its claims just because David was all at once poet, musician, warrior and king.[21] Although *Davideis* may have been written before Cowley had had a chance to read Hobbes on poetry and imagination, or indeed the fourth book of *Leviathan* with its efforts to naturalize the Scriptures and to resist the literary charms of pagan "idols," it shares that book's spirit by accommodating a Biblical story to Greek epic forms and vice-versa (see, for instance, the poem "To Sir William Davenant" which spells the program out).

This was not just a breakthrough volume for Cowley personally but for English literary history and indeed the high humanities generally not so much because it introduced new genres (two of which would be especially influential) but because it involves a broad understanding of the poet as simultaneously critic, intellectual, rhetorician, humanist involved in, but not overcome by, the new natural philosophy (i.e., science) and ethically responsible to the nation as a whole. The poet conceived like that was a prototype for the idea of the contemporary humanities practitioner.

It was at this point that Cowley published a series of essays to which poems and poetic imitations were attached. These discourses are also important literary-historically since they form a bridge between Bacon and Montaigne's humanist essay writing and the eighteenth-century periodical essayists, even though their more immediate impact was to help formulate "Country ideology,"—in other words, a retreat from metropolitan life; attachment to local land, farming, and community;

insistence on virtue and personal independence; contempt for money and decadence. That ideology, too, became part of the humanities' lineage, as we will see.

When Cowley died in 1665, he was given a lavish funeral and buried in Westminster Abbey next to Spenser and Chaucer. His reputation as England's preeminent poet endured for about fifty years after his death but would decline throughout the eighteenth century, and was dealt a decisive blow by T. S. Eliot, who blamed Cowley for absorbing the Baconian/Hobbesian philosophies, and made him the scapegoat for the metaphysical movement's failure to sustain Dante's coherence of sensibility.[22] That seems to have taken Cowley out of curricula in the modern humanities, except maybe in specialist courses.

But, in fact, as this sketch of his career has begun to suggest, Cowley was an extraordinarily innovative and ecumenical literary figure whose early reputation as a *Wunderkind* and whose license to range across genres (whether "creative" or analytic or polemical) helped him play a role in the circles in which Hobbes and Boyle were eminences as well as to contribute to the revolutionary period's hard politics, all the while articulating new literary, intellectual, and political programs, living as a poet who was also a scholar and a social commentator. As I say, it would allow his work to reveal ideas, values and practices that would slowly mutate into today's high humanities, partly just because he was involved in the new philosophy, and yet to do so in ways that have little or nothing to do with Latour's modern constitution.

This is not the place to complete this argument. Rather I want to adumbrate just four positive programs or thought practices which Cowley helped secure and which would be extended after his death to become generative of the liberal arts and then the humanities. I make no claim to adequacy here—this is a very partial view of both Cowley's works and its afterlives. Nor am I making a case either for Cowley's originality or his influence. My point is not that Cowley invented these thought practices or that he personally shaped how later generations would construct the methods and values that came to help constitute the high humanities. My point is a more modest one: I want to show that (1) that Cowley's innovative work broaches certain core humanities' thematics; (2) that Cowley's work as a proto-humanist is not reactive, and (3) that it cannot be properly understood in terms of Latour's paradigm for the "modern."

Creativism

The first of these thought practices I will call "creativism." It describes a particular way of thinking about relations between God, man, and nature. Those relations were, of course, crucial to the Hobbes/Boyle debate as well as to Latour's understanding of the modern. To sum up what will be familiar to many: Hobbes argued for an ontology in which the universe consisted just of bodies in motion. It contains no essences, no spirits, no final causes. Philosophy itself was "the knowledge acquired by reasoning" aimed at showing how the origins of things ("the manner of generation") lead to the "properties" of things and vice versa (as, for instance, motion leads to consciousness). Philosophy's purpose was to aid "human life."[23] The key term here is *reasoning*. For Hobbes (circularly), philosophy required leisure; life and leisure required a secure Erastian sovereign state; and a secure sovereign state required philosophical legitimation, that is legitimation based on indubitable philosophical reasoning proceeding from the elemental given of bodies in motion rather than on divine or spiritual action. This last is why Boyle's claim to have shown that vacuums exist so perturbed Hobbes. As he saw it, a vacuum was a break in universal causality which could interrupt reasoning's power and reach as well as provide a place for spirits to inhabit. Thus, vacuums threatened the sovereign Erastian state's legitimation.

Boyle's interest in natural philosophy was also theo-political. His main purpose was not to problematize Hobbes's insistence on deductive reasoning but rather to rethink the nature of nature. And his primary antagonist here was not plenism and materialism à la Hobbes but a still current scholastic understanding of nature as possessing its own creative and vital powers distinct from God, what some Cambridge Platonists then called "plastic nature."[24] Against this, he wanted to promote an idea of a passive nature that nonetheless *signified* divine authority and creation along with a human world whose piety and devotion to God might take the form of contemplating nature, mediating on it, and reading its signs. In his *Occasional Reflections* (1665), in which he offers a series of meditations on everyday life objects and situations, Boyle went so far as to hope that the practice of occasional meditation would be widely imitated so that Christendom would have "the Satisfaction of making almost the whole World a great *Condave Mnemonium* (i.e., a room furnished with images designed to help the memory) and a well-furnished *Promptuary*, for the service of Piety and Vertue, and may almost under every Creature and Occurrence lay an *Ambuscade* against Sin and Idleness."[25] Boyle's experiments fitted into this scheme because they acted on and revealed nature rather than relying either on reason or on nature's own agential powers. His revision of interactions between God, nature and humanity was ultimately designed to protect a particular relation between Church and State, one in which religion, not now narrowly bound to ritual, tradition and ecclesiastical institutions, would be available in everyday life so as to reduce the confessional struggles that remained politically difficult in late seventeenth-century Britain. Hobbes argued for a (perhaps secretly atheist) radical Erastianism; Boyle for a rational religion, accepting of an inert but signifying nature that might bring the Protestant confessions together.

Cowley welcomed the new thought while placing himself to its side. He declared himself an agnostic in relation to the vacuum pump controversy. The first stanza of his "Ode to Mr. Hobs" addresses the issue head on:

Vast *Bodies* of *Philosophie*
I oft have seen and read,
But all are *Bodies Dead*;
Or *Bodies* by *Art fashioned*;
I never yet the *Living Soul* could see,
But in thy *Books* and *Thee*.
'Tis only *God* can know
Whether the fair *Idea* thou dost show
Agree intirely with his *own* or no.
This I dare boldly tell,
'Tis so *like Truth*, 'twill serve our turn as well.

> *Just*, as in *Nature*, they Proportions be,
>   As full of *Concord* their *Varietie*,
>   As *firm* the parts upon their *Center* rest,
>   And all so *Solid* are that they at least
>   As much as *Nature, Emptiness detest.*[26]

For Cowley, Hobbes and his books represent thought as life, the very opposite of "Bodies Dead" or even "Bodies by Art Fashioned." For all that, only God knows whether Hobbes's supremely solidly structured philosophy and its detestation of vacuity is right or not, but (and this is important to my argument to come) that does not matter as much as it might because Hobbes's philosophy in its deductive rigor, vitality and order is so antithetical to emptiness to serve as well as nature as a basis for truth and reality.

Cowley, then, accepts the philosophic force of what may only be what he elsewhere calls "useful lies."[27] This notion underpins his own literary and poetic account of creativity and thus of relations between God, men and nature. His line of thought proceeds thus: God created the universe, and his creative power is not so much shared by human beings in their "reason" (as the scholastics and Cambridge Platonists believed) but is rather mirrored by writing practices that Cowley often thinks of as "wit." What is this wit? It is not a rhetorical device or driver, nor is it just linguistic play. As Cowley spells it out in a poem addressed precisely to wit:

> In a true piece of *Wit* all things must be,
>   Yet all things there *agree*.
> As in the *Ark*, joyn'd without force or strife,
> All *Creatures* dwelt; all *Creatures* that had *Life*.
>   Or as the *Primitive Forms* of all
>   (If we compare great things with small)
> Which without *Discord* or *Confusion* lie,
>   In that strange *Mirror* of the *Deitie*.
>
> But *Love* that moulds *One Man* up out of *Two*,
>   Makes me forget and injure you.

I took *you* for *my self* sure, when I thought
That you in any thing were to be *Taught*.
Correct my errour with thy Pen;
And if any ask me then,
What thing right *Wit*, and height of *Genius* is,
I'll only show your *Lines*, and say, *'Tis this*.[28]

These stanzas are somewhat ambiguous: the "you" in the second stanza denotes "wit," of course, but also, because wit "mirrors" God, it denotes God too. And the point is that, even though poets (as "geniuses") are vehicles of wit, they cannot actually describe and teach it. Wit appears just in the writing itself: it is a form of life which can be *pointed to* in lines like the poem itself, which are claimed to be written simultaneously by God, wit, and the poet. It is as if Cowley is combining opposites by occupying a point of intersection between Boyle's and the Cambridge Platonists' absolutely different understandings of nature. That is an ambiguity that can allow Cowley to think of the world itself as an exercise in wit, as "God's poem," as he puts it in these lines from *Davideis*, book 1.

As first a various, unform'd *Hint* we find
Rise in some god-like *Poets* fertile *Mind*,
Till all the parts and words their places take,
And with just marches *verse* and *musick* make;
Such was *God's Poem*, this *World's* new *Essay*;
So wild and rude in its first draft it lay;
Th' ungoverned parts no *Correspondence* knew,
An artless *war* from thwarting *Motions* grew;
Till they to *Number* and fixt *Rules* were brought
By the *eternal Mind's Poetique Thought*.[29]

I want to emphasize these lines' affirmation that God and the poet are both creative generators of order and rhythm as well as Cowley's supposition that that order is elaborated simultaneously in historical and in rhythmic time. But I want to insist most that the distinction between the natural and the artificial breaks down here, and, with that collapse, so do distinctions between what is divine, what is human and what is an object. Ultimately the world and all things in it are not to be thought of under the concept of "nature," but rather as "creation," a made object.

This is a way of thinking that is not at all modern but nonetheless will exercise a lasting force on the high humanities, especially after Giambattista Vico turned creativism into a method. Vico focused on that sector of the universe (let us call it "culture") which has been constructed or rearranged by acts of human will and labor (which for him too are also acts of divine will) so as to enable human beings to live in civil society. He conceived of culture as a relation to life's material conditions, especially to agriculture (as property), sex (as marriage), and death (as burial).[30] Conceived thus, culture is universal to, but different among, all gentile peoples, and each particular gentile culture has its own historical temporality.[31] Furthermore, even though cultures are constructed and various they can be *understood* as we understand poems. This way of thinking presses against Latour's "modern constitution" since it knows no hard division between the human and the nonhuman. It is a resistance to Cartesianism and Hobbesianism not an accommodation of it. For Vico in particular, human beings are only human as acculturated into cultures which continuously negotiate between people and things, between needs and desires, between life and death: they are creatures, we might say, of tooled-up lifeworlds. At any rate, to the degree that creativism forms an enduring strut of the humanities, they elude Latour's denunciation of the modern.

Ambiguity and Annahme

Unlike Vico, who was a rhetoric teacher and philologist, Cowley was, of course, a poet. This allows us to pose an intriguing question: what precisely is his relation to his own propositions? Is he being, in Bernard Williams's sense, *truthful* (i.e., making sincere truth claims)?[32] Or is he being *creative*, allowing his imagination loose rein in such a way that it would be inappropriate to examine his statements' truth status too closely? Is he knowingly articulating "useful lies"? Or is he *speculating*, entertaining notions that may or may not be true, who really cares?

Without being able to follow this inquiry as far as it might lead us, let me just say that I accept T. S. Eliot's argument (made in relation to Donne) that poetry of Cowley's kind is presented to be "entertained not precisely believed," that is, it has the epistemological status proper to what Eliot (after Alexis Meinong) called *Annahme*.[33] There are good intellectual historical reasons to stake this claim. It is based in Pyrrhonism, especially as expressed by Montaigne who devotes many pages of his "Apology for Raymond Sebond" to the importance, both moral and epistemological, of "remaining in suspense" in relation to judgments, especially truth judgments.[34] More distantly, it resonates with Hobbes's argument, finally based in Descartes's skepticism, that all ideas exist in the "Imagination" (even our ideas of space and time), with the implication that constitutionally and ontologically they remain at a distance from that absolute truth which belongs, for instance to the deductions which connect propositions, ideas and events rationally and causally. But in Cowley this quality of "being entertainable" is not philosophical, it is literary or humanist.

This matters to us because, *Annahme*, too, are common in the humanities. There, propositions are routinely put forward that are "entertainable" rather than true (or false), even if admittedly they are entertainable within a larger framework committed to truthfulness just because the humanities are finally legitimized by their putative capacity to produce true and useful knowledge. Furthermore, the humanities' attachment to *Annahme* allows us to think that what we as humanities academics do is "interpretation," and that two different interpretations can *both* be true. If creativism leads to understanding as a humanities paradigm, *Annahme* lead to the paradigm of interpretation. In the context of Latour's work, this is pertinent because it highlights the fact that Latour's theories marginalize the role of interpretation along with its epistemological and attitudinal groundings, and thus pass too quickly over a key aspect of the humanities. In this regard, once more, Cowley escapes being "modern" at the same time as he feeds into the humanities, this time just because he was so receptive to the new philosophy's skeptical epistemology, and its impulsion to speculation and interpretation.

Translation

As we have begun to see, Cowley both promulgated and practiced a theory of translation which used "wit" and "invention" to create texts whose effect and spirit, but not their phrasing, was that of the original.[35] On one level, this theory aided the transmission of classical literature into the English tradition by later writers such as Pope and Dryden. But Cowley was also involved in translating and transmitting in a broader sense which, once again, would form a crucial strut of the humanities. To a significant degree the humanities—history, literary studies, art history, philosophy, etc.—are involved in the transmission of heritages, archives, lineages and knowledges across generations. This involves continual revisioning as well as conserving, but also often involves allowing an inherited form, concept or value to shape something quite new, just as Cowley himself transformed some passages of the Old Testament into Pindaric Odes, or turned to the Greek epic genre to tell a Biblical story. This is also true of many of our artifacts—fiction films, for instance, are routinely an archeology of older fictional forms—and it is true as well of our tools and methods, many of which have been transmitted to us from a distant past. Cowley's versatility and adventurousness in translating and transmitting what was new to the culture allowed him to mediate between humanism and the new philosophy. Which is also to say that he stood at a distance from the ancient versus modern debate that opened up just after his death. Like his friend John Evelyn as described by Joseph Levine, he was *both* a modern (a promoter of progress and productivity) *and* an ancient (a conservator of classical culture), a conjunctive possibility that the high humanities will radically expand, if on terms that (once again) have little to do with Latour's own different concept of "the modern."[36]

Retreat and Critique

Toward the end of his life, Cowley retreated from public life to embrace a Horatian/Virgilian persona. Importantly, this retreat also involved a critique of urban commerciality which was itself a *translatio* just because it reframed classic tropes in relation to a new situation, namely early-modern urban commercialism. By the first decades of the next century, Cowley's ethos of retreat would form a background to an established—patriot—critique of what by that time was thought of as Whiggism, and which we today might regard as emergent capitalist modernity.

A slightly less political way of thinking about Cowley's affirmation of retreat with its implication of critique is to note that the *otium* or leisure that was then often recognized as a condition for practicing philosophy and literature was not actually to be *assumed* of a learned gentleman but was a position to be staked out, paid for, and defended against the demands of politics, business and, indeed, against the demands of the Baconian ideology of labor, experiment and productivity that Cowley himself championed. However that might be, it is quite clear that Cowley posed his retreat against what he thought of as the commercial and urbanizing tendencies of his era.

> Whilst this hard Truth [about solitude's ethical value] I teach,
> methinks, I see
> The Monster *London* laugh at me;
> I should at thee too, foolish City,
> If it were fit to laugh at Misery,
> But thy Estate I pity.[37]

At the risk of overstating my case, this, too, makes Cowley a progenitor of the modern high humanities. That is because ever since the mid-seventeenth century the humanities have themselves have had deliberately and *critically* to step away from business, productivity, and partisan practical politics. Critique, we might say, is one of their structuring conditions. It is required to mark out a distinct cultural space for them, even if, of course, the humanities have by no means been consistently critical of dominant social values and institutions or, indeed, uninvolved in commerce and politics.

Nonetheless: here, too, we are at a distance from Latour's account of the modern which famously dissociates itself from criticism. Latour vows to "bring the sword of criticism to criticism itself" on the grounds that critique and criticism has become bound to a generalized, relativist hermeneutic of suspicion which has led to a loss of truth, and is hand in glove with (as we now say) fake news.[38] Worse still, so Latour, in its pursuit of fetishes and hegemony as well as in its obsession with conditions of possibility, contemporary critique devalues and reduces the actual world's solidity and diversity. Further, it stops us generating newness.

But the high humanities, which have been, (1) so focused on what is *made* (as Vico recognized); (2) so concerned to transmit and revision artifacts (textual and otherwise) from the past; and (3) so open to the play of the imagination, are routinely required to conduct those programs at a distance from society's productivity apparatuses—a distance that they maintain only by spelling out what is lacking and lost when you live merely rationally, obediently, productively. That is by being, implicitly even where not explicitly, *critical*.

We might end this (admittedly limited) account of the way in which the humanities developed alongside the new science on terms other than those posited by Latour's modern constitution by asking whether this analysis changes how we might think of Latour's work's own relation to the humanities. I believe it does. For it encourages us exactly to *entertain* Latour's ideas and findings rather than to follow them or to believe them or to take them wholly seriously as, for instance, a correction to the high humanities' characteristic protocols. Some of us may reckon that, as Cowley said of Hobbes, they are "so like Truth" to "serve our turn as well." Others may enjoy and intellectually play with them just as elaborately worked out imaginative projections. Others may wonder whether they represent a swerve from the humanities' central streams and try to think harder what that swerve towards the post-humanities means. I suppose too that some, disclaiming their epistemological status as *Annahme,* might even see them (mistakenly I'd say) as a threat.

Notes

1. For the "politics of nature," see Bruno Latour, *Politics of Nature: How to Bring the Sciences into Democracy*, trans. Catherine Porter. Cambridge, MA: Harvard Univ. Press, 2004. For a good account of the context in which Latour

developed his thought, see Henning Smidgen, *Bruno Latour in Pieces: An Intellectual Biography*, trans. Gloria Custance (New York: Fordham Univ. Press, 2015).

2. For Latour's own admissions on the rather messy status of some of his core concepts, see http://www.bruno-latour.fr/sites/default/files/121-CASTELLS-GB.pdf.

3. See also Bruno Latour, "Why has critique run out of steam? From matters of fact to matters of concern," *Critical Inquiry* 30/2 (2004): 225–38.

4. See, for instance, Geoffrey Galt Harpham, *The Humanities and the Dream of America* (Chicago: Chicago Univ. Press, 2011), 80–90.

5. The relation between the modern humanities and these older formations remains understudied, but there are some suggestive remarks in John Guillory, "Who's Afraid of Marcel Proust? The Failure of General Education in the American Univ.," in *The Humanities and the Dynamics of Inclusion since World War II*, ed. David A. Hollinger (Baltimore: Johns Hopkins Univ. Press, 2006), 25–49.

6. One of the clearest theorizations of the humanities in roughly these terms is to be found in Ernst Cassirer's 1939 essay, "Naturalistische und humanistische Begründung der Kultuphilosophie," published in English translation as the introduction to *The Logic of the Humanities,* trans. Clarence Smith Howe (New Haven: Yale Univ. Press, 1960), 3–38.

7. Bruno Latour, *We Have Never Been Modern,* trans. Catherine Porter (Cambridge, MA: Harvard Univ. Press, 1993), 4–6.

8. *We Have Never Been Modern*, 10 ff.

9. *We Have Never Been Modern*, 67.

10. *We Have Never Been Modern*, 34.

11. The most recent and fullest account of the Reformation in these terms is to be found in Brad Gregory's *The Unintended Reformation: how a religious revolution secularized society* (Cambridge, MA: Harvard Univ. Press, 2012).

12. This argument is especially associated with John Milbank, *Theology and Social Theory: beyond Secular Reason* (London: Blackwell Publishing, 1990); and Michael Allen Gillespie's *The Theological Origins of Modernity* (Chicago: Univ. of Chicago Press, 2008).

13. R. H. Tawney, *Religion and the Rise of Capitalism* (New York: Harcourt, Brace and Co., 1926).

14. Charles Péguy, *Temporal and Eternal*, trans. Alexander Dru (London: The Harvill Press, 1958), 115.

15. Friedrich Schiller, *Über naïve und sentimentalische Dichtung*, ed. Klaus. L. Berghahn (Berlin: Reclam, 2016).

16. Anthony Giddens, *Consequences of Modernity* (Stanford: Stanford Univ. Press, 1990), 20, as described in Thomas Pfau, *Minding the Modern: Human Agency, Intellectual Traditions and Responsible Knowledge* (Notre Dame, IN: Univ. of Notre Dame Press, 2013), 42.

17. Eliot first used the phrase and put the case in a 1921 review of an anthology of seventeenth-century metaphysical poets assembled by Herbert Grierson. See T. S. Eliot, "The Metaphysical Poets," *Times Literary Supplement* 1031 (1921): 669–70. In his lectures given at Harvard in 1926 (and published posthumously as *The Variety of Metaphysical Poetry*), he situated the dissociation of sensibility earlier, in Dante's time.

18. Abraham Cowley, *The Works of Mr Abraham Cowley. Consisting of Those which were formerly Printed; And Those which he Design'd for the Press, Now published out of the Authors Original Copies, the Seventh Edition* (London: Henry Herringman, 1681), 40 (verses written on several occasions). I have used this early edition because from the mid-eighteenth century on editions of Cowley's works are incomplete and riddled with errors. Henceforth Cowley, *Works*. But this book is not paginated sequentially: each of its sections has its own pagination (and "The Preface" is not paginated at all) so where using the volume below I have cited the section in brackets and then the page number, where available.

19. He most likely also attended Gassendi's lectures at the *Collège Royale* which helped inspire Hobbes. Jean-Charles Darmon, "Pierre Gassendi et la République des Lettres: questions liminaires," *XVIIe Siècle*, 2006 (233): 579–585. And Thomas M. Lennon, *The Battle of Gods and Giants: The Legacies of Descartes and Gassendi, 1655–1715* (Princeton: Princeton Univ. Press, 1993). The relationship between Gassendi and Cowley was noticed by Richard Hurd, Cowley's eighteenth-century editor. See Richard Hurd, *Moral and Political Dialogues; with Letters on Chivalry and Romance*, 3 vols. (London: T. Cadell, 1771): 2: 137.

20. Abraham Cowley, *Essays, Plays and Sundry Verses*, ed. A. R. Waller (Cambridge: Cambridge Univ. Press, 1906), 391.

21. Cowley, *Works* ("The Preface").

22. T. S. Eliot, *The Varieties of Metaphysical Poetry*, ed. Ronald Schuchard (London: Faber, 1993), 185–200.

23. These phrases are from Thomas Hobbes, *Leviathan, or the Matter, Forme and Power of a Commonwealth, Ecclesiasticall and Civil,* ed. Michael Oakeshott (Oxford: Basil Blackwell, n.d.), 435.

24. See Ralph Cudworth, *The True Intellectual System of the Universe, Second Edition* 2 vols. (London: J. Walthoe et. al., 1743): 157–168.

25. Robert Boyle, *Occasional Reflections upon Several Subjects. With a Discourse about Such Kind of Thoughts* (Oxford: Alex Ambrose Masson, 1848), xxxi.

26. Cowley, *Works* (*Pindarique Odes*), 26.

27. Cowley, *Works* (*Pindaric Odes*), 23.

28. Cowley, *Works* (*Miscellanies*), 4.

29. Cowley, *Works* (*Davideis*), 13.

30. Giambattista Vico, *The New Science*, trans. Thomas Goddard Bergin and Max Harold Fisch (Ithica, IL: Cornell Univ. Press, 1968): 53, 98–9.

31. See Vico, *The New Science*, 102–6. That Vico's way of thinking has been central to the humanities was an observation made by Auerbach and Berlin among others. See, e.g., Erich Auerbach, "Vico and Aesthetic Historicism," in *Time, History and Literature: Selected Essays*, ed. James I. Porter (Princeton: Princeton Univ. Press, 2016), 36–46; and Isaiah Berlin, *Three Critics of the Enlightenment: Vico, Hamann and Herder*, ed. Henry Hardy (London: Pimlico, 2000), 21–84.

32. Bernard Williams, *Truth and Truthfulness: An Essay in Genealogy* (Princeton: Univ. of Princeton Press, 2002).

33. T. S. Eliot, *The Varieties of Metaphysical Poetry*, ed. Ronald Schuchard (London: Faber, 1993), 88.

34. Michel de Montaigne, *The Complete Works*, trans. Donald Frame (New York: Everyman's Library, 2003), 453.

35. Cowley, *Works (Preface to Pindarique Odes)*.

36. Joseph M. Levine, *Between the Ancients and the Moderns: Baroque Culture in Restoration England* (New Haven: Yale Univ. Press, 1999), 3–35.

37. Cowley, *Works (Several Discourses)*, 95.

38. Bruno Latour, "Why Has Critique Run out of Steam? From Matters of Fact to Matters of Concern," *Critical Inquiry* 30 (2004), 227. Latour's critique of critique has been taken further and in different directions by Rita Felski in *The Limits of Critique* (Chicago: Univ. of Chicago Press, 2015). My argument here that the humanities are defined by their *critical* distance from the ordinary social world of commerce and politics does not, of itself, imply an embrace of critique as it came to be conceived of in the second half of the twentieth century.

CHAPTER TEN

The University of Life

NIGEL THRIFT

Introduction

IN THE SIXTEENTH CENTURY, Henry VIII dissolved the English monasteries. Led by Thomas Cromwell, assessors descended on the monks' abodes, valued them and sold them off, leaving many monks with just a small pension. The monks were thrown onto their own devices and ended up as teachers or priests or apothecaries—or as beggars. To read some commentators nowadays, much the same thing is happening now as the market makes its way into universities. Old contemplative habits are being replaced by an alien attendance to innovation and creativity, attacking what are understood as the university's fundamental values and producing all manner of impoverishment. It's all horrible, except for those who write the laments. At least they have plenty of material—and a pleasing sense of moral superiority.

The problem is that it often seems as if all thinking about universities has been hijacked by this "it's all got worse" narrative. There is a blasted heath of alternative explanation and a palpable sense of mourning as faculty move from a position in the clerisy to a position in the laity, as A. H. Halsey puts it, a progression—or should we say regression—that is now even more far advanced than in Halsey's day and that has brought both bitterness and bewilderment.[1]

There is a reason for this sense of sadness apart from a general erosion of social position. Since the nineteenth century, the idea of universities has often been based around "values" (a word I will subsequently want to question): moral-epistemological honorifics that, as a matter of course, deal in absolutes as a currency and transcendence as a goal. More crepuscular forms of explanation are given short shrift. The result is clear, in any case. Explanatory biodiversity is reaching perilously low levels. It is just assumed that we are—we must be—in the badlands.[2]

But this will no longer do. There is, to begin with, the fact that from many perspectives, universities are, at least in some respects, better places than they were.[3] Women are no longer outliers. Promotion is, generally speaking, on merit. On balance, more really mind-bending research is getting done. And all this is before we get to the fact that many more people are getting a university education than ever before. That said, it is true that universities no longer cleave to such a tight-knit set of values. It is not just the simple empirical fact that no institution can take in just one value.[4] There is also the fact that universities, like most institutions, have accreted new values (and thrown others aside) as they have acquired new strands of activity.

This is where the work of Bruno Latour and a number of other authors can help us out. In this paper, I will use this work to begin to sketch out an alternative account of universities and their future, one in which all variation is not variation from a theme but just, well, variation, and in which domains of explanation are never all-encompassing and never way stations on the road to some kind of utopia of the mind that was once supposedly realized in fact. Universities are and always have been— or so I would assert—coalitions of practices which have put their so-called values in uneasy tension.

To address this tension, I will try to read across from universities to Latour and back again. The paper is in four parts, therefore. In the first part, I will briefly attend to the new strands of activity that have accreted to universities over the last twenty years or so and, at the same time, to the growth in variation in what is called a "university." The second part of the paper alights on the different and seemingly opposed values that are a result of the current much-expanded spectrum of universities and the clashes that these different strands of activity engender. The third part of the paper suggests how universities could be redesigned in the face of this greater diversity, not just as a means of compromise but as a new synthesis based around experimentation and adventure. Universities badly need redesigning because of the process of accretion of new strands of activity. This can either be done as a result of these new strands of activity being given free rein or through conscious intervention so as to produce a more considered and *dutiful* outcome, an intervention that includes connections between each strand as a matter of course.[5] Finally, there is a brief conclusion that raises the prospect of reinstating a certain kind of honor in the lives of universities.

Of course, the experience of universities around the world has been variegated, to put it but mildly. University systems and the institutions they have created vary widely. Some are inherently statist. Others have always had a much greater private influence. Some were born in societies where the pursuit of knowledge was born out of religion as much as Enlightenment. Some were responsible for producing a specific intelligentsia. Others were not. Some were inherently more egalitarian in tone. In others, elitism ran amok until quite recently. And so on. But whereas there is the study of comparative religion, comparative anthropology, comparative sociology, and comparative economics, universities suffer from a lack of comparative analysis of their different institutional machineries and creeds that have often had very different relationships to state and economy and religion. In any case, I will reserve most of my commentary for the university systems I know best: the United Kingdom and the United States. But this more general heterogeneity needs to be kept in mind when talking about values: all universities do not subscribe to the same values and any convergence is often nominal. Until we can get to see this, we will keep on going round in circles.

The Light Is Still On

Universities have changed since the Second World War, as what was unabashedly a pursuit for elites has become a system for educating the mass of the middle class. The number of students and faculty has expanded mightily. Hints of this expansion were already to be seen directly after the War in, for example, the GI Bill of 1944. But the great acceleration took place from the 1960s on. The UK is illustrative. Whereas there were about 250,000 students in 1965, now there are near to two and a half million.[6] This expansion has been mainly in domestically based students, but a significant additional element has been international students flocking from many parts of the world. Since these students pay more, they have become a crucial element in the makeup in the economy of most universities.

Universities have also changed since the Second World War because of the rise of big science research. They have become the main redoubts of Science with a capital S, displacing, in large part, corporate research and research in specialized government institutions.[7] In turn, corporate interests have come to gain greater sway in university research than ever before, shaping what research gets done to an extent that is still argued about, not least because, as Philip Mirowski points out, "the 'commercialization of science' turns out to be a stubbornly heterogeneous phenomenon," built mainly around the evolution of the laboratory under very different economic and military regimes and trends that would gratify popular culture (for example, the quaint idea that start-ups and intellectual property would be a panacea for lack of funding).[8]

But, whatever else, this science is not one bloc. It is striated by different, sometimes radically different, disciplinary ways of proceeding, from the mass engineering projects such as CERN that depend upon agglomerations of scientists through the behemoth that is the medical and life sciences, to formerly single-scholar disciples such as mathematics, statistics, and computer science.

Because of these two changes in particular, universities have become larger and larger: whereas in previous times they might have been compact and college-like, many universities have now become large bureaucracies. In contradistinction to writers such as David Graeber, who seem to think that bureaucracies are automatically reprehensible, it is difficult to know how these expansions could have been underwritten in any other way, whether we are talking about student financial advice or help filling in grant applications or mounting online courses.[9] In particular, they have grown because they have taken on more and more functions that require significant administrative capacity. Whereas the function of a university was once teaching and a bit of research, now universities encompass all manner of activities, only some of which relate to teaching and research: economic growth, regional cohesion, outreach to the community, and all manner of global structures.

And that brings us to the final major change: in lockstep with this tripartite expansion in scale, universities have become economies. That is not, of course, the same thing as universities taking on economic values lock, stock, and barrel, but it is obvious that as more and more people are employed and institutions come to have more and more economic influence simply through the size of their turnover, they will become thought of as economic entities in their own right. Add to the mix governments looking for growth, and universities become framed as major export industries susceptible to government and corporate influence in a way that neither body would have considered before. Universities come to be understood as intellectual property.

In these circumstances, the idea that the bulk of universities would stay mainly as they were—as small-scale craft economies teaching a liberal arts curriculum—seems bizarre. But, for all these changes, the important fiction of continuity still persists. Like the Church, universities tend to think of themselves as inheritors of tradition, even while they have constantly innovated in order to endure. Indeed, even new institutions often play to this tradition, literally in the case of practices such as graduation ceremonies and the like. Just as in the case of the Church, the university is "an almost perfect model of the complexity of the relations between a value and the institution that harbors it: sometimes they coincide, sometimes not at all; sometimes everything has to be reformed, at the risk of a scandalous transformation; sometimes the reforms turn out to consist in dangerous innovations or even betrayals."[10] Equally, the narrative of what is wrong with universities persists. For example, the idea that business is taking over and all is going to rack and ruin—that universities are becoming mere appendages of capital—arises around 1895 in the United States and has been a constant ever since, spreading out to other parts of the world as a ritual academic complaint.[11] In other words, as universities have changed, many of the values that seem to underlie them have stayed pretty much the same. Even those who have done so much to upset the old order still appeal to them.[12] We can argue about whether the proponents of these values entirely believe them, practice them in their daily lives with the appropriate kinds of equipment as backup, or what have you. But what cannot be denied is that they have a hold. In the terms of social psychology, when asked for an account of what they are doing, this is the account many people in universities give—muddled or not, momentary or not. They are well-worn scripts powered by the inertia of repetition and thus difficult to shift to a new plane.

Perhaps the one thing that universities have done pretty consistently has been to act as temples of privilege. With a few exceptions, through the warp and weft of history, they have served the well-off, whether in actuality or through meritocratic systems that produce surprisingly similar results. They have added to the tendency to literally write off the practices of large parts of the population as "ordinary" or "average," most recently through strategies of exclusion, which depend upon the argument that universities try ever so hard to include but are forced to exclude because of merit. It is only comparatively recently that this situation has changed. With the expansion of higher education in many parts of the world, a degree of exclusion has been removed, at least for the middle class. But most states have only been partially willing to pay for this expansion, with the result that universities have had to make up their funding from sources other than government. In many countries they have also turned to the private sector to help. Meanwhile, in a race to prolong social privilege, more and more attention is given to the internal differentiation of the tertiary sector by parents and students.

In summary, what a university is has changed and is now changing again. There is no going back. But the university could be redesigned so that it retains some of the best features of the old and also constructs new features that fit but also extend its activity.

The Clash of Values

One upshot of these changes is that universities now encompass a series of value systems based on different ways of mobilizing and earning a living. Universities take in many modes of existence. They are not just one hieratic mission, involving gestures and hieroglyphics left over from a previous age. One should not make too much of this clash of values—or too little. So, in Latour's terms, a series of different rationalities and means of veridiction are being practiced through different modes of existence: "every instauration implies a value judgement."[13] Each value system has its own forms of specification, evaluation (including calculation), and means of acting out good and bad will. But the difference is that in universities these modes of existence and their respective means of justification are brought into contact on a daily basis rather more starkly than in many other arenas. They *have* to pass through representatives of other modes of existence in order to endure. Indeed, nowadays, *each mode of existence depends on the other to survive to a much greater degree than ever before.* The idea that one academic mode of existence can ride roughshod over the others is a fantasy. Indeed, very often, they are in resonance. Certainly, they are in constant negotiation. Take, for example, science, administration, economics, and community outreach.

Science nowadays depends on a large penumbra of administrators to sustain itself. These administrators are involved in tasks such as aiding with and submitting grant applications, making sure that grants are spent according to profile, liaising with grant funding agencies, maintaining a supply of consumables to laboratories, even making sure that power supplies are reliable, as well as enumerating numerous risk analyses, codes of conduct, dashboards, and all the other paraphernalia that turn administration into a moral enterprise. Equally, administrators depend on scientists to provide the ideas for grants that can be turned into a reliable stream of research income. But, as science has become larger in scale and complexity, so the relationship has become one of very often deft mutual dependence—and not just the parasitic relation that so many academics are intent on portraying, wherein the paraphernalia of administration is a burden imposed by outside forces, especially when things aren't going the way academics would like and they are looking for things to blame.

Again, it is clear that universities now run according to standardized financial procedures. In turn, these procedures have made them more susceptible to thinking about the world in financial terms, not least in terms of the necessity to generate some kind of surplus in order to maintain expenditure on capital and in terms of borrowing (again, usually in order to support capital expenditure) as well as in terms of investment strategy—"no margin, no mission" as the saying goes. This is hardly surprising. Not only do universities routinely pursue activities that are unprofitable that they then need to finance through surplus-generating activities, but they also need to build and rebuild mundane things like research laboratories and halls of residence.[14] Then, universities interact much more closely than in the past with corporate interests such as consultancies and with private enterprise more generally, not least because of a general withdrawal of state funding. In the United States, this has been the case since the end of the nineteenth century (when the first jeremiads about the malign influence of business also started).[15] In other countries, the change came later, but all countries now understand universities as economic growth poles—sometimes, it has to be said, in grossly exaggerated ways that suggest that every university is a hive of staff and student start-ups, that every university is boosting small firms, and that most universities are doing significant research for large firms. Add in all manner of corporate funding, and it is clear that the economic mode of existence has become increasingly influential. This state of affairs has been much decried—as a capitulation to the immediate and worldly, as economic serfdom, as the triumph of an economic logic, as a pathology of productivity.[16] But, notwithstanding all manner of reservations, it is dangerous to generalize to too great a degree. There are good and bad corporate links, after all. What is equally clear is that for large universities to achieve extra growth in income and influence makes it all but inevitable that they will call on business. As they have done so increasingly, so they have become more businesslike. Economentality has become a habit.

This tendency has only been furthered by the growth of development offices seeking out gifts and donations, in part to substitute for state funding. Not too much should be made of it, however. Universities have always been malleable. For example, they have never been as far from the world of commerce as is often made out. From the earliest times, many have curried favor with the wealthy. The early Oxford colleges performed all kinds of pirouettes to get their hands on money, often from sources that could hardly be called respectable.

Finally, there is a university's community influence. Most universities aspire to having an impact on their community. Recently, this influence has widened and has become something approaching an industry. Universities run hospitals, schools, arts centers, massive volunteering programs, sports programs. Staff run all manner of charitable programs and other forms of outreach. Students are similarly active. So are alumni. The exact makeup varies, but one of the aspects of the modern university that is unabashedly to the good is the commitment to service by the institution, by staff, and by students. This has grown substantially in recent years and has also internationalized, so that a university's community may now span many countries.

With each of these modes of existence comes a more general tendency: the importance of measurement. Whether it is the citation mania of parts of science, the hunger of administration for quantitative comparators, the demands of monetary and fiscal discipline, the apparatuses of relationship management, or the need to display community commitment, each mode of existence is measured out, quite literally.

Nearly all of this activity has grown up in a way that attests to the fact that "universities are complex institutions where individuals, groups, and units of various kinds (e.g. academic departments, research centers and institutes, professional schools, support and external relations staff, administrators) compete for resources both inside and outside of their organizational domain while simultaneously trying to coordinate with or at least not intrude on one another."[17] There is, after all, no reason why any institution has to be a neatly trimmed bundle of activities, all working in lockstep. Indeed, most institutions, for all their protestations to the contrary, are not. But, that said, there is now a pressing need to redesign the university or see it beat an increasingly hollow drum. Yet one of the frustrations that accompany the actions of anyone trying to run a university is their colleagues' indifference to becoming involved in any substantial redesign of the university. Colleagues are happy to criticize from afar. They are rarely interested in doing more, with the result that they cede the ground to exactly the tendencies they abhor, having no solutions of their own to the pressing task of reproduction that universities now face. The criticism comes from both Right and Left, although it is probably more virulently argued from the Left because it has a dragon to slay that must clearly be responsible for all the ills of the world—neoliberalism. But there is too much in the world for it to fit together into such an easy explanation,[18] not least because universities have themselves been a part of the unfolding history of rationalization and commodification, not a breed set apart.[19]

Wendy Brown's recent book is a classic example of the genre. Pretty much everything significant is ascribed to an all-encompassing force, neoliberalism, which is in the process of destroying universities via the substitution of economistic for humanist values. The trouble is that while there is certainly something in this critique—few people would deny that a certain "economentality," to deploy Timothy Mitchell's term, has taken hold of universities as of many other institutions—it is by no means the whole story.[20] Brown makes it very difficult to see alternatives.

For Brown, the 1960s United States is the apotheosis of higher education. The decade

> promised not merely literacy, but liberal arts to the masses. It also featured cultivation of a professoriat, and a professional class more generally, from the widest class basis in human history. And it was a time in which a broad, if not deep college education—one inclusive of the arts, letters, and sciences—became an essential element of middle-class membership. No mere instrument for economic advancement, higher education in the liberal arts was the door through which descendants of workers, immigrants, and slaves entered onto the main stage of the society to whose wings they were historically consigned. A basic familiarity with Western history, thought, literature, art, social analysis, and science was integral to middle-class belonging, in many ways more important than a specific profession or income.[21]

One does not have to be a committed follower of Bourdieu to question this narrative. The 1960s were a time in many countries when only a very small elite, with commonly held values based on an education that prepared them to be a cultural as well as economic elite, went to university. Even in the United States, there were only 5.92 million college students in 1965, compared with 20.24 million now (and this is a much smaller increase than in many other countries). This was clearly not the education of or for the many. Indeed, as in many other countries, the working class was and has remained firmly excluded from higher education, consigned to a vocational education on which, except in a very few countries, far fewer resources were spent. Equally, the idea that the liberal arts somehow carried the weight of a common citizenship seems to me to beg the question of who counts as a citizen. Finally, insofar as higher education was not a tool of economic advancement, this was mainly because a graduate at that time was pretty well guaranteed a job. The two went together and, of course, in the higher reaches of US higher education, they still do: as in many other countries, elite universities are not just places of learning; they are also machines for reproducing elites by means of defining merit as "entrenched in applicants' and employers' own upbringings and biographies" in ways that favor their children above others.[22] There is no getting around the fact that elite universities are part and parcel of the class ceiling and opportunity hoarding. To argue that in the US at least public universities were committed for more than fifty years to a mission of "egalitarianism and social mobility . . . as well as for providing depth and enrichment to individuality" is to gild the lily,[23] just as it is an exaggeration to argue that universities have now become simply a means of training for jobs. Rather, as the system has expanded, some have taken on this function whilst others haven't. Equally, in many universities around the world, there is still considerable room to maneuver.

Take the case of the humanities. To listen to many commentators, the humanities, which are often somehow assumed to constitute the soul of the university, are now being crushed in a philistine vice. But the facts do not support this interpretation.[24] In the UK, for example, the absolute number of humanities students, by the narrowest definition, has increased fivefold and by a broader definition tenfold since 1967. Even in the United States, the numbers have doubled, with the only sharp decline in enrollments occurring in the 1970s and 1980s, mainly because of a movement of women into professional courses: "Talk of a crisis triggered by a decline of a per cent or two does seem like an over-reaction that is likely to contribute to rather than ameliorate the alleged problem."[25]

Redesigning the University

Against this background of not just shifts in the substance of what a university is about, but also shifts in thinking about the university *and* thinking about thinking itself, it seems obvious that the idea of the university, as originally conceived by John Henry Newman in religious terms or by Wilhelm von Humboldt in statist terms, needs more than a modest makeover: the heterogeneous and multifaceted universities of today have genuine *design* limitations.[26] But what might replace the idea of a university other than the high-functioning sleepwalking of a "filopietism" that dictates that all universities should want to pursue the same vision, even though that is patently impossible? It seems that universities need to be consistently designed once more. But they need to be designed with an eye to what they have become—large, multifaceted organizations that are part of a diverse ecology including many types of rationality—rather than what they once were, and with an eye to the task of articulating values that are appropriate to these institutions without ever being exactly homologous.

This process of design doesn't have to be seen as a benighted rebirth of central planning or as a playing field for consultancies with their imperatives of constant evaluation and assessment. After all, design is not an alien concept in universities. Virtually each and every system of higher education has been the subject of intervention on a large enough scale to earn the description of design. Clark Kerr's design of the California State system is often cited in this context. In the United Kingdom, the university system has been the subject of numerous government reports that have led to redesign, often on an epic scale.[27] The same applies in many other countries where there has often been wholesale top-down change, as in the agglomeration of several universities into one superuniversity that is evident in Denmark or France. But, equally, the work of redesigning individual institutions has been a constant work in progress. It can consist of something as simple as changes in committee structure or the addition of a new School or department. Or it can be of wider consequence, as in the vogue for wholesale reorganization that was current a few years back—and that mainly failed.

Quite often, these externally and internally generated changes are incorporated into a strategy of some kind that, as an almost obligatory moment, will include a statement of values. Yet these statements often ring curiously hollow. They often seem far removed from the actual business of the university and act more as a convenient prop, rather than any kind of civilizational template. Of course, practices vary. In some elite institutions, a repeated invocation of values, conducted in a high moral tone, may still be the general currency of the common room. But, in most higher education institutions, values are now thin threads of justification to be brought out only when a situation requires them. In each case they are underpinned by a persistent suspicion of change that is not so much a case of cynicism as a self-fulfilling defense mechanism.[28]

Instead of the hieratic values so often espoused on all sides, universities might be better suited to taking on *duties or obligations*, understood as neither total self-interestedness nor pure altruism. Instead of a point, a vector. Instead of moral perfection, an aspiration to virtue. Demands are still made, but within reason and tempered by a certain partiality born out of love for objects that can rightly be perceived as worthy of love. The notion of duty has a long history going back to at least Cicero and passing through philosophers such as Kant on perfect and imperfect duty, but it has fallen out of favor recently, even though moral philosophy still has a hold in Western societies (and even though it is not necessarily the case that morality has to impinge on the idea of duty).[29] The word sounds stuffy and conformist, concerned with slavish adherence to a moral or civil law rather than to thought. But duty was of course also considered as central to political thought until recently, as a practical emendation of value.[30] The notion of duty as that which is owed to others could be recast so that it is no longer subjectivist, but rather is concerned with the cultivation of new types of powers. What duties might universities then have? I will count four.

Let's start with the two most obvious duties—research and teaching. The changes to universities outlined previously give the lie to the idea that universities are closed off from "the real world." Indeed, nowadays at least, they very rarely hesitate to intervene in it. For example, in the UK, for better or worse, "impact" is itself becoming a measure of actual academic verisimilitude, as if everyone could and should be a John Dewey, but a John Dewey updated for times in which everything has to be enumerated. Putting the pros and cons of this particular argument aside, and taking into account the changing nature of research, the university is still the place where we call into question what we know and the sense of the world that has become naturalized over time, where the tradition is meant to be one of breaking with tradition. But there is more.

Universities remain one of the few repositories of long-term wisdom in a time when such wisdom has become an imperative in the face of the intrusion of Gaia, as "something that does not demand a response from us, that is utterly deaf to our repentances," that completely exceeds us. In an era in which several existential threats loom on the horizon—not just climate change but also resource scarcity, plagues, even artificial intelligence, perhaps—and in the absence of other institutions willing or able to take up the baton, universities have moved from commentators at the side of these issues to being centrally involved in them. Equally, they have become almost the only translators of the long-term into the short-term priorities that tend to assail everyday life, especially in consumer-led cultures. They have become temporal arbitrageurs. This is a serious shift of priority and probably a heavier responsibility than has been laid upon them previously. As a result, universities are becoming something akin to planetary watchkeepers, requiring a kind of Eudaimonism that is oriented to caring for the planet and planetary flourishing rather than the individual and human flourishing.[31] Such a task requires constancy certainly, a commitment to accuracy, of course, but, most of all, the kind of adventurousness in research and general enquiry that Isabelle Stengers has repeatedly lauded as representative of the best science,[32] "the creation of a situation enabling what the scientists question to put their questions at risk."[33]

To achieve this imperative of rich "description" will require several actions.[34] One will be cooperation. Even though universities are often set up against each other in various competitions, still they are already doing this quite extensively. Of course, there is the groundswell of international cooperation between academics that has become the norm, over and above the large projects like CERN being sponsored by governments and universities. This grassroots movement is being joined by the explicit global research priorities set up by most research-intensive universities of any note, which are often tied in with burgeoning inter-university cooperative networks of various kinds. There is a lot more to do, of course, but it is the case that a base now exists that is growing daily. Another action will be producing more interdisciplinary work, not just because it is legislated by various national and international research bodies (though that might be part of it). That is happening naturally in the empirical natural sciences, partly because of the need to group around certain core technologies (for example, in the life sciences CRISPR-Cas9), partly because many of these core technologies (from magnets to various forms of atomic microscopy) are very expensive, and partly because modern science often requires large teams. It is also happening in the social sciences and humanities, where increasingly, research takes in large teams. An interesting by-product is a question mark over the whole notion of authorship. We may well be moving from a time when individual authorship was the norm to a time when group authorship is the yardstick of progress, from the novel to Pixar.

The second duty is to transmit knowledge through teaching and mentoring. But I would want to add, to do so adventurously with an eye to the whole of the population and not just a part. Throughout their history, universities have been educators of elites. But, probably for the first time in history, that is no longer necessarily the case so far as many, if not most, institutions are concerned. Elite education is now the preserve of a relatively few institutions. Of course, there is a job to be done to shift the scandalous levels of inequality that these institutions help to perpetuate, since the majority of students from low-income families never apply to them in the first place and are concentrated in other higher education institutions by default—the so-called "undermatching" issue that is really a cover for exclusion.

But there is a larger canvas to paint. All the other higher education institutions are, in effect, vocational, even though they are mainly teaching the scions of an expanded middle class that has made more room at the top but remains fiercely opposed to extending its privileges to the working class. The snobbery around this fact is sometimes quite extraordinary, based on a notion of academic charisma that is deeply embedded in academic selfhood through a host of different means—research papers, seminars, lectures and lecture rooms, grading and examinations, annual meetings, and so on.[35] These other institutions, by contrast, are seen as doing "mere" training. Yet most people do want jobs and are not in the upper middle-class zone where they can automatically expect to get them. Most people concerned with finding work are not drones whose thinking is restricted by extant cultural tramlines, but are searching out a practical life with all of its worries and concerns. Most people do find their way into an understanding that is more than their work, but some people also find satisfaction in their work. Most people participate in a life that can never be tied down to one thing and whose resonances may be unknown but are none the worse for that. Just read the final paragraph of George Eliot's *Middlemarch*.

Yet, ironically, in many countries the higher education system is not particularly good at training students to get and keep jobs. Record numbers of students may be graduating from universities and colleges, but many of them are not finding work, not least because they are not being trained for the jobs available. It is a matter of record that companies do not feel that they can get enough staff with the correct skills. Though these kinds of surveys of industry can be taken with a pinch of salt, they underline a more serious problem: having a job—being in work—is much better than having no job. But many jobs could fade out of existence in the next decades as automation of middle class jobs becomes a reality.[36] Ironically, many commentators argue that an education system that teaches students to think creatively, exactly as a liberal arts education is supposed to do, may be the only way to steal a march on the march of the robots.[37] The alternative is a world in which "0.1 per cent own the machines, the rest of the 1 per cent manage their operation, and the 99 per cent either do the remaining scraps of unautomatable work, or are unemployed."[38]

Whatever the case may be, universities are now twixt and between. They have become increasingly practical and vocational, whilst also relaying humanist values that seem to have been founded on quiet contemplation by a singleton self. There is absolutely no reason, however, why a university education could not mix the contemplative and the practical in interesting ways that would illuminate both aspects of knowledge. Indeed, there is currently a focus on high-quality insurgency in precisely this sphere. I am thinking here of the way in which some universities are trying to refocus the teaching of engineering so as to make it into a process of thinking and making in which the "and" is removed, or how other universities are mounting humanities courses that include a heavy practical element. Yet others are considering how to produce degrees that involve students making their own curriculum in search of solutions to practical issues. Others still are trying to do all of these things, while also introducing a substantial locational variation to courses so that students sample the world in order to understand different cultural perspectives. This is an age of experimentation aided precisely by online developments that expand what can be achieved.

The third duty is drama. Universities need to become more dramaturgical. This will no doubt sound like a strange demand. But insofar as universities are about thinking together in ways that are inevitably fictive in requiring a projection of the imagination into the future, it is not. So many of the issues that universities are concerned with, which require the world to be kneaded and adjusted in various ways, require heightened levels of communication so that they become more likely to press on the passionate interests of the population.[39] At present, while university academics are involved in all kinds of experiments in heightened communication and shared experience, universities themselves do far too little to speak to the concerns of larger publics, except insofar as they are involved with their own survival. That is a pity. The population contains all kinds of unrequited longings that can be turned from lead into gold, all kinds of enunciations that can become collective.

Take one of the most obvious topics—climate change. University academics almost unanimously cleave to the reality of climate change and its dangers but, as Per Espen Stoknes points out, communication about climate change has been an almost unmitigated disaster.[40] Citizens cared more about it twenty-five years ago than they do now. But it would be possible for universities to mount subtle, concerted, and, above all, positive campaigns that would start to change the balance of opinion.

They could do so based on an archive of enunciation and performance that runs across the arts and humanities in an almost seamless fashion and that is intent on conjuring up the feeling that life just walked onto the stage.

And then, last, an accidental duty. By a process somewhere between design and osmosis, universities have become increasingly involved in service to the community since the Second World War. This service takes place across many registers: I am not sure that many people in universities realize just how extensive it now is or have made enough of it. Whenever I hear an academic being cynical, I can direct them to their own backyard where staff and students are busy at work on a myriad of projects, mainly but not only small-scale, mostly but by no means only charitable, which cover an extraordinary variety of forms of reaching out into the world: from teaching in schools around the globe to science outreach, from running hospitals overseas to giving basic health advice, from providing legal aid to pursuing human rights, from religious and other conciliation services to ecumenical pursuits, from the whole gamut of arts and cultural activities to producing film and video, from ecological and environmental projects to battling pollution.

In other words, universities have become, perhaps unwittingly, a mainstay of civil society. This can only be to the good. It provides new connections, new relays, and in a way that gives the lie to the by now raddled idea of any kind of ivory tower.

Conclusions: Standing Tall

Universities need to stand tall by emphasizing the duties that they fulfil. They can no longer be—if they ever could be—detached from the world. But these duties are not akin to the grand moral mountains of the past. They are smaller—peaks and prominences, if you like—but also more achievable, as well as dependent on a certain kind of honor. By deploying this term, I mean to signify not a competition for status but the kind of respect that emanates from an honor code that is an index of a different kind of worth.[41] It seems to me that this is what needs to be reinstated in the lives of universities. Yes, the idea of the university is under threat because economentality has become more prevalent.[42] But it is also under threat for all kinds of other reasons. For example, because members of elite institutions, for all of their protestations to the contrary, continue to regard the world as a status hierarchy that they have the monopoly in interpreting, they do not render sufficient respect for the different institutions universities have become and their different missions and different means of gaining esteem. We need, as Félix Guattari would have put it, a means of answering the question "how can we become united and increasingly different?"

One answer to this conundrum is to take some simple practical steps. For example, elite universities could join with other universities in much more positive ways. In other writing, I have argued that what used to be called vocational education needs to become more fully a part of the university venture, not something corralled off for working-class people who, it is assumed, are better off remaining working-class people. All higher education institutions of whatever stripe could become involved in this mission.

Another is to invest in shared projects that acknowledge that a division of labor doesn't necessarily have to create divisions: difference could also produce unity. The shock of the old could again become the shock of the new. In *Finches of Mars*, Brian W. Aldiss foresees a future in which "the universities of the cultivated world [have] linked themselves together under a charter which in essence represented a great company of the wise, the UU (for United Universities)," an organization that, amongst other things, has colonized Mars.[43] Now this might seem like a form of Platonic colonialism, but perhaps some larger projects like this might serve to bind universities together rather than set them apart.

Part of all this is a problem of the kinds of theory we tend to deploy. Too often they are too big, too ambitious to spot the kinds of practical maneuvers that can give birth to new ways of going on. Equally, this kind of high theory, whether the linguistic-philosophical frameworks popular in the humanities or the totalizing models of a certain kind of sociology, forces the world into positions that it cannot live up to: it is all—or it is nothing. The world has to take sides. But the world is many things and no account can take them all in. We need frameworks more open to contingency, surprise, and empirical variability (which is where actor-network-theory can prove especially helpful). The university can no longer revolve around values arising out of high theory.

What we need instead is to move to "low theory" that can deal in concepts that are modest—but not too modest: "Rather than imagine theory as a policing faculty flying high as a drone over all the others, a low theory is interstitial, its labor communicative rather than controlling. . . . Theory proposes, practice disposes. It does not set its own agenda but detects those emerging in key situations and alerts each field to the agenda of others."[44] Universities are important institutions, of that there is no doubt. But they do not require justification by appeals to universal and essentially hieratic forms of theory and value that cut across or colonize what are now so many different modes of existence. Bloated concepts and values only obscure university missions by appealing to moral ideals that cannot be reached by most institutions unless they convert themselves into religious institutions. But there are more modest, *practical ways of proceeding* that might more readily fit with the range of duties I have outlined and with academic life as it really is lived. For example, terms such as "dialogue" and "judgment" and "connection" might signify one practical means of proceeding. Another way of proceeding might be signified by the terms "curiosity," "growing knowledge," "inspiring new generations." A third might use terms like "sharing" and "held in common" and "care of the possible."

Making such moves could be seen as following on from Latour's attempt to provide a new habitation in which we might establish values, "or better yet . . . install them, in institutions that might finally be designed for them,"[45] institutions that might be bound by certain formalities, but that could never lapse into certainty. In other words, universities could become—to the extent that they are not—houses with many modes of justification, all woven together by experimentation in many registers. They would be founded in multiple duties rather than values. These duties would be embedded through an implicit code of honor that does not have to be written down as a mission statement, and they would be bolstered by practical ways of proceeding that are able to become the subject of genuine diplomacy.

Notes

1. A. H. Halsey, *Decline of Donnish Dominion: The British Academic Professions in the Twentieth Century* (Oxford: Oxford Univ. Press, 1992).

2. See Nigel Thrift, "Universities 2035," *Perspectives, Policy and Practice in Higher Education* 20, no. 1 (2016): 12–16.

3. In most places outside the privileged enclaves from which most views about universities emerge, it may be that the terms and conditions of academic life have been subject to attrition but that has probably been the price that had to be paid for creating so many academic and other jobs with such stretched resources.

4. See Bruno Latour, *An Inquiry into Modes of Existence: An Anthropology of the Moderns* (Cambridge, MA: Harvard Univ. Press, 2013).

5. See Isabelle Stengers, interview by Erik Bordeleau, "The Care of the Possible," trans. Kelly Ladd, *Scapegoat* 1 (2011): 12–13, 16–17.

6. L. Robbins, *Higher Education* (London: Her Majesty's Stationery Office, 1963.)

7. See Elizabeth Popp Berman, *Creating the Market University: How Academic Science Became an Economic Engine* (Princeton, NJ: Princeton Univ. Press, 2012).

8. Philip Mirowski, *Science-Mart: Privatizing American Science* (Cambridge, MA: Harvard Univ. Press, 2011), 88; Christopher C. Morphew and Peter D. Eckel, eds., *Privatizing the Public University: Perspectives from Across the Academy* (Baltimore: Johns Hopkins Univ. Press, 2009).

9. David Graeber, *The Utopia of Rules: On Technology, Stupidity, and the Secret Joys of Bureaucracy* (New York: Melville House, 2015). Of course, many academics argue that administration has now taken over, leaving them as little more than corporate handmaidens with no control over their own affairs. There is an issue but, at its worst, this is no more than a "know your place" argument: administrators need to return to the status of porters.

10. Latour, *An Inquiry into Modes of Existence*, 44.

11. See Richard F. Teichgraeber, introduction to Thorstein Veblen, *The Higher Learning in America: A Memorandum on the Conduct of Universities by Business Men* (Baltimore: Johns Hopkins Univ. Press, 2015), 1–29.

12. Thus, many academics bemoan these changes whilst also being their keenest beneficiaries.

13. Latour, *An Inquiry into Modes of Existence*, 452.

14. See Burton A. Weisbrod, Jeffrey P. Ballou, and Evelyn Diane Asch, *Mission and Money: Understanding the University* (New York: Cambridge Univ. Press, 2008).

15. See Teichgraeber, introduction to *The Higher Learning in America*, 1–29. Jonathan R. Cole, *The Great American University: Its Rise to Prominence, its Indispensable National Role, and Why it Must be Protected* (New York: Public Affairs, 2009).

16. See Berman, *Creating the Market University*; Mirowski, *Science-Mart*, 88.

17. Alessandro Duranti, "On the Future of Anthropology: Fundraising, the Job Market and the Corporate Turn," *Anthropological Theory* 13, no. 3 (2013): 214.

18. This is simultaneously a matter of content and method.

19. See William Clark, *Academic Charisma and the Origins of the Research University* (Chicago: Univ. of Chicago Press, 2006).

20. Timothy Mitchell, "Economentality: How the Future Entered Government," *Critical Inquiry* 40, no. 4 (2014): 479–507.

21. Wendy Brown, *Undoing the Demos: Neoliberalism's Stealth Revolution* (New York: Zone Books, 2015), 180.

22. Lauren A. Rivera, *Pedigree: How Elite Students Get Elite Jobs* (Princeton, NJ: Princeton Univ. Press, 2015), 15–16.

23. Brown, *Undoing the Demos*, 184.

24. None of this is to excuse the more ludicrous manifestations of vocational monocultures as found in states like Texas or Florida. See Matthew Reisz, "Humanities crisis? What crisis?" *Times Higher Education*, July 9, 2015, 10.

25. Peter Mandler, "The 'Crisis in the Humanities' in Comparative Perspective" (lecture, Australian Historical Association Conference, Sydney, July 7, 2015). Usually the idea that there has been a precipitate decline in the student population taking humanities (or liberal arts) subjects is made by using relative rather than absolute figures. See, for example, most recently Fareed Zakaria, *In Defense of a Liberal Education* (New York: Norton, 2015).

26. See Michael M. Crow and William B. Dabars, *Designing the New American University* (Baltimore: Johns Hopkins Univ. Press, 2015).

27. For example, in the 1960s the Robbins Report.

28. See Harry Brighouse and Michael S. McPherson, eds., *The Aims of Higher Education. Problems of Morality and Justice* (Chicago: Univ. of Chicago Press, 2015).

29. Cicero, *On Duties*, ed. M. T. Griffin and E. M. Atkins (Cambridge: Cambridge Univ. Press, 1991).

30. Although it is worth remembering Nietzsche's argument that it is the task of higher education "to turn men into machines," a feat accomplished by means of the concept of duty (Nietzsche). Friedrich Nietzsche, *Twilight of the Idols; and, The Anti-Christ*, trans. R. J. Hollingdale (Harmondsworth: Penguin, 1990), 29.

31. Of course, the term eudaimonia originates with Aristotle but it can be easily related to a planetary notion of the term in that Aristotle can also be understood as the first to both investigate and mount a defense of the living world: "There is something awesome in all natural things." Quoted in Armand Marie Leroi, *The Lagoon: How Aristotle Invented Science* (London: Bloomsbury, 2014), 10.

32. Another will be the fact that as the world becomes more "scientific" in tone and content, so the fiction that there is a stentorian venture called Science with a capital S becomes harder to maintain. With science involved in all parts of popular culture, it becomes a currency that increasingly will be shared in one form or another. This can be freeing. Instead of science with a capital S, there are now adventures in science. Latour and Stengers have been the main advocates of this approach.

See Latour, *An Inquiry into Modes of Existence*; Stengers, interview by Brian Massumi and Erin Manning, "History Through the Middle: Between Macro and Mesopolitics," *SenseLab*, November 25, 2008.

33. Stengers, "Reclaiming Animism," *e-flux* 36 (2012).

34. Henning Schmidgen, *Bruno Latour in Pieces: An Intellectual Biography* (New York: Fordham Univ. Press, 2015).

35. See Clark, *Academic Charisma and the Origins of the Research University*.

36. See Erik Brynjolfsson and Andrew McAfee, *The Second Machine Age: Work, Progress and Prosperity in a Time of Brilliant Technologies* (New York: Norton, 2014).

37. See Zakaria, *In Defense of a Liberal Education*.

38. John Lanchester, "The Robots Are Coming," *London Review of Books* 37, no. 5 (2015): 7.

39. See Ash Amin and Thrift, *Arts of the Political: New Openings for the Left* (Durham, NC: Duke Univ. Press, 2013).

40. Per Espen Stoknes, *What We Think About When We Try Not To Think About Global Warming: Toward a New Psychology of Climate Action* (London: Chelsea Green Publishing, 2015).

41. See Kwame Anthony Appiah, *The Honor Code: How Moral Revolutions Happen* (New York: Norton, 2010).

42. See Gaye Tuchman, *Wannabe U: Inside the Corporate University* (Chicago: Univ. of Chicago Press, 2009).

43. Brian W. Aldiss, *Finches of Mars* (London: HarperCollins, 2013), 7.

44. McKenzie Wark, *Molecular Red: Theory for the Anthropocene* (London: Verso, 2015), 218.

45. Latour, *An Inquiry into Modes of Existence*, 7.

PART TWO

LATOUR AND THE DISCIPLINES

CHAPTER ELEVEN

Critique, Modernity, Society, Agency

Matters of Concern in Literary Studies

DAVID J. ALWORTH

BRUNO LATOUR IS best known within literary studies as an opponent of critique, seeking to cultivate an entirely new ethos of knowledge production. A lively writer with a knack for shaking up received ideas, Latour is one source of inspiration for the robust methodological discussion that currently spans the disciplines of English and Comparative Literature. In his much-cited 2004 article, "Why Has Critique Run Out of Steam?," he argued for an analytical reorientation premised not on skepticism and wariness but on openness and generosity. "The critic is not the one who debunks," he wrote, "but the one who assembles. The critic is not the one who lifts the rugs from under the feet of naïve believers, but the one who offers the participants arenas in which to gather."[1] It did not take long for actual critics, scholars of literature, to warm up to this characterization of the critic. In the years following the publication of his article, Latour's ideas were welcomed by an otherwise diverse array of humanists, casting about for new methods of interpretation: alternatives to the orthodoxies of historicism, post-structuralism, and post-modernism.[2]

This initial embrace was followed by pushback. As in other fields, from sociology and anthropology to philosophy and physics, Latour's claims have provoked strong reactions within the literary humanities. While some scholars applaud his emphasis on "composition" and positive knowledge, others see him as woefully benighted, ensnared by the very trap of critique that he means to destroy. Where some see stimulating arguments cast in delightfully literary prose, others see dangerously misguided ideas, bad politics, and rhetorical self-indulgence.[3] Not that this mixed reception is anything new to Latour himself. Ever since he published his first ethnographic study of scientific practice, *Laboratory Life* (1979), he has been "reviled" by certain thinkers and adored by others.[4] Indeed, the *New York Times* recently labeled him "France's most famous and misunderstood philosopher."[5]

The controversy that has always surrounded Latour's work, however, is not simply the result of his tendency to inflame the passions with bold claims. Rather, it seems to be a specific formal effect of his prose, which fits neatly in no genre whatsoever. Synthesizing the rhetorics of metaphysics, anthropology, sociology, aesthetics, cosmology, and fiction, Latour stylistically inscribes his philosophical conviction that modern epistemology is constitutively mixed, impure, and heterogeneous. Just as his universe of ideas is populated not by subjects and objects but by "conceptual characters" called quasi-subjects and quasi-objects, so too his writing amalgamates forms, tropes, devices, and conventions from a wide range of discourses and genres.[6] As I have argued elsewhere, the polyphony that characterizes Latour's prose—what could be called its *heteroglossia*—makes it decidedly novelistic and thus, like the novel form itself, capacious, voracious, and difficult to pin down.[7] Readers failing to categorize his writing often manifest signs of confusion and even horror, as though it were a "large loose baggy monster" necessitating but resisting confinement.[8] And yet, Latour's deep commitment to clarity nonetheless renders even his most complex arguments amenable to reductive summary—one consequence of which is that the name "Latour" now calls to mind any number of contrarian academic sound bites: critique has run out of steam, we have never been modern, there is no such thing as society.

To be sure, these are not merely tidbits of critical wisdom, but intricately argued claims that present real challenges and opportunities for scholars in the humanities. Still, in assessing the relationship between Latour and literary studies, one must immediately confront the literariness of Latour: the performativity of his work, its exuberance of idiom, the way that his writing novelizes other discourses and tilts toward hyperbole and formal monstrosity.[9] There are a few reasons why it is important to consider stylistic elements in this case. First, and most generally, if literary criticism is premised on the notion that style cannot be disarticulated from substance—that style is, indeed, substantial—then a literary-critical account of a self-consciously literary writer, irrespective of discipline, must address rhetorical idiosyncrasies. Second, Latour's thinking is deeply indebted to both literature and literary theory; his ideas about actants and actors, for instance, have a basis in narratology and semiotics. And third, Latour frequently tests the limits of genre, as in the case of *Aramis, or The Love of Technology*, which he calls a work of "scientifiction" for its combination of science writing, philosophy, and fantasy. Thus, it is not simply that literary critics will want to read Latour precisely *as a writer*. In fact, he demands to be read this way.

Latour's literariness amounts to more than a litany of peculiar tics and tendencies; it serves his polemical attempt to defamiliarize fundamental concepts—such as "society, discourse, knowledge-slash-power, fields of forces, empires, capitalism"—even as it stokes the sort of confusion and controversy that can be the price of making a strong argument.[10] To read almost any paragraph of Latour's writing is to glimpse the workings of a would-be novelist elaborating characters, crafting dialogues, building narrative suspense through complex scenes of conflict and resolution, and gently badgering his reader to pay attention. Take, for instance, the grandiose tone that animates even his most conventionally written book, *Laboratory Life* (1979).[11] Alongside collaborator Steve Woolgar, Latour declares that "the particular branch of philosophy—epistemology—which holds that the only source of knowledge are ideas of reason intrinsic to the mind, is an area whose total extinction is overdue."[12] John Guillory is right to notice both the vaulting "ambition" of this statement ("which seeks nothing less than to 'overcome' the binarisms of all previous philosophies with a few sweeping gestures pronounced alongside the business as usual of ethnographic case studies") and the performative force of its rhetoric (which is "rather like the Louis XIV's declaration that with the revocation of Edict of Nantes all the Protestants in France were Converted to Catholicism").[13]

Latour's literariness, however, is not only a matter of style and rhetoric. It is also evident in his productive engagement with textuality and narrative theory. "We presented the laboratory as a system of literary inscription," explain Latour and Woolgar, "an outcome of which is the occasional conviction of others that something is a fact."[14] Even at this early stage in his career, Latour was emphasizing the production, circulation, and mediation of texts within the social life of knowledge. Having begun his intellectual life as a student in biblical exegesis, he was sensitive to the ways that, as Nietzsche had it, "concepts never leave behind their stylistic embodiment," but are always part of a "complex trajectory of interpretation, rewriting, invention, reprise, fabrication, canon formation, and institutional incorporations that allow statements to gain a meaning." Reflecting back on this time in his career, Latour has confessed, "I was a Derridean for a whole year after reading *Of Grammatology* (a key influence on the later notion of scientific *inscription*)."[15] In *Reassembling the Social: An Introduction to Actor-Network-Theory* (2005) he goes a step further, downplaying the conflict between C. P. Snow's "two cultures" and thereby rejecting any difference "between 'scientific' and 'literary' minds." Instead, he argues that the salient difference appears "between those who write *bad* texts and those who write *good* ones."[16] Like much of Derrida's or Nietzsche's work, a good text is highly self-aware, alert to its own "thickness," which is to say "its pitfalls, its dangers, its awful way to make you say things you don't want to say, its opacity, its resistance, its mutability, its tropism." From Latour's perspective, writing becomes "all the more accurate" when it acknowledges its own artificiality.[17]

If at times Latour sounds like a Derridean poststructuralist, this is not because he has rejected the lessons of structuralism and semiotics. To the contrary, when he explains in *Reassembling the Social* that Actor-Network-Theory (ANT) "uses the technical word *actant* that comes from the study of literature," he is referring specifically to the narratology of A. J. Greimas. "[A]n actant," Greimas writes, "can be thought of as that which accomplishes or undergoes an act, independently of all other determinations," meaning it is "a type of syntactic unit, properly formal in character, which precedes any semantic or ideological investment."[18] According to this analytical model, narrative structure includes three binaries of actants (subject/object, sender/receiver, helper/opponent) that become binaries of "actors" once they are "invested," or figured semantically. Greimas continues, "An actor may be individual (for example, Peter), or collective (for example, a crowd), figurative (anthropomorphic or zoomorphic), or non-figurative (for example, fate)." Thus, Greimasian narratology not only grants agency to nonhumans but also provides two different ways of rendering the capacity for action, either abstractly as an actant or concretely as an actor, which is why Latour tends to use these two terms interchangeably throughout his work. "[A]*ny thing* that modif[ies] a state of affairs by making a difference is an actor," he asserts, "or, if it has no figuration yet, an actant."[19]

One of Latour's most compelling maneuvers is to adapt narratology beyond its conventional use, expanding its purview to account for the empirical world. Through his engagement with Greimas in particular, he develops two interrelated claims about agency. First, he equates agency with effectivity, without privileging subjective intentionality over other sources of action. "An invisible agency," he writes, "that makes no difference, produces no transformation, leaves no trace, and enters no account is *not* an agency. Period. Either it does something or it does not." Second, he differentiates agency (an abstract capacity for action) from its figuration (an empirical manifestation of that capacity). With typical literary panache, for example, he describes "four ways to figure out the same actant":

> "Imperialism strives for unilateralism"; "The United States wishes to withdraw from the UN"; "Bush Junior wishes to withdraw from the UN"; "Many officers from the Army and two dozen neo-con leaders want to withdraw from the UN." That the first is a structural trait, the second a corporate body, the third an individual, the fourth a loose aggregate of individuals makes a big difference of course to the account, but they all provide different figurations of the same actions.[20]

A greater emphasis on "figuration," Latour contends, would enable sociologists "to gain as much inventiveness as that of the actors they try to follow—also because actors, too, read a lot of novels and watch a lot of TV!" His point is that, if sociologists want to understand "complex repertoires of action," then they "need as much variety in 'drawing' actors as there are debates about figuration in modern and contemporary art." In this respect, "sociologists have a lot to learn from artists" as well as from fiction writers, since "it is only through some continuous familiarity with literature that ANT sociologists might become less wooden, less rigid, less stiff in their definition of what sort of agencies populate the world."[21]

In examining Latour's literariness, then, we find a constellation of themes, or "matters of concern," that are central to his thinking— critique, modernity, society, and agency—all of which he has developed through a productive engagement with literary theory. At a time when literary studies is a net importer of ideas and methods from other disciplines, it is refreshing to see how Latour "has borrowed from narrative theories" to develop his most influential arguments.[22] We are now in a position to explore what happens when intellectual traffic flows in the other direction. To be sure, the contact zone between Latour and Literature is a busy one, and it is governed by shared matters of concern.

I. Critique

"My argument," Latour asserted in 2004, "is that a certain form of critical spirit has sent us down the wrong path."[23] So begins his effort to dismantle critique, an effort that was welcomed by literary critics eager to find alternatives to symptomatic reading and the hermeneutics of suspicion. In the humanities, arguments against critique have taken multiple forms over the past decade or so, but in general they tend to disparage the reflexive operation of debunking or unmasking—along with the attitude of superiority and the ethos of negativity associated with this operation. "The task of the social critic," as Rita Felski puts it, "is now to expose the hidden truths and draw out unflattering and counterintuitive meanings that others fail to see."[24] She contends that this task not only defines the object of study as a tissue of ideological determinants but also elevates the critic to a position of power over the text. Rewriting the text "in terms" of what Fredric Jameson once called "a particular interpretive master code," the critic aims to disclose its "political unconscious": the veiled source of its form and meaning.[25] Once this mode of analysis had attained prominence in literary studies, it assumed a kind of hegemony. "Why is critique," Felski wonders, "so frequently feted as the most serious and scrupulous form of thought? What intellectual and imaginative alternatives does it overshadow, obscure, or overrule?"[26]

Latour offers some answers to these questions. As an alternative to critique, he suggests *compositionism*: a term that connotes positive knowledge, critical making, gathering, and assembly, rather than standing back, holding forth, and unveiling. While "critique did a wonderful job of debunking prejudices, enlightening nations, and prodding minds," it created "a massive gap between what was felt and what was real," precisely because "it was predicated on the discovery of a true world of realities lying behind a veil of appearances." For the compositionist, by contrast, there is no dialectic of appearance and reality: "there is no world of the beyond. It is all about *immanence*." This commitment to immanence places Latour in a specific philosophical tradition, namely that of pragmatism or radical empiricism, that still has much to offer literary studies. In fact, Latour frequently cites William James as an important predecessor. Instead of "remaining faithful to . . . the philosophy of Immanuel Kant," with its basis in critique, Latour seeks to foster what James called "a stubbornly realist attitude" alongside a renewed commitment to empiricism.[27] And instead of moving away from the empirical world in order to puzzle over "the world *beyond*," he argues that we must move closer to the flux of experience—that is, to apprehend "matters of fact" as "only very partial" and "very polemical, very political renderings of matters of concern, and only a subset of what could also be called *states of affairs*."[28] Facts are real, but they are not just out there, awaiting discovery; they are, in Latour's view, very fragile figurations of experience produced and nurtured by the everyday business of science. Experience is therefore the master category from which facts derive. "James felt and captured something about the originality of this mode of existence," Latour stated in an interview with *The Los Angeles Review of Books*. "Nothing except the experience but no less than an experience": this is a great line by James and I completely agree with him. I'm a Jamesian."[29]

What does Latour mean to suggest with this act of self-definition? Looking back on his work with Latour over several decades, Antoine Hennion describes *An Inquiry into Modes of Existence* (2013) as "an explicitly pluralist and pragmatist reformulation of ANT methodology." He goes on to explain that their pragmatism has little in common with "enunciative pragmatics, analytical philosophy, and theories of action," which is to say the pragmatism of "the 1980s" that is most familiar to scholars of literature.[30] Rather, ANT takes pragmatism "back to its founding principles, *pragmata,* the *agency* of things, and is not used as an alias to shift out of critical sociology without reconsidering the very narrow arena it had reserved for objects." Hennion suggests, in other words, that he and Latour found retrospective confirmation of their ideas in foundational texts by James and John Dewey. "It was as if Dewey were confronted by contemporary problems, such as the environment, development, energy, sexuality . . . all of this in a world without exteriority, but plural and open, an expanding tissue of heterogeneous realities, but 'still in the process of making,' as James nicely put it."[31]

In addition to rejecting "the world *beyond*" that James also rejected, Latour adapts two main ideas from his philosophical forebear that are pertinent to literary studies. The first idea is that relations are real. "To be radical," James argued, "an empiricism must neither admit in its constructions any element that is not directly experienced, nor exclude from them any element that is directly experienced." For both James and Latour, then, empiricism demands a posture of openness toward the world, one that excludes nothing and that resists the urge to cast certain experiences as unrealistic. James continues, "For such a philosophy, *the relations that connect experiences must themselves be experienced relations, and any kind of relation experienced must be counted as 'real' as anything else in the system*" (1912, italics in original), If traditional empiricism ascribes reality only to concrete phenomena such as chairs and billiard balls, then James's empiricism is radical to the degree that it ascribes reality to the conjunctive or disjunctive states between such phenomena. As James writes, "in actual experience the more substantive and the more transitive parts run into each other continuously, there is in general no separateness needing to be overcome by an external cement; and whatever separateness is actually experienced is not over-come, it stays and counts as separateness to the end."[32]

Inspired by this claim and its import for Latour, Brad Evans provides a compelling example of how emphasizing relationality might reorient literary criticism away from the familiar protocols of critique. "I take it," Evans contends, "that the project of the humanities in its most ambitious formulation is that of tracing new relations to older ones." Drawing on Latour's *Reassembling the Social,* he puts this project into practice through a reading of Henry James that enlists the novelist as a fellow traveler, a kind of Actor-Network-Theorist *avant la lettre.* Evans argues against the "emphasis on consciousness" that "misdirects us from what the late James stories and novels are all about: the failure of social rela-tions to apprehend themselves." Rather, he wants us to read James for what the novelist might teach us about "the impossible-to-describe art of associating" through his experiments with the novel form, a genre of prose fiction that is, from Evans's perspective, all about sociality: the liminal, transitive flux of persons, things, and systems. Whereas a con-ventional historicist reading of James would presuppose a social ground (e.g., the transatlantic world, an upper-class expatriate milieu) against which his fictions emerge and through which his fictions attain the sort of meaning that can be divulged through critique, Evans seeks "to rei-magine the sociality of literature" by "open[ing] a link to James as a theorist of the artwork of networks, for ultimately his novels are neither the residue of consciousness or mere objects. They are, rather, part of the ever-swarming assemblage of relations" that humanists are in the busi-ness of tracing and apprehending.[33]

The second idea that Latour adapts from James is that of the "pluri-verse," which James elaborates most fully in *A Pluralistic Universe,* a book that defines reality as all flux, change, and contingency. "The rul-ing tradition in philosophy," James argued, "has always been the Pla-tonic and Aristotelian belief that fixity is a nobler and worthier thing than change. Reality must be one and unalterable. Concepts, being themselves fixities, agree best with this fixed nature of truth, so that for any knowledge of ours to be quite true it must be knowledge by univer-sal concepts rather than by particular experiences, for these notoriously are mutable and corruptible."[34] Against this "ruling tradition," James held that "What really *exists* is not things made but things in the making . . . Reality falls in passing into conceptual analysis; it *mounts* in living its own undivided life—it buds, bourgeons, changes, and creates."[35] For his part, Latour echoes this claim throughout the *Inquiry*, often citing his predecessor directly. "William James," he writes, "asserts that there exists in the world no domain of 'with,' 'after,' or 'between' as there exists a domain of chairs, heat, microbes, doormats, or cats. And yet each of these prepositions plays a decisive role in the understanding of what is to follow, by offering the type of relation needed to grasp the experience of the world in question."[36]

This engagement with James enables Latour to shift the emphasis from the ontological question of what things are to the experiential question of what things do.[37] He then goes on to develop the implications of the latter question, which animates much of his work, through a familiar literary conceit. "If you find yourself in a bookstore," Latour writes, "and you browse through books identified in the front matter as 'novels,' 'documents,' 'inquiries,' 'docufiction,' 'memoirs,' or 'essays,' these notices play the role of prepositions." While such genre labels "don't amount to much," they play a "decisive" role in shaping your experience of the text at hand. "Everyone can see that it would be a category mistake to read a 'document' while believing all the way through that the book was a 'novel,' or vice versa."[38] While this might seem like a basic point, emphasized in genre theory from Gérard Genette to Wai Chee Dimock, it exemplifies how Latour approaches literary art more generally. Unlike a traditional literary critic or literary historian, and very much in the spirit of pragmatism, he asks not what texts mean but what they do—how they function as actants or matters of concern.[39] For him, literature is not a primary source, whose meanings must be excavated through strenuous critique, but a pedagogical or conceptual resource. Literature, in this sense, is a report on experience that facilitates the process of apprehending reality. Just as Latour encourages sociologists to attend closely to "[n]ovels, plays, and films from classical tragedy to comics," so too he suggests that, "the resource of fiction" can facilitate our understanding of experience, reality, and social life.[40]

II. Modernity & Society

For many literary critics, abandoning critique entails rethinking the relationship between literature and society. If, conventionally speaking, literary critics presuppose society as the ground or context for interpreting texts, then Latour suggests a decidedly unconventional approach to reading and thinking about literature, for he argues that society cannot be posited as a stable given. His position is not unlike that of Manuel DeLanda, an assemblage theorist who contends that because "we never reach a point at which we may coherently speak of 'society as a whole,' the very term 'society' should be regarded as a convenient expression lacking a referent."[41] In agreement with this claim, Latour rejects the traditional understanding of "society" or "the social" as a special domain of reality (distinct from "the material" or "the natural") governed by abstract laws, structures, or functions. Rather, he considers the social to be "the act and the fact of association," as Felski puts it, "the coming together of phenomena to create multiple assemblages, affinities, and networks."[42] In this analytical model, the social is not a fixed container where anything can be situated, but "a process of assembling" whereby persons, things, texts, ideas, images, and other entities (all of which are considered actors or actants) form contingent networks of association. As its title implies, therefore, the goal of *Reassembling the Social* is to articulate an "alternative social theory" that concentrates on associative processes: the coming together of actors into networks that must be traced in order to be understood.[43] In this sense, society is not presupposed as a cause, but defined as an effect of how actors assemble, disassemble, and reassemble anew.

This new definition of society has emerged because Latour and other social theorists have become increasingly convinced that the traditional definition, as Patrick Joyce explains, "does not describe the world very well, neither the world of the present nor the world of the past."[44] Thinking in more general terms, Latour sees the prevailing conception of the social as a fault of modern epistemology *tout court*. In *We Have Never Been Modern*, a short book that serves as a kind of primer on his philosophical polemic, he argues that modernism is defined by two mutually reinforcing dynamics: translation and purification. The first "creates mixtures between entirely new types of beings, hybrids of nature and culture," whereas the second "creates two entirely distinct ontological zones: that of human beings on the one hand; that of nonhumans on the other. Through purification, foundational modern binaries (human vs. nonhuman, subject vs. object, society vs. nature, mind vs. matter, fact vs. fetish) are scrupulously maintained, despite the widespread proliferation of hybrids, such as commodities and cyborgs. Thus, when Latour asserts that "we have never been modern," he does not mean to deny the fact of modernization but to suggest that it relies on the mutual reinforcement of purification and translation: one force that establishes and shores up boundaries, another that transgresses them.[45] "I've worked a lot on the question of modernism," Latour sighed in a recent interview. "People never really understood what I was saying. I was just arguing that humans and nonhumans are now mixing together more and more intimately."[46]

Émile Durkheim, Latour insists, failed to apprehend such mixing when he conceptualized society as an autonomous, *sui generis* and, specifically, human domain of interactions. And just as Latour seeks to displace Kant with James in the philosophical tradition, so too he defines Gabriel Tarde as the true founder of sociology and "an early ancestor of ANT."[47] This move complements his broader effort to sideline critique. "What if explanations resorting automatically to power, society, discourse," Latour asks, "had outlived their usefulness and deteriorated to the point of now feeding the most gullible sort of critique?" His point is that both "serious" and "popularized" versions of this type of thinking "have the defect of using society as an already existing cause instead of as a possible consequence" and that "probably the whole notion of *social* and *society* that is responsible for the weakening of critique."[48]

Such claims have an oblique but nonetheless powerful impact on literary analysis, for they disrupt long-held assumptions and entrenched protocols. If society is not conceptualized as a preexisting container— "that total genus," as Durkheim calls it, "beyond which nothing else exists,' or "that whole which includes all things"—then the basic (historicist) procedure of "placing a text in social context" makes no sense, since the latter would not be prior to and broader than the former.[49] Moreover, if society is defined as an effect rather than a cause, then the trajectory of a typical "symptomatic reading" must be inverted: instead of asking how society and its attendant ideology cause literature to assume a certain form, the critic asks how literature imagines society and social relations. These two questions are not mutually exclusive, but they do inflect literature in different ways. While the former configures the text as an epiphenomenon of a given social order, the latter seeks to identify something like a radically literary sociology. Indeed, to ask how literature imagines society and social relations is not to see it merely as a reflection of or response to the material and ideological conditions of its production, but to assume that literary work, to borrow a phrase from C. Wright Mills, expresses a "sociological imagination" all its own.[50] Which is also to assume, as Latour himself would put it, that sociologists have a lot to learn from artists.

III. Agency

"[M]odernism," writes Bill Brown, "always knew that we have never been modern."[51] In dialogue with Latour, Brown argues that the art and culture of modernity—from surrealist collage to cinematic montage, from the boxes of Joseph Cornell to the novels of Virginia Woolf—regularly disrupts modern epistemology, undoing the binarisms of human and nonhuman, subject and object, society and nature, exposing such concepts as intellectual tools that, however convenient, fail to capture what Latour calls the "many metaphysical shades between full causality and sheer inexistence."[52] In general, such disruption is an effect of how art mediates agency (understood, in the most basic sense, as a capacity for action). For instance, when poems defamiliarize the palpable world, when narratives enlist objects as part of the plot, or when films alter your perception of space and place, Brown suggests, they disclose agency as both promiscuous (shared with nonhumans) and utterly irreducible to subjective intentionality. Many literary critics seem to agree.[53]

Latour's claims about agency have had a wide impact, both in literary studies and beyond. They are at once intuitive and controversial.

Indeed, when he drafts off Greimas to conceptualize an agent or actant "as that which accomplishes or undergoes an act, independently of all other determinations," he is not only eschewing a conception of agency based on identity or subject-position within a hierarchized society but also decentering the human species, a maneuver that feels increasingly urgent in a time of climate crisis. As Donna Haraway explains, "Latour passionately understands the need to change the story, to learn somehow to narrate—to think—outside the prick tale of Humans in History. Searching for compositionist practices capable of building effective new collectives, Latour argues that we must learn to tell 'Gaïa stories.'"[54] Acknowledging how agency is shared between humans and nonhumans, and thereby undoing a central tenet of modern thought, serves as a kind of prolegomena to this new mode of storytelling.

Still, as Hennion points out, Latour's ideas about agency are not exactly new. They have their roots in what Hennion calls "American extensions of action theory," which itself is indebted to the philosophy of James and other pragmatists. "In contrast to our experience of Bourdieu," Hennion explains, "we had not been immersed from the beginning of our training in the language of *affordances*, those orientations of objects toward new uses whose possibility they nevertheless suggest."[55] When they found that language, however, it opened up new conceptual possibilities for them. The concept of "affordances" was invented by psychologist James J. Gibson and applied to industrial design by Don Norman, two thinkers seeking to answer the question of whether perception entails interpretation. Gibson, who pioneered what is known as the ecological approach to perception, claimed that our sense organs, without any intervention from the brain, constantly gather information about the world around us. Norman argued that the brain is always involved in the perceptual process, and without the brain to organize all the perceptual data that we acquire through sight, smell, touch, and other sensations, we could not attain a coherent picture of the world.

For Norman, this debate was a breakthrough. As he writes in *The Design of Everyday Things*:

We live in a world filled with objects, many natural, the rest artificial. Every day we encounter thousands of objects, many of them new to us. Many of the new objects are similar to ones we already know, but many are unique, yet we manage quite well. How do we do this? Why is it that when we encounter many unusual natural objects, we know how to interact with them? Why is this true with many of the artificial, human-made objects we encounter?[56]

The answer is that we are always learning from the objects in our midst. Through a process that Gibson calls "information pickup," we sense that balls are for throwing, knobs are for turning, and chairs are for sitting. These sensations tell us about an object's affordances, or lack thereof. A hammer affords hitting, but it does not afford a reduction in bad cholesterol. You wouldn't use a fire hose to fix a flat tire. But there is a crucial qualification: affordances are not simply stored in a given object. Rather, they are defined, as Gibson puts it, "across the dichotomy of subjective-objective."[57] An affordance is, therefore, a relation between two entities, not a property of any single entity. This means, for instance, that some chairs afford both sitting and lifting for some human agents, whereas others, as Norman explains, "can only be lifted by a strong person or a team of people. If young or relatively weak people cannot lift a chair, then for these people, the chair does not have that affordance."[58] Still, the concept of affordance is not synonymous with prescribed use. Many affordances add to what Brown calls the "misuse value" of a given object, as when you treat a chair like a ladder.[59]

If the notion of distributed agency, of a capacity for action that resides "across the dichotomy of subjective-objective," is central to Latour's thinking, then it is no less important to the literary and cultural criticism that his work has inspired. While Brown and other critics associated with "thing theory" have introduced a "new materialism" into current debates over literary-critical methodology, other scholars, also influenced by Latour, have argued for a "new formalism" that tracks the agency of forms—asking, in the idiom of ANT, how received patterns and orders function as actants that define the shape of social life.

Intended as "a methodological starting point," Caroline Levine's *Forms* (2016), for instance, "proposes a way to understand the relations among forms—forms aesthetic and social, spatial and temporal, ancient and modern, major and minor, like and unlike, punitive and narrative, material and metrical. Its method of tracking shapes and arrangements is not confined to the literary text or to the aesthetic, but it does involve a kind of close reading, a careful attention to the forms that organize texts, bodies, and institutions." This method, Levine continues, "builds on what literary critics have traditionally done best—reading for complex interrelationships and multiple overlapping arrangements." But she also urges these same critics "to *export* those practices, to take [their] traditional skills to new objects—the social structures and institutions that are among the most crucial sites of political efficacy." In the end, then, "new formalism" is not just a method for thinking transhistorically about the "shapes and arrangements" that govern both art and life; it is also a way of doing political work, for Levine contends that "the most strategic political action" begins with understanding forms and ends with the reordering of things.[60]

In addition to stimulating both new materialism and new formalism, ANT offers a few other provocations to literary and cultural criticism. As Felski has argued, it enables critics to conceptualize "networks within texts" as well as "texts within networks."[61] Instead of separating characters from their settings, or emphasizing what Georg Lukács called "the mesh of human destinies" over the fate of the nonhuman, an ANT-inspired literary criticism takes as its object of analysis the motley assemblages of actants that appear in novels and other literary works.[62] As in real life, assemblages in literature are understood to have agency: precisely as assemblages, that is, they can do things that none of their constituent parts could accomplish on its own.[63] It becomes the task of the critic, therefore, to track what this agency does, and perhaps also how it means. At the same time, though, ANT-inspired criticism pulls in a different direction, not inward to look more closely at the text itself but outward, toward the links between text and context, for such criticism conceptualizes any individual literary work as part of a larger network, refusing to reify the text as an autonomous artifact that is somehow divorced from the "relations" supporting its mode of existence.

At its most fundamental, then, all of the criticism that I have in mind heeds Latour's call to "follow the actors," whoever or whatever they are and wherever they may lead.[64] From his perspective, the most important scholarship in the humanities and the social sciences begins from this premise; it does not start from the top down, with a received idea about *agency*, *modernity*, or *society*, but from the bottom up, with the particular, minute, ordinary, and often unremarkable actions of individual actors, human or otherwise. Moreover, even if "follow the actors" has replaced "Always historicize!" as an imperative slogan for many literary critics, this imperative would seem to have a special purchase on scholarship addressing contemporary literatures and cultures, since "the contemporary" names an historical period *in process*. The very term *contemporary* has a least three meanings (immediate, contemporaneous, and cotemporal) and it designates a time characterized by "the coming together of different but equally 'present' temporalities or 'times,' a temporal unity in disjunction, or a disjunctive unity of present times."[65] The uneven developments of global capitalism, in other words, are now experienced as "noncontemporaneous contemporaneity," but the object toward which they aspire—a fully globalized world, or what Marshall McLuhan, in a different context, called the global village—is already longed-for in the present because lost in the future, since this object can have no earthly existence.[66] "[Y]ou will remind me," Latour laments, "that there exist many groups of scholars calculating the number of additional planets necessary for the development of all eight billion humans—from two to five virtual planets depending on calculations and expected level of development—when we have only one planet."[67]

Given this particular problem of scale, Latour's effort to tell "Gaïa stories," which is to say his commitment to rethinking the question of agency in the Anthropocene, has assumed a special urgency for scholars in the humanities, including those who seek to understand what role (if any) literature and literary criticism might play in registering and ultimately disrupting the forces of global warming.[68] But Latour's import for scholarship on contemporary literature is even broader than this. For one thing, the human actors in question—the authors, editors, agents, publishers, and other professionals who work in the book trade—are very much *in action,* and their labor results in "the occasional conviction," as Latour and Woolgar might put it, "that something is" in fact literature: a lasting work of art, a soon-to-be canonical piece of fiction, a future classic. To work in the field of contemporary studies without following these actors (without interviewing them, for instance, or scrutinizing their sites of interaction as Latour and Woolgar studied the scientific laboratory) would be a missed opportunity to apprehend how literature actually gets made in the twenty-first century. In this sense, Latour's imperative to follow the actors becomes an invitation for literary critics to learn the basic techniques of qualitative research in the social sciences (e.g., ethnography, in-depth interviewing, observation, &c.) as a way of bolstering their arguments about literature's current mode of existence.[69]

IV. Conclusion: Controversies

At the end of the eighteenth century, in the context of Prussian totalitarianism, Kant wrote about "the conflict of the faculties" in the research university. Latour, for his part, inhabits what Marshall Sahlins calls "the conflicts of the Faculty": the skirmishes between different thinkers, employing different disciplinary tools, to answer the same questions.[70] What is modernity? How might we conceptualize society, reality, or experience? Has critique outlived its time? From the beginning of his career to the present day, Latour has taught us to "feed off" the controversies that emerge at the interstices between disciplines.[71] To put the point in literary terms, if every academic discipline offers a specific point of view on the world, then Latour operates as a kind of *omniscient narrator*, roving between disciplines, inhabiting their arguments, and drafting off their methods to develop a body of work whose *variable focalization* is among its most prominent formal features. Of course, the figure of the omniscient narrator is controversial within literary criticism, even today, which is one reason why Latour regularly ironizes his own authority.[72] "I am not very disciplined," he recently confessed. "I'm at heart a philosopher but I've been accepted as an anthropologist by anthropologists because I work on modernity, and now modernity is turning into the Anthropocene question. So work I had done many years ago—saying we have never been modern—is now being vindicated. Everyone agrees we will never modernize the planet. Something else is happening. It's called the Anthropocene. And despite the fact that no sociologist really recognized me as one of them, I'm a sociologist."[73]

Being undisciplined can cause problems, a few of which I want to address here by way of conclusion. First of all, Latour's reach beyond the inchoate field of science studies in the 1970s has turned him into an academic superstar, the heir to the throne once occupied by Derrida and Foucault. This means that his name alone can be used as a weapon in the academic battle that Sahlins dubs "competitive xenophilia," which involves "trumping one's disciplinary mates by importing prestigious ideas from unrelated disciplines" and thereby "marking off the enterprising professor from colleagues not yet in the know."[74] Within literary studies specifically, Latour's name itself is so powerful that its use, as Lytle Shaw has pointed out, can eclipse the authority of the literary text that ANT is being called upon to analyze.[75] In such cases, the aura of the new Master Theorist overshadows the object of study to such an extent that the object no longer matters for the inquiry at hand. We know this is a "Latourian" analysis, irrespective of what is being analyzed. From Latour's own perspective, however, this would seem to be an unfortunate side-effect of his influence, since he would urge us all to "follow the actors," which in the case of literary studies means forestalling critique in order to give primary sources the opportunity to make their own impact.

A second objection to Latour evident in literary studies stems from what could be called the materialist critique of new materialism. Unconvinced by ANT, some critics of Latour consider his thinking insufficiently attuned to the power dynamics of capitalism, to exploitation, ideology, reification, and class struggle. Others go even further, suggesting that ANT simply replicates the new spirit of capitalism. Alexander Galloway makes the point bluntly when he asks, "Why do these philosophers," referring to Latour, Alain Badiou, and their legatees, "when holding up a mirror to nature, see the mode of production reflected back at them. Why, in short, is there a coincidence between todays ontologies and the software of big business?" It is a question that marks the difference between ANT and, on the one hand, the materialist tradition within Western philosophy and, on the other hand, the historical materialism of Marx. But it is not fair to say that Latour's "philosophy mimics the infrastructure of contemporary capitalism" simply because it employs the "network" metaphor, nor is it accurate to say that "Latour's actants" come from "systems theory" and are therefore sullied by their link to "big business."[76] Never mind that ANT's notion of actants and agency actually derives from narrative theory: a more careful reading of Latour would acknowledge that his claims (some of which are explicitly in dialogue with Marx) stubbornly resist the omnipotence of capitalism by developing an analytic for "the enchantment of the object world" that does not "depend on the structure of the commodity."[77]

Finally, ANT has been chided for failing to be a diachronic rather than synchronic enterprise, which is to say for failing to be a kind of historiography. Without a robust conception of "the social" as a durable structure, ANT's critics suggest, it has no way of accounting for either stasis or change over time.[78] As Michael Lucey and Tom McEnaney contend, "So many things perdure—hierarchies of value, forms of inequality, relations of domination, language as organized hierarchically into registers—and precede any instance of language-in-use, or any structured social encounter."[79] That "things perdure," however, is never doubted by ANT; rather, what ANT aims to disclose is how and why they do so. It is only by following the actors, regardless of which "ontological zone" they occupy, that we can understand how complex abstractions such as "relations of domination" persist through concrete scenes of encounter and exchange. This is why Levine, in explaining the political stakes of her formalism, turns from Marx to Latour, from demystification and critique to observation and compositionism. Instead of employing "traditional ideology critique, which aims to expose the false and seductive discourses and cultural practices that prevent us from recognizing human unfreedom," she urges us to focus on "the many different and often disconnected arrangements that govern social experience" as a first step toward reassembling the social. It is only through such focus and its rhetorical counterpart, what Heather Love calls "thin description," that we can discern "which forms of protest or resistance actually succeed at dismantling unjust, entrenched arrangements."[80]

Although competitive xenophilia and disciplinary envy can be found in the reception of Latour in literary studies, as they can be found whenever ideas and arguments move across disciplinary lines, at its best this reception suggests an ambitious effort to reset the terms and methods of literary criticism for the twenty-first century. Partly inspired by Latour, and exemplified by critics such as Levine, Felski, Evans, and Love, the vibrant methodological discussion of the past decade reveals not a "new modesty" but a new analytical audacity.[81] In addition to rethinking key matters of concern such as critique, modernity, society, and agency, scholars of literature are asking what, how, and even whether to read. As these questions become increasingly urgent in a time of political turbulence, ecological crisis, and media revolution, we will continue to need Latour's original and contrarian voice.

Notes

1. Bruno Latour, "Why Has Critique Run Out of Steam? From Matters of Fact to Matters of Concern," *Critical Inquiry* 30 (Winter 2004): 246.

2. See, for example, Stephen Best and Sharon Marcus, "Surface Reading: An Introduction," *Representations* 108.1 (Fall 2009): 1–21; and Heather Love, "Close but not Deep: Literary Ethics and the Descriptive Turn," *New Literary History* 41.2 (Spring 2010): 371–91.

3. For a highly influential positive engagement with Latour, see Rita Felski, "Context Stinks!" *New Literary History* 42.2 (Autumn 2011): 573–91. For a critique of Latour from the purview of sociology, see Charles Turner, "'Travels without a Donkey': The Adventures of Bruno Latour," *History of the Human Sciences* 28.1 (2015): 118–138. For a critique from the purview of art criticism, see Hal Foster, "Post-Critical," *October* 139 (2012): 3–8. I discuss Latour's detractors below.

4. On this reception history, see Ian Hacking, *The Social Construction of What?* (Cambridge, MA: Harvard Univ. Press, 1999), ch. 3.

5. Ava Kofman, "Bruno Latour, the Post-Truth Philosopher, Mounts a Defense of Science," *New York Times Magazine*, October 25, 2018.

6. Bruno Latour, "Life Among Conceptual Characters," *New Literary History* 47, no. 2–3 (Spring & Summer 2016): 463–76.

7. David J. Alworth, "Latour & Literature," in *Theory Matters: The Place of Theory in Literary and Cultural Studies Today*, ed. Martin Middeke and Christoph Reinfandt (London: Palgrave Macmillan, 2016), ch. 22. On "heteroglossia," see M. M. Bakhtin, *The Dialogic Imagination: Four Essays*, trans. Caryl Emerson and Michael Holquist (Austin, TX: Univ. of Texas Press, 1981).

8. The phrase "large loose baggy monster" was used, famously, by Henry James to describe the novel form (Henry James, *The Art of the Novel* [Chicago, IL: Univ. of Chicago Press, 2011], 84). For an excellent example of confusion and horror in the reception of Latour, see Alexander Galloway, "The Poverty of Philosophy: Realism and Post-Fordism," *Critical Inquiry* 39.2 (Winter 2013): 347–66.

9. Felski, "Context Stinks!" 575; Alworth, "Latour & Literature," 307–308.

10. Latour, "Why Has Critique Run Out of Steam?" 229.

11. For a less conventional and more experimental example, see Bruno Latour, *Aramis; or The Love of Technology* (Cambridge, MA: Harvard Univ. Press, 1996).

12. Bruno Latour and Steve Woolgar, *Laboratory Life: The Construction of Scientific Facts,* 2nd ed. (Princeton, NJ: Princeton Univ. Press, 1986), 280.

13. John Guillory, "The Sokal Affair and the History of Criticism," *Critical Inquiry* 28.2 (Winter 2002): 491.

14. Latour and Woolgar, *Laboratory Life*, 105.

15. Latour, "Conceptual Characters," 466.

16. Bruno Latour, *Reassembling the Social: An Introduction to Actor-Network-Theory* (New York: Oxford Univ. Press, 2005), 124–25.

17. Latour, *Reassembling the Social*, 54–55.

18. A. J. Greimas and J. Coutés, *Semiotics and Language: An Analytical Dictionary* (Bloomington, IN: Indiana Univ. Press, 1982), 5–7.

19. Latour, *Reassembling the Social*, 7. Latour also cites other literary theorists, such as Louis Marin and Thomas Pavel, but Greimas seems to be the most important to his thinking. "It would be fairly accurate," he writes in a footnote, "to describe ANT as being half [Harold] Garfinkel and half [A.J.] Greimas" (54). For a thorough account of the term *actant* in literary theory, see David Herman, "Existential Roots of Narrative Actants," *Studies in Twentieth Century Literature* 24.2 (2000): 257–70.

20. Latour, *Reassembling the Social*, 54–55.

21. Latour, *Reassembling the Social*, 54–55.

22. Latour, *Reassembling the Social*, 54–55.

23. Latour, "Why Has Critique Run Out of Steam?" 231.

24. Rita Felski, *The Limits of Critique* (Chicago, IL: Univ. of Chicago Press, 2015), 1.

25. Fredric Jameson, *The Political Unconscious: Narrative as a Socially Symbolic Act* (Ithaca, NY: Cornell Univ. Press, 1981), 10.

26. Felski, *Limits of Critique*, 5.

27. Latour, "An Attempt at a 'Compositionist Manifesto,'" *New Literary History* 41, no. 3 (2010): 475.

28. Latour, "Why Has Critique Run Out of Steam?" 231–32.

29. Steve Paulson, "The Critical Zone of Science and Politics: An Interview with Bruno Latour," *The Los Angeles Review of Books*, 23 February 2018.

30. On the reception of pragmatism within literary studies, see Nicholas Gaskill, "Experience and Signs: Towards a Pragmatist Literary Criticism," *New Literary History* 39.1 (Winter 2008): 165–83.

31. Antoine Hennion, and Stephen Muecke, "From ANT to Pragmatism: A Journey with Bruno Latour at the CSI," *New Literary History* 47.2–3 (Spring & Summer 2016): 289, 301–302.

32. William James, *Essays in Radical Empiricism* (New York: Longmans, 1912), 42–43. Emphasis in the original.

33. Brad Evans, "Relating in Henry James (The Artwork of Networks)," *The Henry James Review* 36.1 (Winter 2015): 1–23.

34. William James, *A Pluralistic Universe* (New York: Longmans, Green and Co., 1920), 237.

35. James, *Pluralistic Universe*, 264.

36. Bruno Latour, *An Inquiry into Modes of Existence: An Anthropology of the Moderns*, trans. Catherine Porter (Cambridge, MA: Harvard Univ. Press, 2013), 57.

37. Clifford Geertz, the anthropologist who inspired the New Historicism of the 1980s, once complained that "social scientific theory" had "been virtually untouched" by literary criticism. Conceptualizing culture as a text, Geertz considered such criticism to be part of the toolkit for understanding what human practices, rituals, and behaviors mean (Clifford Geertz, *The Interpretation of Cultures* [New York: Basic Books, 1973], 208–209). Latour, by contrast, turns to literature and literary criticism for help in thinking not about what practices mean but about what practitioners do—how specific actants capacitate or constrain the repertoires of action that we call "culture" and "society."

38. Latour, *Inquiry*, 57.

39. On the pragmatist emphasis of *doing* over *meaning*, see Gaskill.

40. Latour, *Reassembling the Social*, 82.

41. Manuel DeLanda, *Assemblage Theory* (Edinburgh, UK: Edinburgh Univ. Press, 2016), 38.

42. Felski, "Context Stinks!" 578.

43. Latour, *Reassembling the Social*, 1.

44. Patrick Joyce, ed., *The Social in Question: New Bearings in History and the Social Sciences* (London and New York: Routledge, 2002), 2.

45. Bruno Latour, *We Have Never Been Modern*, trans. Catherine Porter (Cambridge, MA: Harvard Univ. Press, 1993).

46. Paulson, "Critical Zone."

47. Bruno Latour, "Gabriel Tarde and the End of the Social," in *The Social in Question: New Bearings in History and the Social Sciences*, ed. Patrick Joyce, 117–33. London and New York: Routledge, 2002.

48. Latour, "Why Has Critique Run Out of Steam?" 230.

49. Émile Durkheim, *The Elementary Forms of Religious Life*, translated by Karen E. Fields (New York: Free Press, 1995), 443. Fredric Jameson quotes this claim of Durkheim's in the epigraph to *The Political Unconscious*, thereby suggesting that the pages that follow will exemplify a kind of literary criticism that takes a Durkheimian conception of "society" as its premise. I discuss this further in David J. Alworth, *Site Reading: Fiction, Art, Social Form* (Princeton, NJ: Princeton Univ. Press, 2016).

50. C. Wright Mills, *The Sociological Imagination* (New York: Oxford Univ. Press, 1959), 14.

51. Bill Brown, *Other Things* (Chicago, IL: Univ. of Chicago Press, 2015), 169.

52. Latour, *Reassembling the Social*, 72.

53. The range of work inspired by "thing theory" is both vast and varied. For an overview, see Sarah Wasserman, ed., "Thing Theory 2017: A Forum," *Arcade: Literature, Humanities, & the World*, https://arcade.stanford.edu/content/thing-theory-2017-forum.

54. Donna Haraway, *Staying with the Trouble: Making Kin in the Chthulucene* (Durham, NC: Duke Univ. Press, 2016), 40.

55. Hennion, "From ANT to Pragmatism," 300.

56. Don Norman, *The Design of Everyday Things: Revised and Expanded Edition* (New York: Basic Books, 2013), 10.

57. James J. Gibson, *The Ecological Approach to Visual Perception* (New York: Psychology Press, 2015), 121.

58. Norman, *Design of Everyday Things*, 10–11.

59. Bill Brown, *Other Things*, 373.

60. Caroline Levine, *Forms: Whole, Rhythm, Hierarchy, Network* (Princeton, NJ: Princeton Univ. Press, 2015), 1–24.

61. Rita Felski, "Being Diplomatic: On ANT and Literary Studies," lecture, SDU, Odense, Denmark, September 21, 2017.

62. Georg Lukács, "Narrate or Describe? A Preliminary Discussion of Naturalism and Formalism," in *Writer and Critic and Other Essays*, trans. Arthur D. Kahn (New York: Merlin Press, 1970), 137.

63. See Jane Bennett, *Vibrant Matter: A Political Ecology of Things* (Durham, NC: Duke Univ. Press, 2010), ch. 2.

64. Latour, *Reassembling the Social*, 68.

65. Peter Osborne, *Anywhere or Not at All: Philosophy of Contemporary Art* (London: Verso, 2013), 17.

66. Harry Harootunian, "Remembering the Historical Present," *Critical Inquiry* 33 (Spring 2007): 475.

67. Bruno Latour, "On a Possible Triangulation of Some Present Political Positions," *Critical Inquiry* 44 (Winter 2018): 214.

68. See, for example, Kate Marshall, "What Are the Novels of the Anthropocene? American Fiction in Geological Time," *American Literary History* 27.3 (2015): 523–38.

69. Introducing "new sociologies of literature" in 2010, James F. English observed that scholars of literature tend to borrow claims and conclusions rather than methods from the social sciences (James F. English, "Everywhere and Nowhere: The Sociology of Literature After 'the Sociology of Literature,'" *New Literary History* 41.2 [2010]: v–xxiii). While this is true of Latour's reception within literary studies, his imperative to follow the actors is a methodological prompt.

70. Marshall Sahlins, "The Conflicts of the Faculty," *Critical Inquiry* 35.4 (2009): 997–1017.

71. Latour, *Reassembling the Social*, 21–27.

72. For a concise account of this topic, see Paul Dawson, "The Return of Omniscience in Contemporary Fiction," *Narrative* 17.2 (May 2009): 143–61.

73. Paulson, "Critical Zone."

74. Sahlins, "Conflicts," 1014–15.

75. Lytle Shaw, "lowercase theory and the site specific turn," *ASAP/Journal* 2.3 (2017): 653–76.

76. Galloway, "Poverty," 347, 348, 362.

77. Brown, *Other Things,* 171.

78. This objection approaches the broader theoretical problem of how different social scientific disciplines conceptualize time. For a thorough account of this problem, see William H. Sewell, *Logics of History: Social Theory and Social Transformation* (Chicago, IL: Univ. of Chicago Press, 2009).

79. Michael Lucey and Tom McEnaney, "Introduction: Language-in-Use and Literary Fieldwork," *Representations* 137.1 (2017): 8–9.

80. Levine, *Forms*, 17–19.

81. Jeffrey J. Williams, "The New Modesty in Literary Criticism," *Chronicle of Higher Education*, January 5, 2015, https://www.chronicle.com/article/The -New-Modesty-in-Literary/150993.

CHAPTER TWELVE

Cinematic Assemblies

Latour and Film Studies

CLAUDIA BREGER

WHAT IS THE RELEVANCE and promise of Bruno Latour's philosophy for film (or cinema) studies?[1] Exploratory answers to this question have been coming in more slowly than in literary studies on the one hand, and the larger, in part social science-inflected field of media studies, on the other hand.[2] Of course, as Latour himself might put it, the borders between these domains indicate "less a dividing line between two homogenous sets than an intensification of crossborder traffic" in the contemporary processes of trans/disciplinary recomposition that form the occasion for this volume. The emerging body of Latourian scholarship specifically on cinema certainly reflects such traffic, even where it intends "to remain within the traditions of film studies."[3] While it is fascinating to trace—and for the purpose of scholarly diplomacy crucial to respect—relatively distinct disciplinary trajectories, this chapter situates itself against the broader backdrop of Latour's reception across disciplinary domains, to specifically outline his relevance for film studies at the intersections of such domains. That is, I hope to show that Latour's philosophy gains its promise for this field not least by how it speaks to, and allows us to productively remap, the complex position of film studies in the ongoing process of de- and recomposing the humanities. Some of these complexities reflect familiar external pressures: like other humanities disciplines, film studies has been targeted by neoliberal administrative efforts to confine scholarly inquiry within professional degree programs focused on media advertising or management.[4] At the same time, and with more interesting implications, the field has been internally de- and recomposed by heterogeneous disciplinary trajectories, as inquiries into cinematic form and aesthetic response have long coexisted and been amalgamated not only with cultural studies approaches but also with interests in the material technologies and institutions of production, distribution, and exhibition in a shifting landscape of old and new media. In 2018, the domain border between *film* or *cinema* and *media studies* has arguably become anachronistic.[5]

In reviewing existing scholarship and proposing further developments for Latour reception in film studies, the following responds to this situation with a twofold—but I claim, consistent—move. On the one hand, I emphatically affirm the significance of film studies in and for the humanities. My Latourian account of the creative processes of filmmaking and film viewing underlines the aesthetic and cultural productivity of cinematic worldmaking: its forceful contributions to ethically and politically urgent concerns about living together, or (in Latour's own words) to the task of reassembling more inclusive (part-human) collectives.[6] On the other hand, I develop this argument precisely by taking seriously the challenge of Latour's transdisciplinary provocation, as I connect my own interests in film aesthetics to a more comprehensive account of the material-semiotic networks of production, circulation, and reception that facilitate cinematic composition and reading. In other words, this chapter joins other contributors to this volume by arguing that Latour's philosophy facilitates a "defense" of the humanities not against the sciences (or for that matter by plainly allying with traditional scientific method), but within a transdisciplinary continuum of perception and knowledge practices that imbricate epistemological realism and reflexivity, empiricism and (critical, imaginative, formalist) techniques of reading.

Latour himself has referenced the world building processes of film and engaged in film and multimedia projects, inviting the adoptions, and adaptations, through which film scholars have begun to deploy his philosophy.[7] I propose a somewhat more encompassing translation of Actor-Network-Theory (and its modifications in the *AIME* project) into film studies, one that bridges the gap between competing disciplinary approaches focused on film technology, production, exhibition, and aesthetics respectively. I will develop this proposal by connecting the emerging reception of Latour in film studies to longer-standing discussions in the field, ranging from a whole legacy of interest in nonhuman agency to early explorations of postcritical reading techniques.[8] From my own angle, re-reading Latour for film studies provides an opportunity to link many of the field's heterogeneous methodologies into a layered, multi-faceted approach to cinematic worldmaking. The two main sections of this piece unfold such a layered approach starting from the major thematic clusters of Latour reception in film studies to date: (1) nonhuman and distributed agency and (2) documentary and other realist genres. As I proceed, I will introduce Valeska Grisebach's film *Western* (Germany, Bulgaria, Austria 2017) as an example that substantiates my claims to theoretical productivity.

Nonhuman (and Non-Sovereign Human) Agency: Film's Actor-Networks

An early essay linking Latour's philosophy to cinema highlights how film studies have long overlapped with media studies, itself a heterogeneous field that can be characterized as part transdisciplinary cultural studies endeavor, part "harder" history of technology. In a 2008 German media studies anthology on distributed agency and collectivity, media philosopher Lorenz Engell investigated cinema's multiple actors, with a focus on technology. His chapter collected representational and aesthetic evidence for how historical and contemporary films have dramatized the agency of collectives, things, apparatuses, and light, and linked these observations to media-theoretical reflections on the nature of the film image itself as an animated "thing"; film's foundation in mechanical recording; its historical lineage in scientific measurement techniques; and its indexical qua photographic quality.[9] Through this lens of history of technology, cinema presents itself as the medium par excellence for concretizing Latour's claims about nonhuman agency. Classical film theory, Engell reminds us, preempted the call to accept nonhumans as "full-blown actors": early and mid-century theorists regularly emphasized "the instrumentality of a nonliving agent" (in André Bazin's classical formulation) and cast film as showing "the world from the perspective of a thing among things."[10]

This return to an earlier moment of medium theory is a delicate move in twenty-first-century scholarship. Arguably, the theoretical controversy around the strong version of Bazin's classical notion has been stabilized: contemporary film theory no longer encourages us to foreground film's automatic, "objective"—or object lens-driven—image-making to the point of ignoring either the human actors assembled with the camera or the cultural scripts shaping their framing activity.[11] Engell's own preoccupation with the agencies of technology makes him forego an explicit critique of the classical trope; however, he does indicate a fuller account of, in fact, *distributed* cinematic agency. This account links to contemporary film studies positions, which have, in some respects, reflexively renewed the earlier interests in cinema's "thing"-like look(s), in response to the sovereignty presumptions that characterized dominant mid- and later twentieth-century film theory. These sovereignty presumptions had taken different forms: auteurist conceptions of cinema deployed them in foregrounding the more or less sovereign human agency of the artist, whereas the psychoanalytic, ideology-critical apparatus theories of the 1970s (and beyond) postulated the film's similarly powerful, "disembodied and godlike vision."[12] In the early 1990s, Vivian Sobchack was influential in critiquing this notion of "tyrannical technology," and conceptualizing the spectator, along with the film itself, as a non-sovereign "*viewing* subject."[13] Sobchack's phenomenological account resonates with Engell's evocation of the film image as an animated thing endowed with active looking power, as it gives weight to the "film's body" as both "*instrumental mediation*" and "situated, finite" embodiment of "perception and expression."[14]

It seems that Sobchack's film, as a material "node" of mediation to be taken seriously as a full-fledged actor, could find a place in Latour's "actor-network." As he has it, this network is populated by "*overtaken*" or "*other-taken*" participants; Latour's rethinking of action in general as a "conglomerate of many surprising sets of agencies" displaces fantasies of human along with nonhuman sovereignty (*RS* 44–45; Latour's emphasis). To be sure, both Engell and Latour himself might object to the anthropomorphizing undertones of Sobchack's rapprochement of nonhuman and human actors. As the "mediation of camera and projector" perceptually aligns "filmmaker and spectator . . . with each other" as well as "a world that is their *mutual intentional object*," Sobchack's emphatically materialist account of cinema's loop of distributed agency remains inflected by classical phenomenology's emphases on consciousness and intention, and perhaps closer to (non-distributive) auteurist theories than she intends.[15] In contrast, Engell underlines how intention is "*overtaken*" in Latour's sense as he more fully dissembles cinematic agency: what he wants taken into account are not only the empirical plurality of humans involved in making a film (as producers, actors, cinematographers, editors, and so forth) and the "varied and complicated instruments" foregrounded by his own approach, but also—with a nod to cultural studies approaches—filmmaking "conventions, rules, and styles," along with the "expectations, viewing habits, iconographies and ideologies" that shape reception processes.[16]

In more comprehensive theoretical terms, a Latourian model of collective cinematic agency is developed in Ilana Gershon and Joshua Malitsky's 2010 "Actor-network theory and documentary studies," a key contribution to the project of bringing Latour to film studies.[17] Introducing ANT as "fundamentally a theory of relationality," Gershon and Malitsky detail "four major conceptual consequences": "First, everyone and everything contributes" to the cinematic network—which, as Gershon and Malitsky stress, encompasses processes from pre-production to reception—including the "strength and directionality of the sun's light" or the variously cooperative or resistant microphones of the Fidel Castro documentaries they introduce as examples.[18] Second, however, "not all actants are the same": "each actant's physicality," such as "the materiality of the pro-filmic space and the cinematic apparatus," matters along with the "different social and historical trajectories" of objects and people in the network, trajectories constituted by power asymmetries, norms of gender, or genre (67, see 68). Third, relations in the cinematic network are "performative": "instability" is the given condition, while any orientation to the "classical" cinematic norms of "consistency and coherence" always requires labor.[19] Fourth (as perhaps implied by the first three principles, but worth underlining), "all" human and nonhuman "actants" are "network effects"; their mediating agency always remains contingent on the forces with which they are entangled (68).

In my own work, I have developed these principles into a syncretic framework that conceptualizes cinematic worldmaking as a collective process of assembling heterogeneous but entangled elements—including affects, associations, bodies, gestures, matter, memories, perceptions, sensations, things, topoi and tropes—via images, words and sounds in the communicative networks of composition, production, and spectatorship.[20] My accent on (networked) communication processes aims to balance prevalent film and media studies emphases on technology (as evident in Engell's contribution and to a lesser degree in Gershon and Malitsky's examples) by foregrounding the constitutively entangled, *non-sovereign human, humanlike and part-human actors* in film's networks. In my definition, human-like and part-human actors include, for example, characters on the diegetic level and camera-cinematographer assemblages on the level of narration.[21] This proposal goes against the grain of radically post-humanist interpretations of Latour. Hoping to mediate between competing humanist and post-humanist approaches in the larger contemporary humanities landscape, it aims to answer concerns about ethical and political orientations in the network.[22] In particular, I explore unexpected resonances between Latour's methodological ethos of deploying "the actors' own world-making abilities" and a postclassical phenomenology no longer focused on intentionality. A key principle here is that the networked, or (in Deleuzian language) "*divisible*," nature of individuals does not preclude us from "respecting . . . what is 'given into' their 'experience.'"[23] At the intersection with affect studies, I underline how Latour's network creatures are moved and (re-)constituted by their attachments [ATT] to nonhumans along with other humans (see *AIME,* 423, 425): by sensations, affectations, memories, and fantasies that are variously mediated by technology and history but are not therefore necessarily any less intense. The ethos of carefully tracing this non-sovereign, often nonconscious activity of human, humanlike, and part-human participants facilitates a forcefully egalitarian conceptualization of cinematic network operations. Thus, the empirical voices of extras, costume designers, and regular audience members are, in principle, as worthy of attention as the interpretations of directors and professional critics, even where access asymmetries tilt the scholarly collection of voices in the direction of the most privileged actors. Respectfully listening to the director's self-interpretation in yet another interview, I don't have to conclude that I have "been dispatched thanks to the flesh-and-blood author" (Latour, *AIME* 247). After all, she sometimes "doesn't know very well what she has done" and is, again, herself as much a product of the cinematic communication loop as her "admirers" (Latour, *AIME,* 247)—potentially no less powerfully seduced or repulsed by the characters and objects populating the collectively assembled cinematic worlds for which she gets credit.

This proposed framework connects to other domains, including literary studies as shaped by debates on postcritical reading techniques over the course of the past fifteen years. Simultaneously, the framework resonates with earlier discussions on (multisensory) spectatorship within film studies, such as Sobchack's call to overcome 1970s film theory's "paranoia," Steven Shaviro's challenge to embrace "film viewing" as "pleasure and more than pleasure," and cognitive scholars' calls to displace modernist *distanciation* mandates with reconsiderations of empathy and sympathy.[24] Integrating this range of heterogeneous impulses with Latour's methodology of following the actors, I argue, facilitates a multifaceted, flexible approach to cinematic composition, production, distribution and reception processes.[25] In my iteration, this approach does not entail a full-fledged displacement of critique but foregrounds an ethos of "[c]ritical proximity" and practices of reconfigurative rather than iconoclastic critique.[26] With a layer of patient phenomenological description, this kind of reconfigurative critique modulates distance and approximation as it cautiously assembles diverging "local," first-person perspectives into collective accounts that afford a degree of contextualization and historicization.[27] In different corners of cinema's communicative loop, I can, for example, trace the circulation of (any combination of) passionate feeling, conditional empathy, intellectual curiosity, quiet reflection or loud anger in the relations between diegetic beings, the relations of editors to the worlds emerging from their image and sound material (mediated by the performance of actors and the light technician's semiconscious orientations towards classical or avant-garde norms), and the engagement of audiences with any of these circulations. Of course, audience engagements are further shaped by genre templates, deeply personal (but, therefore, no less political) experiences and cranked-up Cineplex loudspeakers or miniscule smartphone screens. In line with Gershon and Malitsky's reminder that instability is the given condition, I underline that these relations rarely add up to the "aesthetically unified feeling" and narrative coherence (rewarding audience empathy with the hero vis-à-vis the villain) that have been conceptually privileged by scholars in philosophical aesthetics and cognitive theory.[28] Instead, I emphasize the productivity, for example, of unexpected bursts of audience affect in relation to characters whose identity we may despise, and more generally of diverging or layered affective vectors in the distributed assembly of cinematic worlds.

A look at Valeska Grisebach's *Western* allows me to concretize some of these ideas of re-thinking cinematic agency through actor-network-theory, including the accent on following non-sovereign human, human-like and part-human actors. The German director's work became known as part of the so-called Berlin School of the 2000s, a loose network of filmmakers interested in renewing experimental takes and, specifically, forms of a phenomenological realism privileging everyday spaces and movements over tight plots.[29] More recently, however, several of these filmmakers have turned to creative explorations of genre, such as the Western reflexively featured (*as* a genre) in the title of Grisebach's 2017 film. In line with the expectations thereby stirred, *Western* sets up a classical scenario of human agency embedded in—and in tension with—nonhuman agency, along with that of other humans. A team of German construction workers with the job of building a hydro-electric power station with large machinery moves into the rural landscape close to Bulgaria's Southern border with Greece. Again and again, the film's camera (a powerful new generation Arri Alexa, held mostly by Bernhard Keller) makes time for—and thus gives weight to—the impact of the mountainous landscape, often in the morning or evening light. The film's relatively classical cinematography and editing notwithstanding, most of these takes emphasize less the cognitive orientation functions of establishing shots than the affective ones of grounding the action (and the prospective film audience) in a world not exclusively centered in human presence.[30] Unlike other recent returns to the Western, to be sure, Grisebach's film foregoes the widescreen format that invites minimizing this human presence altogether. Instead, the director emphasizes her joint attempt, with the editor Bettina Böhler, to balance the weight of the landscape with the forces of technologically-fortified human interference.[31]

On the diegetic level, the water of the river floods and temporarily silences the power shovel's motor when the construction crew decides to change the course of the river because it is in the way of their design. However, the protagonist Meinhard, who operates the shovel, is less flustered by this incident than his supervisor (and prospective antagonist) Vincent; the film seems to assert Meinhard's (networked) agency in showing how he calmly steps into the water and succeeds in restarting the shovel. Meanwhile, the absence of a shared language initially allows the German workers to ignore the locals' reminders that their undertaking does not happen on "virgin" territory, but within a complex local network of human and nonhuman actors. Thanks to subtitles for both the German and the Bulgarian dialogue, the film audience knows more and understands, for example, that the white horse befriended— or ambiguously "domesticated"—by Meinhard does have an owner. Interhuman conflict interrupts more fully over Vincent's disrespectful flirtation (bordering on assault) with a local woman. Later, he sneaks out of the camp with the horse to manipulate the water network designed to alternate supply under the prevailing conditions of scarcity; in turning the lever, Vincent organizes water for the construction project at the expense of the local tobacco harvest. During the ride back across steep terrain he has a fall, caused by both his incompetence and carelessness. He abandons the badly injured horse, which is found only days later and has to be shot. (The end credits reassure the film audience that no animals were hurt during production.)

The diegetic parable of human actors exploiting nonhuman along with less powerful human actors in the unequal competition for limited resources is hard to miss, and certainly intended by the director. (Emphasizing the non-sovereignty of agency does not necessarily imply disregarding intentional dimensions of art making altogether.) In interviews, Grisebach discusses her interest in quasi-neocolonial power asymmetries in contemporary Europe, including those facilitated by complicated application procedures for European Union infrastructure funding that put experienced and liquid West European companies at an advantage. The "Western" (or rather "Eastern") film setting dramatizing these institutional conditions, Grisebach comments further, allowed her to tackle questions of "diffuse xenophobia" without tapping into a neo-Nazi genre, as a German setting would have made her do.[32] A German reviewer underlines this connection between cinematic genre and contemporary political context by praising Grisebach's "masterful" play with the prejudices of the Western, the "elementary values" of which have "a comeback" in today's "populism."[33] I agree that the film presents a highly intriguing artistic accomplishment; the notion of (playful) mastery, however, is a very misleading characterization of its collective worldmaking project. If the explicit recourse to a classical genre combined with critical directorial intent makes us expect a biting parody or similar distance format, *Western* thoroughly fails to deliver. Pointers towards its different mode of engagement are, perhaps, in the director's admissions of her own affectability: her "big longing for the Western genre," the "genre of her childhood," and her fascination for the construction workers' "old-fashioned form of masculinity."[34] Instead of an iconoclastic critique, *Western* diegetically and extradiegetically probes *engagement* with the culture of xenophobic masculinity it explores.

On the level of composition and production, this different mode of engagement is enabled—or perhaps demanded—by the director's collaborative ethos. Grisebach describes not only the extensive exchanges with her editor and camera man as a form of "surrendering control" resulting in a "joint transformation" of the ideas at hand, but also talks about creating a "shared world of experience" in a "long joint process" with the lay actors, most of them real-life construction workers, whose tenderness for each other and humorous, imaginative language use she admired.[35] In the film they made together, the camera follows the protagonist Meinhard most of the time. In line with the Berlin School's stylistic signature, it operates mostly unobtrusively, at enough of a distance to anchor him in the landscape and to facilitate imagined audience co-presence rather than penetrating analysis, including many shots from behind. At the same time, the (non-parodic) exploration of genre facilitates a higher degree of character relatability than many early Berlin School films; close-ups and medium close-ups on Meinhard's expressive, wrinkled face let us gather his emotions at key moments of the action. The actor's mature, slender yet strong physique and his—and/or the character's—quiet, sometimes shy, but usually competent demeanor are suited to facilitate more spontaneous sympathy than the film audience's initial encounters with some of the Bulgarian locals. While the store owner refuses to sell Meinhard cigarettes in the wake of Vincent's improper behavior, her boycott is quickly undercut by the male bystanders admiring the Germans who occupied the country before. This does not mean that the film downplays the racism of the German crew (who, in turn, joke about returning after seventy years), or for that matter offloads it on the unpleasant antagonist. While Meinhard is characteristically silent or soft-spoken rather than loudly inappropriate, it is he who (with another colleague) mounts a German flag on the deck of the worker's camp early on, prompting a third colleague to ask, in line with their usual rough banter, whether they have gone nuts: "ever thought of the locals?"[36]

This reply is crucial for how critique works in *Western*: approaching "the experience" of the actors, the film invites us to "*modify* the account" by deploying a different voice on the same level, with principally analogous authority both within the diegesis and in the process of narration (Latour, *AIME* 8; Latour's emphasis). Later in the film, when Meinhard increasingly spends time with the locals in response to the escalating conflict with Vincent, his Bulgarian friend Adrian will correct his behavior. Making room for the affective complexities unfolded in these exchanges, the film makes high demands of its audiences also, as it deploys its seemingly straightforward Western tropes towards a layered engagement.[37] Rather than merely deciphering clear-cut political allegories, we are invited to forego easy resolutions by respectfully, at moments tenderly, attending to the clashing worldmaking orientations of highly flawed actors. With its ethos of respectfully collaborative worldmaking, *Western* programmatically underlines the processes of distributed agency, which, as I suggested in this section, characterize cinematic worldmaking more generally. Latour's conceptualization of networked, "overtaken" action offers a nuanced framework that allows us to integrate competing theoretical emphases in film studies into a layered account: it makes room for the role of technology, production teams and viewers, for interests in the details of aesthetic composition and audience engagement; and it invites us to unfold the complexity of these processes co-shaped by the forces of affect and discourse, fantasy and genre, materiality and performance. Grisebach's reflexive interest in all of these forces activates the egalitarian bent of Latour's proposal: film is exemplarily tasked with "following the actors" here.

Documentary Matters, Or, Transformations Across Genre

The focus on non-sovereign agency is only part of what actor-network-theory (and AIME) have to offer to film studies. After first viewing *Western* (on the big screen at the *Film Society of Lincoln Center*) and reading a couple of reviews that briefly referenced its production conditions, I was haunted by the question: how much of a documentary is this film exactly? The urgent desire to know more about the modes, and limits, of *Western*'s grounding in contemporary life worlds opens onto larger epistemological and aesthetic questions that have formed a second focus of Latour's reception in film studies to date. In their co-authored article quoted above, Gershon and Malitsky initially turn to Latour for help with the dilemma posed by postmodern critiques of representation: "documentary scholars have been struggling to retain the political purchase of claiming the real while acknowledging the postmodern recognition that truth is socially constructed" ("ANT" 65). Latour himself might be inclined to raise an eyebrow at their use of the notion of "social" construction, but he, of course, annuls the problem at stake. Distinguishing his philosophy from the deconstructive project for which ANT's initial approach to the sciences was often mistaken, Latour positions his own onto-epistemological "constructivism" as an *"increase in realism (RS* 92; Latour's emphasis)." More specifically, he advocates a non-positivist "realism dealing with . . . *matters of concern, not matters of fact."*[38] While "highly uncertain and loudly disputed," Latour explains, matters of concern are nonetheless "real" and "objective"; they should be taken as *"gatherings"* rather than "objects" by a "more talkative, active, pluralistic, and more mediated" empiricism.[39]

In embracing the process of construction, Latour's more mediated realism challenges the critical habit of grounding claims to the real in notions of immediacy and direct contact, in the tradition of Bazin's above-quoted comments on the "essentially objective character of photography," which, Bazin adds, forces us "to accept as real the existence of the object reproduced."[40] Even while few film scholars or critics would unconditionally subscribe to this argument today, motifs of immediacy and contact have continued to shape contemporary documentary norms as well as their critique.[41] They also keep resurfacing in film-theoretical debates on the indexical sign, which seems to have attained new promise as a media-specific anchor for those yearning for referentiality in today's "post-medium" condition.[42] Whether critical or affirmative, the emphasis on immediacy and contact has fueled the dilemma diagnosed by Gershon and Malitsky, inhibiting a non-naïve but positive conceptualization of how documentary "claims the real"—or gathers matters of concern. As Gershon and Malitsky emphasize, a Latourian perspective instead demands from documentary scholars a "reflexive engagement with the labor involved in producing effective documentary facts" ("ANT" 69). In other words, it draws our attention to the technological and compositional choices that ground documentary practices: ANT helps "scholars to re-figure how documentary films function as a genre—as well as to distinguish among different documentary subgenres" ("ANT" 75–76). Thus, Malitsky's work on post-revolutionary Russian cinema amends the dominant historical narrative of how montage theorists retracted their avant-garde affiliations under Stalinist pressure through a nuanced comparative account of their differently modernist practices. This grounding of documentary forms in aesthetic process (broadly understood) does not assimilate them to fiction in the tradition of postmodern critique. Drawing on Latour's discussion of scientific indexicality in *Pandora's Hope*, Malitsky argues that the "realist" turns beyond "rapid juxtapositions and complex narrative" and towards a "descriptive" aesthetics of "accumulation" achieved higher "indexical stability" in Latour's sense of making things increasingly real.[43]

Malitsky's argument builds on the ways in which Latour complexifies concepts of indexicality, as he—again—acknowledges the transformative impact of both Bazin's instrumental actor and the human hand holding the camera, along with the many other agencies involved. In my own work, I develop these forays into a Latourian aesthetics of realism by more fully disentangling the controversial promise of indexicality from its associations with the specific ontology of film as a historically photographic medium. I integrate indexicality's operations of physical contact into a broader spectrum of material-semiotic processes that are principally—although differently—in effect across disciplines, media, and genres. The argument engages Latour's more recent work: whereas *Pandora's Hope* foregrounds indexicality as the paradigm establishing scientific reference, *Reassembling the Social* turns our attention to transdisciplinary processes of composition (including scholarly writing) and emphasizes the productive role of discontinuity and change, or "mediation," "displacement," "transformation," and "translation" in all network processes (*RS* 45, 64, see 106–109). In returning to the question of institutionalized domains, and specifying different modes of existence and verification, *AIME* continues to break down the "barrier between questions of ontology and questions of language," embedding indexicality in a spectrum of "articulation" processes that constitute both "words" and "the world" (144). In [REF], the mode of scientific reference, the general procedure of achieving "a certain *continuity* of action" through a "series of small *discontinuities*" is concretized as that of maintaining "constants" across material discontinuity by breaking "at every step with the temptation of resemblance"; etymologically, Latour underlines, "to refer" is "to *report, to bring back*" (*AIME* 33, 78–9; Latour's emphasis). [FIC] is the mode of existence through which Latour approaches the language side of the dominant disciplinary bifurcation—although he underlines that [FIC] is more than an ensemble of genres.

[FIC]'s constitutive "alteration" is in a particular "way of folding existents so as to make them the blueprint for a kind of expression," that is, in how "raw materials" produce "FORMS, or better, FIGURES."[44] Given these resonant operations, fiction can certainly not be opposed to reality. But *AIME* further foregrounds the "commerce, crossings, misunderstandings, amalgams, hybrids, compromises" between modes, and, in particular, the "very fertile" crossing "[FIC • REF]": "no chain of reference . . . without *narrative*," no discussion of "DNA" without "*characters*" (146, 250; Latour's emphases).

Cinematic documentary inhabits this intersection of fictive formation and productive referentiality in complex ways. On the level of medial affordances, its power "of returning to reality" through mediation (*AIME* 141) is not only in the indexical quality of some of its images, but in the full spectrum of its layered—visual, aural and linguistic—material-semiotic operations, including techniques notoriously faulted by documentary scholars for flagrantly violating presumptions to immediacy, such as voice-over and montage. Of course, not every use of voice-over or montage contributes equally to such a return to reality: once we admit that different types of signs, materials and operations contribute to the making of scientific accounts as well as cinematic documentaries, the key question at stake is whether something "is *well* constructed" (155; Latour's emphasis). In *Reassembling the Social*, Latour defines a "good account" as one that "*traces a network*," that is, treats each participant "as a full-blown mediator" in tracing matters of concern from different angles: "objectivity, or rather '*objectfullness*'" is achieved through an ethics of carefully assembling complexity in multiplying associations and perspectives (128, 133; Latour's emphasis). In *AIME*, this ethics is layered with the qualified return to differences of domain, including genre. "Factual narratives," Latour spells out, "do not differ from fictional ones as objectivity differs from imagination. They are made of the same material, the same figures" (251). What differs is "the treatment to which we subject" these figures and materials: while we "authorize beings of fiction to . . . 'carry us away' . . . into another world," those of documentary are "domesticated" and "disciplined by chains of reference" (251).

In other words, the discussion of documentary standards is not rendered moot by the acceptance of an underlying condition of construction, but re-activated in a nuanced way. For example, some film critics and scholars have reservations about accepting the use of reenactments as documentary, such as in the 2007 film about the Tarde-Durkheim debate in which Latour starred in the role of Gabriel Tarde.[45] If "physical portrayal," to many, attains a higher degree of authenticity than such "nominal" representation, *Western* offers a strong documentary feel with its lay actors, many of whom are introduced by their real-life names.[46] But while the film provides a forceful impression of these lay actors' physicality, of bodily gestures, voices, dialects and linguistic quirks, we are not necessarily seeing their "natural" movements and speech: Grisebach spells out in interviews that she did extensive acting exercises with the cast before the actual shoot, and encouraged them to develop their characters.[47] She also came to them with a script. To be sure, this script was not handed out as a straightforward template, and would be modified in the collaborative process, but it did present a fictional framework that had itself grown out of a combination of interview research with construction workers and Grisebach's above-mentioned genre fascinations (see "AT," "WAM"). Appropriately labeled a "drama," the film certainly does not qualify as a documentary in the sense that its materials and figures referentially "bring back" a full-fledged scenario that actually unfolded as such on Bulgaria's mountainous border with Greece. This "negative" answer does not, however, settle my question about the film's documentary dimensions in all respects. A positive answer can be attained by approaching the intersection [FIC • REF] from the other side.

In my own work, I first turned to Latour's reconceptualization of realism because it enables a different approach to realist fiction (here in the sense of genre), specifically the ways in which the emphasis on "objectfull" assembly facilitates less suspicious readings of filmic constructions that aim to investigate historical actualities via fictional scenarios.[48] Other cinema scholars have undertaken comparable inquiries in the last few years by bringing Latour to a broader range of cinematic genres. Jerome Schaefer's *An Edgy Realism* tackles the film-theoretical difficulties of developing a positive, non-naïve conceptualization of realism in the context of contemporary "shaky cam" horror films such as *The Blair Witch Project*. Schaefer argues that the existing descriptions of these films in the negative language of "mockumentary" or "simulated realism" are inadequate to describe the extraordinary "experiences" of "immersion" facilitated by the genre with its lack of "ontological segregation between the images and the story world" (*ER* 3–4, 8). While Schaefer locates the shaky cam aesthetics of edgy realism as a specific product of contemporary new media culture, he also deploys it as a case study towards a broader film-theoretical intervention. Schaefer's key concept is that of *transformation:* a Latourian "film theory of transformations," he argues, allows us to overcome the "false dichotomy between the textual and the material" in understanding film images as "material-semiotic phenomena in the making" and products "of the networks they are embedded in" (*ER* 12–13, 15). Schaefer's own study focuses on "the intricate network made up of looks and gazes, viewfinders and images, invisible observers and visual narrators, cameras and operators" in his genre (*ER* 18). Meanwhile, Eric Herhuth approaches a different genre and section of the cinematic network—animation— through an analogous conceptual focus on transformation. Starting from how Latour develops many of his arguments about modernity, agency and politics through "figures of puppetry," Herhuth deploys the apparently "minor aesthetic form" of animation to correct film theory's historical focus "on photo-indexicality rather than movement."[49] If "more direct forms of animation" do "align with notions of indexicality," a broader focus on the "real transformations" animation can facilitate allows one to resituate its practices as forms of "realism (or materialism)" in mediation: not just a minor modernist aesthetics, but "philosophical expression of the world" in a "longer view."[50]

I want to expand these arguments: understanding cinematic worldmaking across genres in terms of transformation allows us to conceptualize how both fiction and nonfiction engage (with) surrounding worlds as aesthetic practices folding heterogeneous materials into forms. The outlined critiques of traditional concepts of realism have shaped debates around fiction as well. Outmoded notions of fiction as (immediately) mirroring or indexing the world have prompted the poststructuralist ban on *all* discussion of its real-world referentiality as theoretically (quasi-)inappropriate, often resulting in unacknowledged tensions for cultural studies scholarship with an interest, for example, in cinematic gender roles, legacies of racism or practices of queer intervention.[51] Latour's model of material-semiotic transformations, however, allows us to conceptualize how the composition and reception of fiction—"made," again, "of the same material, the same figures" as nonfiction—imaginatively draws on the life worlds surrounding it, tackling matters of concern through rich folds of mediation.[52] As an ensemble of "styles and trajectories" rather than a "domain" opposed to reality, fiction makes not just worlds, but, I would stress, ontologically remakes "the world."[53] As fiction films creatively suspend physical laws, political probabilities, or simply documentary standards of reassembling events, they present not "falsehood" but the imaginative "exploration of traces."[54] Reference in fictional genres, as I have proposed in extending Latour's account, is less disciplined than in factual genres. Never absent, it operates in piecemeal—in the sense of both unsystematic and fragmentary—ways. If there is no chain of reference without narrative (as Latour highlights from the science angle), there is also no narrative without links, however fragile or wildly transformative, to bits and pieces of the real worlds surrounding it (in an encompassing sense): affects, associations, fantasies, objects or mountain settings, discourse scraps, historical counterparts, and more.[55]

In evaluating "what is well constructed" in fiction (*AIME* 453), many of us will want to negotiate criteria that underline its transformative license rather than disciplining it by recourse to some established standard of realism: the "most powerful fictions," Patrice Maniglier suggests in his contribution to this volume, "are those that 'send' us the furthest" from the world as is.[56] I would like to qualify this by saying *sometimes* and *in some respects*, arguing for a plurality of criteria for a plurality of fictional genres and unique works of art in different life world contexts. Without curtailing fiction's imaginative license, we can allow some criteria that evaluate how (forcefully, thoughtfully, shockingly, playfully, empathetically, convincingly, outrageously, sensitively, complexly, clearly, ethically . . .) a film engages matters of concern through the audiovisual medium's powers of "mobilizing" our "affects" and minds, and the myriad different aesthetic techniques accomplishing these transformations.[57]

Thus conceptualizing the genres of fiction in terms of material-semiotic transformation furnishes a positive answer to the question of how *Western* documents contemporary life world realities: in a piecemeal fashion, and through the imaginative lens of a particular genre. The bodies, gestures, voices, dialects and world views of the construction workers acting in the film become mediants in a process of cinematic worldmaking in which, as indicated above, "fictional idea[s]" and "reality," "research," "casting," "fantasy," and "life experiences and biographies" were complexly entangled from the start (Grisebach, "AT"). During the shooting, the director reports, she sometimes let the Arri Alexa (a camera model that has facilitated some of the most striking long-take experiments in contemporary cinema)[58] run while she took a break to later "reconnect again and intervene in the scene as it was already playing out," transforming the interactions that had developed in her absence ("WAM"). The cinematic effect of these procedures is not smoothly realist by "classical" standards of unobtrusiveness: a reviewer comments on how the film's loose assembly of "deliberately offhand detail"— "snatches of landscape and sunshine; snatches of conversation . . . and group dynamics"—is "rendered unstable by the occasional, obviously pre-written lines of dialogue Grisebach surreptitiously places in her characters' mouths" (Lattimer, "AF").

Arguably, the film title indexes this grafting activity, announcing the two-way transformative encounter between the actualities of twenty-first-century life worlds and the topoi of a historical genre that, in turn, "tells so much about the construction of society" (Grisebach, "WAM"). The film audience probes these narrative possibilities in the associative space opened up by *Western*'s noticeable acts of linking contexts: as Meinhard begins to prefer the locals over his crew, we map the action via the well-known Western topoi of "going native" and intergroup "brotherhood." The fictional building blocks serve as a means for exploring a process of connection and change in Grisebach's twenty-first-century European contemporary scenario that is, perhaps, realist insofar as the change is neither radical, nor definite, nor free of affective tensions and contradictions. Thus, the main actor (as/or character) describes the world as a place ruled by force and aggression ("Fressen and Gefressenwerden," literally "feed and being eaten") while he shares food, drink, work and personal vulnerabilities with his new Bulgarian buddy Adrian. With his inconsistent gestures throughout the film, Meinhard's character elicits less full-fledged empathy or sympathy than a combination of irritation and affective curiosity about his backstory (which is never revealed). He begins a slow, tender romance with a local of his own age before seizing the opportunity to make out with a younger Bulgarian woman who is also unsuccessfully pursued by Vincent; and he claims that violence is not his "thing" right before preparing to use his knife, which he later tries to gift to the young local boy he has befriended, against the explicit instructions of the boy's relatives. The film's undramatic, open ending preserves these affective incongruities. Except for the horse, no one dies; Meinhard does get the beating he arguably deserves from some locals but, as far as we can see, continues to hang out with them—as do the other Germans at this point. In the concluding scene, Adrian corrects Meinhard's inappropriate attempt to rid himself of the knife by returning it, displacing the promise of closure with the prospect of ongoing violent as well as friendly negotiations.

It is precisely this lack of closure, however, that underscores how cinematic transformations engage larger life worlds at the porous borders of their fictional, documentary or, in the case of *Western*, docufictional cosmoi. In terms of ethos, the film's realist project of giving a "good" account of the world through "objectfull" assembly (Latour, as quoted above) thus dovetails with the project of respectfully "following the actors" spelled out in the previous section. As any preconceptions about racist working-class (East) German men are broken down into complex relations with the complicated protagonist and the real-life actor embodying him, the film's reconfigurative critique invites an ongoing engagement that may encompass interest, feeling for, anger, disconcertment and more, without excusing the documented attitudes of cultural superiority. In conclusion, let me underline that the proposed Latourian account of cinematic worldmaking in general—as an onto-epistemological process of aesthetic transformation undertaken in a network of non-sovereign actors—does not commit us to embracing this ethos of cautious reconfiguration as the only valid or regularly preferred way of making cinematic and other worlds. In fact, cautious reconfiguration has its limits. *Western* stops short of imagining a radical dissolving of either the nationalist and racist attitudes, or the structures of inequality it depicts: the film's (both aesthetically and politically) realist gesture of respecting the complexity of people's orientations along with the unlikelihood of rapid change fails to imagine full-fledged alternatives to the status quo or alignments "around new ideas and causes."[59] Encouraged by Latour's own admission that political "rages, too, have to be respected" (*RS* 249), our plural criteria for successful cinematic fiction might include some that valorize forms of loud defiance not offered by Grisebach's cautious take.

More generally, I hope to have shown that reading (and potentially making) films with Latour, positions their media, institutions, and communicative circuits in the lively center of the contemporary landscape of a recomposed humanities, whose curious investigations, urgent concerns and layered methodologies are no longer tied down by old critical habits or stand-offish self-enclosures in the ivory towers of art and criticism. (As materialized conditions of asymmetrically distributed research, teaching, and learning opportunities, to be sure, these towers are very real parts of our networks.) The recomposed humanities I am prepared to defend are egalitarian in their orientation. They respect the nonhuman actors we are entangled with along with less powerful, vulnerable, utterly non-sovereign human actors; they are fueled by ongoing revisions of our democratic standards about who, and what, can be more fully included in a future planetary collective. While I know that such inclusiveness will never be complete, the stubborn clinging to it as a normative horizon contributes to my reluctance to let go of the old and politically compromised terminology of the human(ist) in the humanities. After all, this terminology's history overlaps not *only* with the ruthless exploitation of less powerful humans and nonhumans, but also with efforts to push back against the nationalist, religious, and racist enclosures that are, in 2020, once more threatening to take over our contemporary life-worlds. In this spirit, I locate the contributions of film and film reading to the recomposed humanities as a "[p]ractical ontology" and "[c]osmopolitics" (*AIME* 481), or a worldmaking practice of "philosophy" that has the distinct advantage of being far less abstract and elitist in its audiovisual, imaginative explorations than the old discipline associated with this name, and also more evidently connected to a mesh of modes of existence.[60] At the intersections of [FIC • REF], for example, with [MET]amorphosis, [HAB]it, [POL]itics, [ATT]achment, [MOR]ality, and others, film is a powerful actor in our collective efforts to reassemble a (more) "common world" (*RS* 250).

Notes

1. The methodological alignments indicated by a scholar's preference for either of these two terms can include foci on aesthetics-vs.-networks of production and reception; or the medium-vs.-the institution (see below for detail). If "cinema" long resonated as the more inclusive term, the discussion on the contemporary "post-cinematic" condition has complicated matters. See, e.g., Steven Shaviro, *Post-Cinematic Affect* (Washington: O-Books, 2010). In line with the syncretic proposal developed in this piece, I use the notion of "film" in an encompassing sense, including production and reception networks, and both analog and digital technologies.

2. On the popularity of ANT in media as compared to film studies, see Jerome P. Schaefer, *An Edgy Realism: Film Theoretical Encounters with Dogma 95, New French Extremity, and the Shaky-Cam Horror Film* (Cambridge: Cambridge Scholars Publishing, 2015), 21 (hereafter cited as *ER*). To be sure, this popularity also reflects a relatively recent trend, as indicated by Nick Couldry's skeptical "Actor Network Theory and Media: Do They Connect and on What Terms?" in *Connectivity, Networks and Flows: Conceptualizing Contemporary Communications,* ed. A. Hepp, F. Krotz, S. Moores, and C. Winter (Cresskill: Hampton, 2008), 93–110. See also Tristan Thielmann's comments on how Latour reception in German media studies was slowed by the dominance of technological determinism. "Der ETAK Navigator. Tour de Latour durch die Mediengeschichte der Autonavigationssysteme," in *Bruno Latour's Kollektive,* ed. Georg Kneer, Marcus Schroer, and Erhard Schüttpelz (Frankfurt: Suhrkamp, 2008), 180–218.

3. Bruno Latour, *An Inquiry into Modes of Existence* (Cambridge: Harvard Univ. Press, 2013), 30 (hereafter cited as *AIME*); Schaefer, *ER* 21.

4. Perhaps this footnote can serve as a gesture of mourning for the vibrant intellectual environment of the interdisciplinary "Communication and Culture" department sacrificed to the new "Media School" at my previous institution, Indiana University, Bloomington.

5. The members of SCMS (the *Society of Cinema and Media Studies*, thus renamed in the 1990s) voted to change the name of the organization's *Cinema Journal*, the leading US venue in the field, to *Journal of Cinema and Media Studies (JCMS)*, effective fall 2018.

6. On the ethical orientation at a new collective, see *Reassembling the Social: An Introduction to Actor-Network-Theory* (Oxford: Oxford Univ. Press, 2005, hereafter cited as *RS*). The proposal for cinema I detail resonates with the twofold methodological goal Rita Felski outlines for the field of literature: to balance broader network descriptions with traditional foci on "advanced techniques of reading." See *The Limits of Critique* (Chicago: Univ. of Chicago Press, 2015), 184 (hereafter cited as *LC*).

7. See, for example, *RS* 55, 89; and for detail on Latour's own work with film and filmmakers, David D. [no full last name given], "Bruno Latour's Artistic Practices: Writing, Products, and Influence." *Toronto Film Review* 5 February 2016. On Latour's aesthetic affinities and writing practices more generally see Francis Halsall, "Actor-Network Aesthetics," this volume.

8. Seminal early texts for the film studies discussion on postcritical viewing include Vivian Sobchack, *The Address of the Eye: A Phenomenology of Film Experience* (Princeton, NJ: Princeton Univ. Press, 1992, hereafter cited as *AE*) and Steven Shaviro, *The Cinematic Body* (Minneapolis: Univ. of Minnesota Press, 1993).

9. Lorenz Engell, "Eyes Wide Shut. Die Agentur des Lichts—Szenen kinematographischer verteilter Handlungsmacht." *Unmenge—Wie verteilt sich Handlungsmacht?*, eds., Ilka Becker, Michael Kuntz, and Astrid Kusser (Munich: Fink, 2008), 75–92, here 75, see 77.

10. Latour, *RS* 69; Bazin, André, *What is Cinema? Vol 1*. Essays selected and trans. Hugh Gray (New edition Berkeley: Univ. of California Press, 2005), 13; Engell, "Eyes Wide Shut" 80 ("die Welt aus der Perspektive eines Dings unter Dingen"; all translations are my own) with reference also to Bela Balász, Gilles Deleuze, and Siegfried Kracauer.

11. Bazin, *What Is Cinema?* 13. After decades of embarrassment, recent scholarly reappraisals of Bazin have emphasized that his theory cannot be reduced to the quoted filmontological claims. See *Opening Bazin: Postwar Film Theory & its Afterlife*, eds., D.A. with H. J.-L. (New York: Oxford Univ. Press, 2011).

12. Sobchack, *AE* 263. See Engell, "Eyes Wide Shut," 84 on auteurist theories.

13. Sobchack, *AE* 265 (quoting Baudry), 22; her emphasis.

14. Sobchack, *AE* 167–8; her emphases; see Engell "Eyes Wide Shut," 75.

15. Sobchack, *AE* 173; her emphasis. Latour has distanced himself from phenomenology for these classical foci. See *Pandora's Hope: Essays on the Reality of Science Studies* (Cambridge: Harvard Univ. Press, 1999), 9. As I underline below, however, there is nonetheless significant resonance between Latour and (postclassical) phenomenology. From a contemporary auteurist perspective, see Daniel Yacavone on Merleau-Ponty and Sobchack's reception of

Merleau-Ponty: "Film and the Phenomenology of Art: Reppraising Merleau-Ponty on Cinema as Form, Medium, and Expression." *New Literary History* 47 (2016): 159–186. For the medium of literature, Felski develops a more fully and explicitly Latourian sketch of distributed agency with an analogous accent on the agency of the artwork itself (*LC*).

16. Latour, *RS* 45 (his emphasis); Engell, "Eyes Wide Shut," 84–85: "vielgestaltige and komplizierte Gerätschaften," "Konventionen, Regeln, and Stile[n]," "Erwartungen, Sehgewohnheiten, Ikonographien und Ideologien" (85).

17. Ilana Gershon, and Joshua Malitsky, "Actor-Network-Theory and Documentary Studies," *Studies in Documentary Film* 4.1 (2010): 65–78, here 66, see 72–73 (hereafter cited as "ANT").

18. Gershon and Malitsky, "ANT," 66, see 69. More recent contributions have begun to explore the agency of things and other nonhuman actors in other films. See, for example, Lars Kristensen, "Bicycle Cinema: Machine Identity and the Moving Image," *Thesis Eleven* 138.I (2017): 65–80; Catherine Lord. "Only Connect: ecology between 'late' Latour and Werner Herzog's Cave of Forgotten Dreams," *Global Discourse* 6:1–2 (119–132), and Brad Prager's response in the same volume, "How Herzog remembers images past: a response to Catherine Lord" (133–5).

19. Gershon and Malitsky, "ANT," 68. Although highly precarious even in the historical practice of mid-century Hollywood cinema, these norms remain the focal point of contemporary cognitive film theories. See, for example, Noël Carroll, *The Philosophy of Motion Pictures* (Malden: Blackwell, 2008).

20. See Breger, *Making Worlds: Affect and Collectivity in Contemporary European Cinema* (New York: Columbia Univ. Press, 2020).

21. Felski's discussion of literature includes characters and narrators as nonhuman actors (*LC* 163–5). According to the principle that not all actors are alike, I opt for increased specificity here. Fictional beings are certainly not reducible to human experience, as Stephen Muecke cautions ("An Ecology of Institutions," this volume), but many of them, including some on-screen aliens and unreliable thing narrators, invite engagements based on partial recognition.

22. Whereas cognitive along with classical phenomenological perspectives have privileged the conscious and intentional layers of character, author and audience activity (see, e.g., Carroll, *Philosophy* 157), Latour has been placed in the opposite (Deleuzian and new materialist) camp, where subjective human experiences and actions tend to be discarded altogether (see, e.g., Brian Massumi, *Parables for the Virtual: Movement, Affect, Sensation* [Durham, NC: Duke Univ. Press, 2002]). For the resulting concerns about "ethics, accountability, normativity, and political critique" see Arjun Appadurai, "Mediants, Materiality, Normativity," *Public Culture* 27:2 (2015): 221–37, here 221. My proposal presents a less categorical (and, I hope, more fully Latourian) variation on Appadurai's solution to conceptually foreground "mediants," that is, in a nutshell, human-as-networked actors (222).

23. Latour, *RS* 161, 236; *AIME* 401; Latour's emphasis. For contemporary perspectives inspired by postclassical phenomenology see, for example, Laura Marks, *The Skin of the Film: Intercultural Cinema, Embodiment, and the Senses* (Durham, NC: Duke Univ. Press, 2000); Eve K. Sedgwick, *Touching Feeling: Affect, Pedagogy, Performativity* (Durham, NC: Duke Univ. Press, 2003); Sara Ahmed, *The Cultural Politics of Emotion* (Edinburgh: Edinburgh Univ. Press, 2004); Felski, *The Uses of Literature* (Malden: Wiley-Blackwell, 2008).

24. Sobchack, *AE* 263, xviii; Shaviro, *The Cinematic Body* 10; see Murray Smith, "Altered States: Character and Emotional Response in the Cinema," *Cinema Journal* 33.4 (1994): 34–56; Carl Plantinga, *Moving Viewers: American Film and the Spectator's Experience* (Berkeley: Univ. of California Press, 2009).

25. Operations on both sides of this loop are analogous insofar as we understand "*interpretation*" as a "*co-production*" (Felski, *LC* 174; her emphasis) and composition as a creative activity of world reading (see below on the latter point).

26. Latour, *RS* 253 (modifying his critique of critique). While my insistence on less suspicious modes of critique departs from Felski's terminology, I substantially draw on her intervention, including the Latourian emphasis on (re-)configuration over deconstruction (Felski, *LC* 17).

27. On phenomenological description see also Heather Love, "Close but not Deep: Literary Ethics and the Descriptive Turn," *New Literary History* 41 (2010): 371–91. On the intertwined moves of "Localizing the Global" and "Redistributing the Local" see Latour, *RS* 173, 191; on "[c]are and [c]aution" *Pandora's Hope,* 288. While Latour problematizes contextualization as a mode of large-scale reductive social explanation (e.g., *RS* 173; see Felski, *LC* 152), Love defends the category with Donna Haraway: "The Temptations: Donna Haraway, Feminist Objectivity, and the Problem of Critique," *Critique and Postcritique*, eds., Elizabeth S. Anker and Rita Felski (Durham, NC: Duke Univ. Press, 2017), 50–72, here 56–57.

28. Daniel Yacavone. *Film Worlds: A Philosophical Aesthetics of Cinema* (New York: Columbia Univ. Press, 2015) 196; see, again, Carroll, *The Philosophy*.

29. On the overall "school," see Marco Abel, *The Counter-Cinema of the Berlin School* (Rochester: Camden House, 2013); on their phenomenological aesthetics, see chapter six of Claudia Breger's *An Aesthetics of Narrative Performance: Transnational Theater, Literature and Film in Contemporary Germany* (Columbus: Ohio State Univ. Press, 2012).

30. On reading for affect beyond character and plot see Robert Sinnerbrink, "*Stimmung*: Exploring the Aesthetics of Mood," *Screen* 53, no. 2 (2012): 148–163.

31. Böhler's work with several Berlin School directors has made her into a major name in the contemporary German cinema scene. See Grisebach in James

Lattimer, "At the Frontier: Valeska Grisebach on Western," *Cinema Scope* (n.d.) (hereafter cited as "AF"), available at http://cinema-scope.com/spotlight/at-the-frontier-valeska-grisebach-on-western/.

32. Grisebach in Cumming, Jesse: "'Where Are the Men I Can Imagine on a Horse?': Valeska Grisebach on *Western*," *Filmmaker Magazine*, March 12, 2018 (hereafter cited as "WAM").

33. ". . . spielt Grisebach meisterhaft mit jenen Vorurteilen, ohne die kaum ein Western auskäme. Und deren elementare Werte heute im Populismus ein Comeback erleben" (Daniel Kothenschulte, "'Western.' Der Westen im Osten," *Frankfurter Rundschau*, August 23, 2017).

34. "Ich hatte eine große Sehnsucht nach dem Western-Genre, da es das Genre meiner Kindheit ist" (interview with Toby Ashraf, "Wessen Recht gilt bei diesen Typen?" *die tageszeitung* 24 August 2017); "WAM."

35. "AF"; "ein langer gemeinsamer Prozess"; "eine . . . gemeinsame Erfahrungswelt" (interview with Ashraf, "Wessen Recht").

36. "Schon mal an die Einheimischen gedacht?"

37. The only *amazon.de* audience review of the DVD so far concludes that "der Film bedarf geduldiger, aufgeschlossener Betrachter" ("requires patient, open-minded viewers"), attesting perhaps to the limitations of my example: with all its genre inflections, it remains an art film unlikely to reach mass audiences. However, the audience score on *Rotten Tomatoes* is very respectable (if lower than the critical score at 84% vs. 95%, based on 242 audience ratings), and the—few but strong—audience reviews on *IMDB* highlight the ways in which the film invites affective engagement with its characters.

38. "Why Has Critique," 231 (Latour's emphasis); see *RS* 114.

39. *RS* 114–5. Latour quotes Heidegger's notion of "gathering" here.

40. Bazin, *What Is Cinema?* 13.

41. See, for example, Pooja Rangan, *Immediations: The Humanitarian Impulse in Documentary* (Durham, NC: Duke Univ. Press, 2017).

42. Mary Anne Doane, "Indexicality: Trace and Sign: Introduction," *Differences: A Journal of Feminist Cultural Studies* 18, no. 1 (2007): 1–6, here 2; see her "The Indexical and the Concept of Medium Specificity" in the same volume, 128–52. In addition to Bazin, these discussions reference Roland Barthes's theory of photography and Charles Sander Peirce's semiotics, where the index is the type of sign that works through a direct physical connection with its object (*Peirce on Signs: Writings on Semiotic by Charles Sanders Peirce,* ed. James Hoopes (Chapel Hill: Univ. of North Carolina P, 1991), 239–40).

43. Joshua Malitsky, "A Certain Explicitness: Objectivity, History, and the Documentary Self," *Cinema Journal* 50, no. 3 (2011): 26–44, here 28–29, 41; "Ideologies in Fact: Still and Moving-Image Documentary in the Soviet Union, 1927–1932," *Journal of Linguistic Anthropology* 20, no. 2 (2010): 352–71, here 358.

44. *AIME* 243; Latour's emphases. See also the definition of [FIC] in the online AIME portal: it "designates not the field of art, culture, works of art, but the particular mode awkwardly designated by the adverb 'fictionally,' which indicates that we require that [*fic*] beings be grasped according to a particular relationship between materials and figures which cannot be detached." The argument in the [FIC] chapter in the *AIME* book does shift back and forth between the general mode of figuration (as associated with the symbolic), the domain of art and genres of fiction. However, all of these categories are relevant, as long as we specify usage at any given moment. See Maniglier, "Art as Fiction," this volume, for a proposal that distinguishes art within in the more general category of fiction.

45. See David D., "Bruno Latour's Artistic Practices."

46. On nominal versus physical portrayal see Noël Carroll, *Theorizing the Moving Image* (Cambridge: Cambridge Univ. Press, 1996), 240–2.

47. See "AT"; and in Ashraf, "Wessen Recht."

48. Breger, "Cruel Attachments, Tender Counterpoints: Configuring the Collective in Michael Haneke's *The White Ribbon*," *Discourse: Journal for Theoretical Studies in Media and Culture* 38.2 (2016): 142–172.

49. Eric Herhuth, "The Politics of Animation and the Animation of Politics," *Animation: an interdisciplinary journal* 11.1 (2016): 4–22, here 6–7.

50. Herhuth, "The Politics of Animation," 10–11.

51. In narrative theory, otherwise diverging contemporary conceptualizations of fiction(ality) dovetail by asserting that "fictional texts do not share their reference worlds with other texts" (Marie-Laure Ryan, "Postmodernism and the Doctrine of Panfictionality," *Narrative* 5.2 (1997): 165–87, here 167), or that reading a statement as fictive implies assuming "that it is not making referential claims" (Henrik Skov Nielsen, James Phelan, and Richard Walsh, "Ten Theses about Fictionality," *Narrative* 23:1 (2015): 61–73, here 68).

52. Latour, *AIME* 251, see also 254; and in this book on hermeneutics moving "to the world" (see Afterword, "Life Among Conceptual Characters").

53. The quotes are from Latour, "three versions of the crossing [fic • ref]." *AIME* online 14 August 2014 (http://modesofexistence.org/crossings/#/en/fic -ref). Despite the singular wording here, I would underline with Muecke that Latour overall facilitates refusing "the conceptual singularity of 'the world'" (see chapter 1, "An Ecology of Institutions").

54. Latour, "The Denier's Process," *AIME* online, August 14, 2014 (http:// modesofexistence.org/crossings/#/en/fic-ref).

55. See Latour, *AIME* 249, on the "fragility" of the beings of fiction. For detail on the argument (for the domain of literature) see Claudia Breger, "Affects in Configuration: A New Approach to Narrative Worldmaking." *Narrative* 25.2 (May 2017): 227–51. See also Maniglier on how artistic fiction "organizes relations with the worlds from which it is detached" (see chapter 16, "Art as Fiction").

56. See chapter 16, "Art as Fiction."

57. I quote Citton here (see chapter 8, "Fictional Attachments"), who makes the argument about literary fiction. My cinematic twist to the idea is not to engage in a competition of media but to flesh out the point: the affective powers of art are in the imaginative powers of fiction (across media) as layered with the differently affective capacities of its specific material-semiotic pathways via images, sounds and words.

58. For example, Alejandro G. Iñárritu's 2014 *Birdman* was shot on Ari Alexas, too.

59. Herfurth, *Politics of Animation* 17 (with reference to Galloway).

60. See Afterword, "Life Among Conceptual Characters."

CHAPTER THIRTEEN

Latour, the Digital Humanities, and the Divided Kingdom of Knowledge

MICHAEL WITMORE

TALK ABOUT THE HUMANITIES today tends to focus on their perceived decline at the expense of other, more technical modes of inquiry. The big S "sciences" of nature, we are told, are winning out against the more reflexive modes of humanistic inquiry that encompass the study of literature, history, philosophy, and the arts. This decline narrative suggests we live in a divided kingdom of disciplines, one composed of two provinces, each governed by its own set of laws.[1] In the first province we find the humanities, concerned as they are with a distinct class of objects— with the things humans make or do. These objects must be known in an equally distinct way: reflexively, historically, and in a context that acknowledges that "things could have been otherwise." The other, more technical kingdom is populated with things that humans find already to be the case, things we can know but never negotiate because they are not, at least prospectively, the product of human deliberation. In this province we find the objects of the "scientific gaze," which has for centuries been scanning its part of the kingdom for all that can be measured, registered, and impersonally described. This gaze recognizes proteins and parabolic curves, but passes over peace treaties, epic poems, and piano sonatas.

Enter Bruno Latour. Like an impertinent Kent confronting an aging King Lear, Latour looks at the division of the kingdom and declares it misguided, even disastrous. Latour's narrative of the modern bifurcation of knowledge sits in provocative parallel with the narrative of humanities-in-decline: what humanists are trying to save (that is, reflexive inquiry directed at artifacts) was never a distinct form of knowledge. It is a province without borders, one that may be impossible to defend. We are now in the midst of a further plot turn with the arrival of digital methods in the humanities, methods that seem to have strayed into our province from the sciences.[2] As this new player weaves in and out of the two plots I have just described, some interesting questions start to emerge. Does the use of digital methods in the humanities represent an incursion across battle lines that demands countermeasures, a defense of humanistic inquiry from the reductive methods of the natural or social sciences? Will humanists lose something precious by hybridizing with a strain of knowledge that sits on the far side of the modern divide? What is this precious thing that might be lost, and whose is it to lose?

Readers of this book will recognize the significance of these entwined narratives and the need to explore their conjunction. Whether we are talking about the crisis of modernity, the crisis of the humanities, or the incursion of the digital, the narratives I am describing depend upon some prior understanding of what it means for something to be made or composed, and what follows from that presumption. In his "Compositionist Manifesto," Latour calls attention to the broad sweep of human practices (including science and politics) that involve the composition or immanent reality of "objects of concern."[3] Within this field, no approach to knowledge, whether mediated by remote sensors or human empathy, can trump another on the grounds that it has direct access to a transcendent reality—an access that puts its objects and concerns beyond debate. Everything can potentially be an object of concern, because everything can be made to concern us. On the face of it, this seems good for the humanities. But Latour's reorientation of the debate and emphasis on composition makes this humanities-friendly conclusion less likely. The point is not to preserve the domain of human-generated artifacts—meaning, art, expression—from the encroachments of a value-free science. Rather, Latour urges readers to see that the protocols for knowing nature and knowing ourselves are both bound up in a process of composition: elements of the world must be set up alongside elements of our thought, as it were, in a single line of type. Latour's outlook has a leveling effect, since neither the products of human thinking nor the "facts thrown up by the natural world" can be counted on to settle debates about what is really real or "what is to be done." Neither the sciences nor the humanities, to reiterate a tired version of the distinction, will ever have the last word.

Latour is, to my knowledge, the first person to point out that the modernist divide forces a proliferation of hybrids—which is his word for developments such as anthropogenic climate change that cannot be kept to one side of that divide.[4] Many of these hybrid objects of knowledge appear at the moment a human motive or concern is placed in a productive relation to (is composed alongside) the constraints of a physical system. The speed bump, for example, aligns the desire to regulate traffic speeds with the inertial properties of bodies moving in cars.[5] The Berlin Key takes the unyielding affordances of metal keys and aligns them with the desire to keep strangers out of apartments.[6] The printed hiking map, not to mention the GPS satellite, synchronizes the movements of the smartphone-wielding hiker to the inch-by-inch deviations of topography, making the landscape a calculable interval of human movement.[7] If we speak of hybridization as the point where constraints cease to be either intellectual or physical, where changes in the earth's mean temperature follow just as inevitably from the "political choices of human beings" as they do from the "laws of nature," we get a sense of how rich and productive the modernist divide has been. Hybrids have proliferated. Indeed, they seem inexhaustible.

But what of the hybrids proliferating *within* the humanities, within just one province of the modernist kingdom that appears to be the subject of Latour's play? We may speak of a sonnet's being composed according to the laws of sonnet-making, but those laws are conventional—a motivated artifact of human practice. There may come a day when humanists have the resources to link our neurological apparatus to the formal properties of poems, proving that caesuras are the prosodic equivalent of speed bumps. This type of finding sits somewhere on the horizon of humanities inquiry and may strike some as a cause for despair. Such despair would be misplaced. A degree of freedom remains in the *alignment* of biological "musts" and poetic "mights," an alignment or composition that is itself contingent and available to be fashioned otherwise. There are plenty of ways of making one moment in a poem recall another without exceeding the constraints of our short-term memory, whatever those limits turn out to be. In any event, we do not need to wait for English professors to get their hands on unlimited FMRI budgets to begin asking what such a hybrid object of knowledge can teach us.[8]

At least one example of this type of hybrid already exists in the "digital humanities." The Latourian theory of hybrids provides a useful starting point for thinking about a field of inquiry in which interpretive claims are supported by evidence obtained via the exhaustive, enumerative resources of computing. The range of practices named in the phrase "digital humanities" is broad, perhaps too broad for the term to mean anything. To narrow that definition, I plan to examine one version of digital humanities practice that I know and feel comfortable explaining: the computational analysis of corpora of texts.[9] In order to say what type of hybrid this sort of research is, and how it fits into the larger narratives I have just described, it will be helpful to anatomize some of this practice's key argumentative moves and assumptions. To that end, I will be introducing a claim about the plays of Shakespeare in a form that best shows the moving parts of the argument. Having introduced this "toy example" and shown where the welds or joins between humanities practice and technical inquiry occur, I will then ask what this particular practice adds to the debate about the humanities and, perhaps, to the still-unfolding drama of Latour's modernist divide.

I. Massive Addressability, Feature Explicitness, and Corpus Closure

The computational analysis of texts finds its origins in traditional philological practices—indexing, concordance making, enumeration of forms, and the like.[10] What the computer has added to these practices is the ability to work indefatigably toward a desired outcome, but only once certain conditions have been met. Those conditions can be described with precision, even if they are difficult to satisfy in any general way.

The first condition is easily met if our object of analysis is *texts*. That condition is massive addressability.[11] For a text to be a text, it must be addressable at an indefinite number of levels of abstraction or scales. We can understand this condition by thinking about the different units to which we can direct our attention when working with texts. For example, when Western writing began to introduce breaks between words (instead of leaving them as a continuous flow of glyphs), a reader could more easily address him or herself to the first, second, or third word in a sequence.[12] The division of pages into paragraphs, or in the case of editions of Plato, into the Stephanus divisions, represents another spatial protocol for address. I can refer to this section of the *Meno* because someone bounded it in a unit and gave it a name: *Meno* 70a.[13] But massive addressability implies that a text can be addressed in an almost limitless number of ways. I can refer myself to all of the verbs spoken by the character Hamlet in a certain printed edition of Shakespeare's play. (The fact that the edition matters, and that we can debate what a verb is and what counts as one in Elizabethan English, only proves that there are many ways of specifying what is to be addressed.) In the spirit of Borges, I might distinguish between phrases that the Emperor likes, and those he doesn't. If I can list those phrases in two columns, I have stabilized the unit address. It is important to note that levels of address are arbitrary or stipulated: there is nothing natural or inevitable about them. In 1528, Erasmus mocked Renaissance imitators of Cicero for their tendency to divide Latin words into those that were used by the ancient rhetorician and those that were not. (The latter were to be avoided.)[14] For words to be grouped in this way, people had to assume that the individual word is an addressable "level" or unit within a text. Today we make reference to entities known as "n-grams"—sequences of words of variable length. If modern-day Ciceronians began avoiding "all three-word sequences that were not used by Cicero himself," which is to say, all non-Ciceronian 3-grams, they would be working at yet another level of address.

To the extent that any level of address can be stabilized, features—things that can be searched and counted in a text—become discernable. Features are made and not born; they too are arrived at interpretively, even if they are tallied mechanically. Suppose I identify "all of the verbs in *Hamlet*" as a feature I want to explore. Clearly there are different ways of establishing what might count as a verb in any text considered in its linguistic and cultural context. The Danish Prince, for example, makes reference to that "fell Sergeant, death." In this instance, "fell" is used as an adjective rather than a verb. To correctly exclude "fell" from "all of the verbs in *Hamlet*," I must understand Renaissance usage, which I may not do. And the difficulties will only increase if I move from grammar to rhetoric, choosing as my feature "all examples of metonymy in *Hamlet*." Literary critics make reference to features all the time, and there is usually some formal understanding of what counts as that feature. (Literary criticism and writings about grammar and rhetoric have, traditionally, been the place where features are defined.) But for computer-assisted analysis of texts, the definition must be explicit enough to link an implied understanding (metonymy) to actual words in the text ("my *crown*, my own ambition, my queen . . ."). A computer has to be given instructions about what to count, and will always follow the instructions.

There can be no computationally derived evidence from texts without explicitly defined features, and the simplest way to establish a feature is to enumerate its every example. "When I say metonymies in *Hamlet*, I mean the following passages in acts 1, 2, 3, 4, and 5. "If a computer is involved in textual analysis, some version of this explicit listing is likely (but not always) taking place.[15] But that list could go on forever if another condition is not met. To actually count things in a text, even if one is doing so with pencil and paper, one must have a closed corpus within which to count. All the metonymies in *Hamlet* occur in something we can agree is *Hamlet*, just as the designation "all the verbs used in early modern texts" assumes an explicit definition of what texts fall within the boundaries of early modernity. Like definitions of features, definitions of the corpus must also be explicit and enumerative. One must actually be able to state *which texts* are part of the corpus, and *which words exactly* are in those texts. (There is no way around the kinds of interpretive judgments we associate with periodization and with the editorial establishment of the text.) We cannot automate these particular judgments about scope and contents, just as there is no entirely value-free way of choosing what features will be counted in texts that, by their nature, allow huge leeway in their modes of address.[16] All of this interpretive work must be done in advance of the analysis, whether the work is made explicit or not.

Once the features have been defined and enumerated and the corpus closed, something special happens: the analysis is in a position to halt, and halting is a critical outcome when one is working with algorithms and computers. Readers may be familiar with the "halting problem" as it was encountered by Alan Turing, the computer scientist whose foundational papers in computer science helped establish the discipline.[17] (The "halting problem" refers to the difficulty of establishing whether a computer program, working on a known set of inputs, will "halt" or will continue running indefinitely without a result.) Part of what makes computer science such an interesting field is that one can think abstractly—without actually running any programs—about the conditions under which a computer acting on a finite set of instructions would halt in the course of acting on those instructions. "How many of the integers between 1 and 1000 are prime?" is a question I might want to answer with the help of an algorithm. To the extent that "prime" is a well-defined feature and limits have been set for candidate numbers (i.e., the corpus has been closed), the feature in question can be counted. The program, in other words, will stop. Something similar happens when computers work with words. Indeed, from a computational standpoint, words and numbers are being treated in the same way. If I am interested in how many times the string of glyphs "fell" is used as an adjective in Shakespeare's plays, I need to define the feature explicitly and close the corpus so that I have put some manageable bounds on the text's high potential for address.

II. A Sample Claim: If, And, or But and Shakespeare's Genres

Now to an example of a claim. I establish and close a corpus of texts—the modern-spelling Folger Editions of Shakespeare's plays, stripped of all words that are not spoken onstage—and provide you with an address where you can examine and download that corpus for yourself.[18] Next I define three features, which in this case will be the words "if" "and" and "but." These features are attractive for the purposes of my exposition because (a) three discrete words with unique spellings are easy to enumerate, and (b) readers will likely remember them. There are other ways of deriving features, but this brute force list of three words is easy to grasp and interesting in what it suggests. Each of my three words, "if," "and," "but," is now a feature to be counted.

I instruct a counting algorithm to identify all of these words in the texts of thirty-eight plays, saying, essentially, show me every time this word is mentioned by a character.[19] By turns, I arrive at a score for each word as a percentage of all words used in the play, and these scores are then treated as "observations" about the plays in an empirical trial. Like rainfall, sunlight, and temperature measurements in a rainforest, "if" "and," and "but" are features that we have counted because we believe they may be distributed across the domain or corpus in an interesting way.[20] Continuing with the example, imagine now that I create a three-dimensional space—an if-and-or-but space—within which I want to locate all thirty-eight plays based on how much of each feature they possess. Each axis in the diagram below represents a word, so that if a play is further from the wall on the front left of the cube, it has more "if"; if it is further away from the back-facing left wall, it has more "but"; finally, when it is further away from the floor of the cube, that play has more "and," It is not necessary to be able to accurately gauge depth in the diagram below for the purposes of present discussion: the space is what I want to emphasize (see figure 1).

Because this space has well-understood mathematical properties, we can make systematic comparisons among the items it contains. Dots clustered near one another are more alike in their relative proportions of the three features (i.e., having or not having any of the three words), and degrees of similarity can be assessed. Let me now add a few labels in order to make a comparison (see figure 2).

Scanning the diagram, I see that *Twelfth Night* is much closer to *Two Gentlemen of Verona* than it is to *Henry V*. Is this an accident? I have to know at least a little bit about Shakespearean drama to answer this question, since to explain why two plays are closer to one another than a third, I have to generalize about all three of them. This is another way of saying that I cannot get literary significance out of the diagram without first putting some in, and I am probably on unstable ground if I start

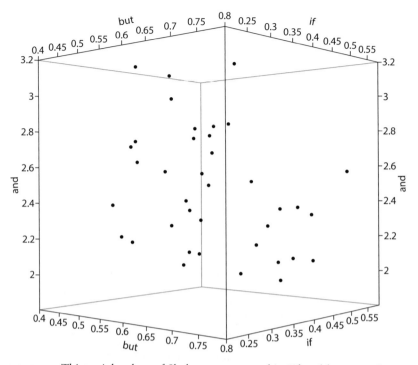

FIGURE 1. Thirty-eight plays of Shakespeare arrayed in "if-and-but space." The axes on these three dimensions represent occurrences of the target word expressed as a percentage of all tokens in the play (words and pieces of punctuation).

saying things like, "*Henry V* is a story about one man's struggle to say the word 'and.'"

As a lifelong reader of Shakespeare's plays, I go ahead and make the likely inference: "*Twelfth Night* is closer to *The Two Gentlemen of Verona* because they are both comedies, whereas *Henry V* is a history play and so sits somewhere else." Names such as "comedy" and "history play" can now begin to motivate the comparisons, providing a shorthand way of summarizing a complex set of relationships among different types of narratives I am trained to recognize.[21] One might say that, like dots arrayed in space, technical genre terms such as "comedy" function as caricatures of the rich and complex set of reactions we have to plays that cannot themselves be fully stated in a natural language. Both the dot-in-space and the generic name are translating the complexity of

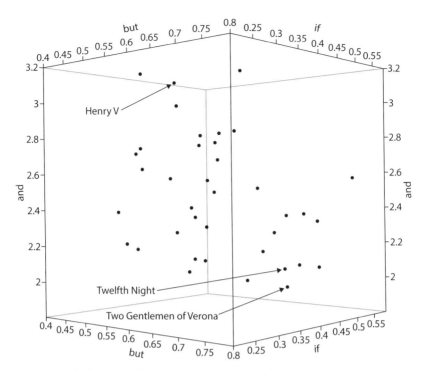

FIGURE 2. Shakespeare plays in three dimensions, with three plays labeled.

what is being discussed, and to the extent that this translation enables a meaningful set of contrasts, the motivation it expresses will seem sound.

"Intriguing idea you have about the distances between these three plays," you say. "But what about a more ambitious comparison?" You continue: "If differences in genre account for where three plays show up in the diagram, shouldn't that difference be visible in the distribution of *all* of them?" This is a much more interesting question. To answer, we begin by replacing the simple dots above with symbols that correspond with the four genres into which these plays are often assigned: comedies, histories, tragedies, and late plays.[22] It is worth pausing now to say that we have just put *a lot* of literary knowledge into the space with this one move. Figures 3a, 3b, and 3c offer three different views of the same cloud, in the same space depicted above, with a short explanation of what they show.

As the captions for each figure indicate, we are now using this three-dimensional feature space to assess the generalizations implicit in the

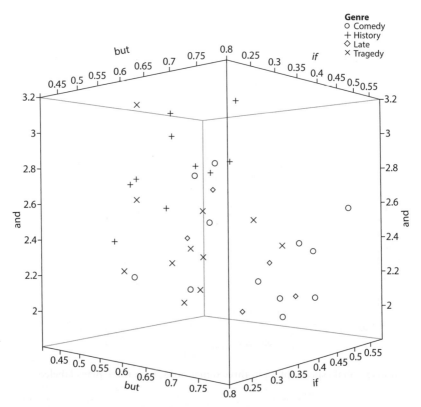

FIGURE 3A. Thirty-eight Shakespeare plays in three dimensions with genre names applied to all. Along the vertical axis, we see that the history plays (+ symbol) generally have a higher percentage of the word "and" in comparison with plays designated comedies (O symbol). The difference in the frequencies of this feature across plays qualifying as either comedies or histories is statistically significant. (P values are < .05 when group means for these two genres are compared with Student's *t*-test.) The same threshold of statistical significance attaches to the other two features, "if" and "but," depicted in figures 3b and 3c.

practice of sorting plays into multiple genres.[23] We have put literary knowledge "back into the diagram" and are now seeing it translate into a distribution of features. Based on observed distributions, comedies and histories use these features in opposite ways. The pattern, verifiable statistically but visible even in these diagrams, is that comedies generally favor "if" and "but" at the expense of "and," while histories do the opposite. An attentive reader will ask how likely it is that this unevenness

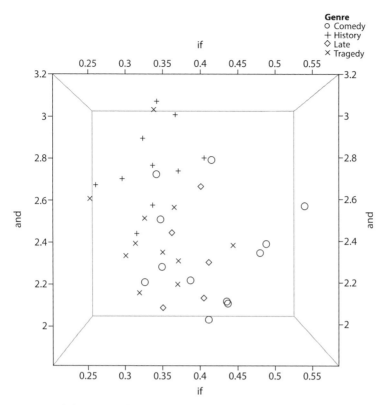

FIGURE 3B. Shakespeare plays in two dimensions: head-on view of the "and–if" axis. Plays designated comedies (O symbol) tend to have more "if" in comparison with plays designated histories (+ symbol).

occurs by chance. Using the same statistical techniques we would use for data observed in rainforests, I could show just how confident I am that chance is not involved (p-values are $< .05$). Now a much more interesting set of questions comes into view. What kind of "force" obliges certain words (the logical contrastives, "if" and "but"; the conjunction "and") to distribute in these specific ways? Surely plays are not ecosystems. Do certain words, like rainfall and daylight, have to occur in inverse proportions? What kind of environment is a play? What kind of social system is a bundle of words?

A play is an immersive, physical mode of storytelling; it is perhaps the oldest interactive art form we have. When we begin asking questions like

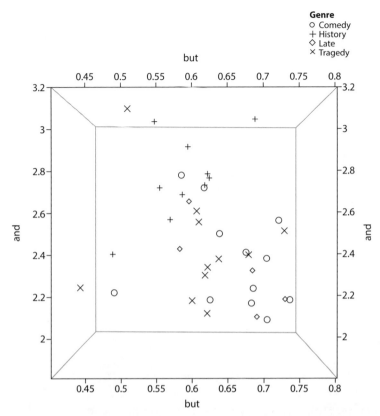

FIGURE 3C. Shakespeare plays in two dimensions: head-on view of the "and–but" axis. Plays with more "but" tend to be comedies (O symbol). Ignoring all depth, one could draw a diagonal line from the lower left corner to the upper right corner in order to separate comedies from histories (+ symbol). The angle of this line would represent a proportion of "ingredients" characterizing items on either side of that partition: items in the lower right have both more "but" and less "and," in comparison with items in the top left. The same line could be drawn in three dimensions as a plane cutting across figure 3a, separating out items in the lower right that are simultaneously high in "if" and "but" while being low in "and"—in other words, "comedies."

the ones I have just listed, the modern distinction between the natural and the artifactual—the found and the made—begins to blur, since we are now forced to engage the physical affordances of the theatrical art form (and its institutionalization in print); the social needs and aspirations that the art form meets; and the intellectual or imaginative limits of its various modes of representation. The discussion is also about to open out on the history of theatrical conventions and the terms we use to describe our experience of different narratives. We are at an interesting conjunction, the kind one hopes to arrive at in this kind of work. The corpus has been closed, features have been chosen, and all the counts made. I am now ready to connect what I know of the constructedness of plays with my sense of why they, and the words they contain, behave as they do.

III. Features and Proxies

My task now is to explain, through what I already know of Shakespeare's plays, why *these three features* seem to track differences in genres, specifically the difference between his comedies and his history plays.[24] This is the most important move in the discussion, since I am saying why my chosen features serve as a *proxy* for something else, in this case, a generic difference commonly accepted by literary critics and historians. What I would like to call the feature-proxy distinction is basic to all computationally assisted criticism of texts, and is unavoidable. No matter how distinctive or statistically sound the pattern discovered may be, one has done nothing to explain it until one provides a motivated link between the features in question, their patterned distribution, and something else that this pattern is a proxy *for*. Here are a few feature-proxy arguments that take, as their motive, the genre terms we have been discussing. The patterns are first stated, then the proxy claims made in argument form:

> STATEMENT 1: Shakespeare's histories have more "and" alongside a corresponding lack of "if" and "but" relative to his comedies. Here are my observations. . . .
>
> STATEMENT 2: Shakespeare's comedies have more "if" and "but" alongside a corresponding lack of "and" relative to his histories. Here are my observations. . . .
>
> ARGUMENT 1: Stories about dynastic conflict involve armies of soldiers and large aristocratic families, not all of whom can be represented on stage by an acting company such as Shakespeare's. The action in

such stories cannot be shown, but must rather be recounted onstage by individual characters who describe a series of events. Such narrative sequences of succeeding events must use "and," which is why being a history entails using this word more frequently.

ARGUMENT 2: If you are telling a story about a frustrated courtship, you are staging a drama of conditions, where an erotic obstacle must be overcome or an erotic wish must be acknowledged. This drama of conditions often involves statements of reticence or confusion ("if you said you love me, I would agree, but you didn't.") or statements of longing and frustration ("If this is true, so is that, but . . ."). "If" and "but" are words that advance this type of action, which is why being a comedy entails using these two words at a higher frequency.

You will notice immediately that both of these arguments involve a familiar form of humanities thinking. I am arguing that certain features are worth paying attention to because, through them, I gain access to a complicated phenomenon that rewards critical attention. When an interested argument transforms a feature into a proxy in this fashion, that argument creates what Latour would call an "object of concern," which means that any feature-proxy argument can, at least potentially, be placed on *either side* of Latour's modern division. It is important to note, however, that even in my overtly quantitative discussion, statistical evidence does not compel me to choose any particular proxy—does not compel me to assert that courtship is a negotiation of conditions, or that dynastic struggles between armies and families must be reported rather than directly seen. These things might be true regardless of what the numbers say. What is significant here, and in a decidedly nonstatistical sense, is the *link* I am making between the features and the conditions that entail their frequent (or infrequent) use.

That link must now be investigated by returning to the texts themselves to see whether or not it holds, and whatever argument is made in support of that link will probably succeed on the basis of those examples. We really are on familiar humanities ground at this point, since most argumentation in literary studies revolves around exemplary passages. I might direct your attention to this exchange from *As You Like It*, for example:

ROSALIND *[to Duke]* To you I give myself, for I am yours.
    [To Orlando.] To you I give myself, for I am yours.
DUKE SENIOR If there be truth in sight, you are my daughter.
ORLANDO If there be truth in sight, you are my Rosalind.
PHOEBE If sight and shape be true,
    Why then, my love adieu.

As You Like It 5.4.120–25[25]

Readers familiar with the play will recognize the competing erotic conflicts that converge here, encapsulated as they are by the conditional ifs that elegantly reason out the recognition scene. Four marriages will soon follow. One would need to go through and inspect many, many more instances of the word "if" in context to decide whether the conditional twists of a comic plot are indeed carried along by this feature. ("Much virtue in if," as Touchstone says.) A similar procedure would be necessary to strengthen the claim about "and" in the history plays.

You may ask why we even need statistics to identify patterns among features that we could ultimately link to proxies by serial reading alone. After all, the humanities is full of nonquantitative arguments in which features are identified and linked to some more complex phenomenon or distinction. Consider, for example, Laura Mulvey's landmark study, "Visual Pleasure and Narrative Cinema."[26] This elegant analysis of scopophelia in classic Hollywood cinema identifies an important feature—the shot that aligns the camera with the gaze of a male character, looking at a woman—and shows how it serves as proxy for the psychic needs of male spectators that organize this narrative practice. There is no more elegant example of a feature-proxy argument; no explicit counting was required to advance it.[27] But counting things will confirm statistically that a feature maps consistently onto some larger generalization, and there are many such generalizations in humanities work that could be further elaborated in this fashion.

Since at least Aristotle's *Poetics*, readers have invented names to describe differences they experience among various cultural forms. Old terms such as "comedy" and "tragedy" have been joined by newer ones—"Shakespearean," "Baroque," "scopophilic," "logocentric." Such terms orient readers and critics in conversation, highlighting certain distinctions for discussion while pushing others out of bounds. The difference computational evidence makes to such a process is that it turns our attention to features of texts that we might not ever have noticed, expanding our range of address. As a *quantitative redescription* of claims made in the humanities, a feature-proxy argument such as the one above shows that a critical commonplace like "*Twelfth Night* is a comedy" entails things we never imagined or would have thought to describe. Who, prior to this kind of work, was ready to say that "being a comedy" entails "having more 'if' and 'but,' with less 'and' than history plays"? Who would want to, you might ask? Well, anyone who wants to know what *other* descriptions we might give of the sources of our experience when it comes to plays and other literary or dramatic forms.

Drawing an analogy with photography, the computational address of, and comparison across, features within a closed corpus offers humanists a version of Eadweard Muybridge's photographs of a horse in full gallop. Because a computer can count things that we cannot attend to in our reading at speed, we learn the literary equivalent of the fact that horses really do lift all four feet off the ground mid-stride.

FIGURE 4. Eadweard Muybridge, Study of Motion, 1870s. Library of Congress Prints and Photographs Division.

Such things are worth knowing, if only because they gesture at a depth of integration that holds between the highly functional parts of our language (conjunctions) and the genre effects we want to understand. Are there other features in our language—semantic, syntactical, rhetorical—that distribute according to differences in genre, cultural milieu, performance conditions, or even human attention span? Like someone who studies ice melt as a proxy for anthropogenic climate change, the literary critic ought be interested in learning what a "deep sampling" of our corpus tells us about the circumstances that constrain, for example, what Shakespeare's characters get to say onstage. Even if a play is a confection of the human imagination, a Pegasus, one can be curious about what this "horse's hooves" are doing, what a sentence- or word-level redescription of our critical distinctions looks like.

Like a prosthetic, computationally assisted comparisons focus our attention on dynamics that remain underexplored in the texts we study; such comparisons can also recontextualize prior judgments about those texts, leading to different descriptions of things we think we know well. Consider once again the diagram showing the relative proportions of "if" and "and" in the thirty-eight plays (see figure 5).

As the earlier caption noted, we see a general separation between the comedies (red o) and histories (green +), with comedies clustering below and to the right and histories above and to the left. But while the generalization holds statistically, it is not honored in every instance. The larger circle sitting in the middle of the field of + symbols is Shakespeare's comedy *A Midsummer Night's Dream*, which puts it in the neighborhood of the history plays (again, only according to the features measured). In almost every study I have undertaken with my collaborator Jonathan Hope, *A Midsummer Night's Dream* has appeared unusual; using very different features, we too have found that this play looks, linguistically, like a history.[28] Why might this be a productive angle from which to view the play? I would want to look more closely at more passages, but I suspect that *Midsummer* is like the history plays in the following respect: both advance via action that cannot be shown on stage, only told. In the case of *Henry V*, it will be armies (or multiple agents) whose movements need to be recounted. (The Chorus asks in the opening

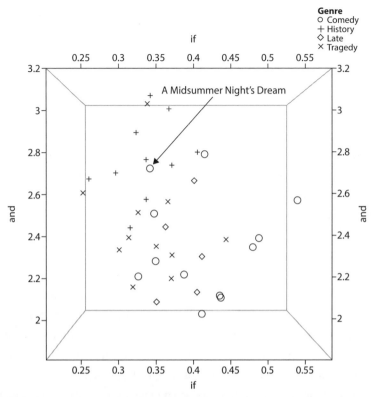

FIGURE 5. Shakespeare plays in two dimensions: head-on view of the "and–if" axis, with *A Midsummer Night's Dream* identified.

scene, "Can this cockpit hold / the vasty fields of France?") In the case of *Midsummer*, it is the action performed or seen by fairies, action that involves movements "swifter than the moon's sphere" that simply cannot be effected with actors moving physically on stage. Such superhuman movement is not necessary, however. Shakespeare conjures these effects through the magic of narrative recitation. When he does this, he is using resources he perfected while writing the early history plays, and this is something we may never have thought to explore without the exhaustive comparisons performed here. He is also responding, more immediately, to the material constraints or affordances of the early modern professional theater, which could only make use of a limited number of actors to depict historic battles and fairy antics.

Feature-proxy arguments that include statistics may also lead us to view "traditional" humanities interpretive practices from a different angle. As I mentioned above, when Shakespeare critics discuss the plays, they inevitably cite passages when they want to illustrate a point. The passage or memorable phrase is the received unit of address in current literary critical argumentation. Most gatherings of literary scholars support a brisk trade in such examples, which acquire the status of currency at professional meetings and in publications. (What would a Latourian anthropologist make of the near ritual invocation of the "To be or not to be" speech in the history of literary criticism, for example?)[29] As readers, as critics, we cannot fully enumerate the vast set of comparisons that informs our reactions to what we read, much less the grounds on which the comparisons are made. When I am writing as a literary historian or critic, I deal with that challenge by choosing passages or phrases that exemplify the underlying contrasts I want to discuss, hoping that these examples will gesture at a vast field of implicit knowledge and experiences that must remain in the background. Hamlet, I may argue, is the first truly modern individual because he has an inwardness that cannot be expressed. "I have that within which passes show," he says in act 1, scene 2, a remark I then go on to claim is exemplary of the Shakespearean treatment of inwardness.[30] The process of making and assessing exemplarity claims is staggeringly complex: it is also foundational to humanities inquiry. Indeed, I would argue that making and assessing exemplarity claims is what humanists do best. At a minimum, the fact that these claims can be recast or described statistically offers proof, if we ever needed it, of the depth and complexity of this basic interpretive activity.[31]

IV. We Have Never Been Humanists.
We Have Always Been Humanists.

We moderns may still live within a divided kingdom of knowledge. The kings who drew those borders, however, are now dead. Latour is right when he asks whether the division of valleys and rivers created by the likes of René Descartes and Robert Boyle is necessary. Strictly speaking, it is not. The times are still many, however, when the humanities are asked to demonstrate the distinctive role they play in our lives, and I see no reason to stop advocating for searching, interpretive work in our respective humanities fields. Nor do I see any reason to stop pointing out the distinctive nature of our methods, as, for example, Rens Bod has done in his excellent *A New History of the Humanities*.[32]

But there are other ways of understanding the distinction between the knowledge practices on our map, and we lose nothing by describing humanistic inquiry as one moment in a series of encounters with the world and the things that concern us—as a species, as a culture, as stakeholders in what Earl Louis has called our "yet to be perfected future."[33] It probably takes a figure like Latour to convince humanists, as he does in greater numbers, that we need not refuse certain types of allies in the work we do; that our insights will not perish under the weight of technoscience, sociology, or mathematics. Indeed, I have tried to show that the kind of complexities we negotiate, whether they stem from our humanity, history, or a locally acting nature, will only become more important as the feature set that is "human culture" receives the attention that only an algorithm can give. If what I have been arguing is correct, someone will always need to say how the features in a corpus become proxies for an object of concern. Someone will always need to put knowledge and motive *back into the diagram*, and that someone will likely come from the humanities.[34]

I would like to end with a thought experiment. Suppose, as sometimes happens in this kind of work, a computer-assisted comparison suggests that a generally accepted "truth" about a piece of literature or drama might be "wrong." In my work with Hope, for example, we found that *Othello* shares linguistic features common to Shakespeare's comedies, particularly during exchanges between the hero and his nemesis, Iago.[35] The dialogue between these two characters is full of allusion and feints; there is an indefiniteness to it—even on the level of articles—that makes these scenes read as a form of courtship. Would it make sense, then, to call *Othello* a comedy? Yes, inasmuch as the play shares linguistic features with a mode of comic dramaturgy that thrives on allusive dialogical exchange, a type of exchange that, in the case of *Othello*, is being put to very different ends. Several decades ago, a particularly good reader of

Shakespeare named Susan Snyder noticed the similarity, and this now *quantitative* redescription of her insight helps us understand how genre rules can be bent, stretched, or decisively broken.[36] When we put this piece of humanities knowledge "back into the diagram," the position of *Othello* begins to make sense. I invite you now, knowing what you know of *Othello* and Snyder's insights, to engage in the following thought experiment. Imagine that I have no access to Snyder's reading of the play. No human being has ever noticed the resemblance of *Othello* to Shakespeare's comedies, either linguistically or on the level of plot. Based solely on computational evidence, I now turn to you and say, "*Othello* looks to me like a comedy."

Would you cease to be a humanist if you agreed? If this question seems unsettling, it is because it gets at the heart of our continuing anxieties about the modernist divide and the perceived dignity of human experience as a pathway to knowledge—an experience that may indeed be increasingly marginalized by the success of technoscience. That version of self-collected experience is, ultimately, what many defenders of the humanities want to locate on one side of the divided kingdom, perhaps on the grounds that no one sees or feels a statistical pattern. (Who among us can actively attend to all of the indefinite articles in a play, poem, or novel?) But any of us can *reflect* on the interconnection between such broadly dispersed features and our own deeply personal or, conversely, deeply social experience of cultural forms. I do not think we need to police what prompts those reflections like border guards, just as I do not think literary scholars should be prohibited from using concordances or other tools that help them gather evidence about what they read. Returning to the example in my thought experiment, I would argue that it is our concern with *Othello*, not our means of understanding it, that makes the play worthy of humanistic inquiry and knowledge claims. To those who object that "we are humanists; we know only what we experience first hand," I respond: we have never been humanists, at least not if that means banishing all thoughts and perceptions that did not first originate in what Hamlet calls "the mind's eye." Or rather, we have *always* been humanists, alive to the unstateable fullness of our experience and engaged in the project of naming it in criticism.

At the current conjuncture, it is the job of the humanities to call attention to that fullness, to sustain its dignity as an object of concern, and to advance the critical project of giving it a name, by whatever means this entails.

Notes

The author wishes to thank the participants at the 2015 conference, "Recomposing the Humanities with Bruno Latour," for their responses to this work when it was presented at that conference.

1. Such divisions are not new; they were first touched on by Giambattista Vico and were well worn by the time they were popularized by C. P. Snow's *The Two Cultures and the Scientific Revolution* (New York: Cambridge Univ. Press, 1959). See Vico, *On the Study Methods of Our Time*, trans. Elio Gianturco and Donald Phillip Verene (Ithaca, NY: Cornell Univ. Press, 1990).

2. An exception to this narrative is linguistics, which has been using computational techniques for decades. See Henry Kučera, W. Nelson Francis, and John B. Carroll, *Computational Analysis of Present-Day American English* (Providence, RI: Brown Univ. Press, 1967) and Randolph Quirk's "Towards a Description of English Usage," *Transactions of the Philological Society* 59, no. 1 (1960): 40–61.

3. Bruno Latour, "An Attempt at a 'Compositionist Manifesto,'" *New Literary History* 41, no. 3 (2010): 471–90.

4. Latour, *We Have Never Been Modern*, trans. Catherine Porter (Cambridge, MA: Harvard Univ. Press, 1993).

5. Latour, "On Technical Mediation—Philosophy, Sociology, Genealogy," *Common Knowledge* 3, no. 2 (1994): 29–64.

6. Latour, "The Berlin Key or How to Do Words with Things," in *Matter, Materiality and Modern Culture*, ed. P. M. Graves–Brown, trans. Lydia Davis (London: Routledge, 2000).

7. Latour, *An Inquiry into the Modes of Existence: An Anthropology of the Moderns*, trans. Porter (Cambridge, MA: Harvard Univ. Press, 2013).

8. Some do have access to such budgets and are publishing results. See James L. Keidel, Philip M. Davis, Victorina Gonzalez-Diaz, Clara D. Martin, and Guillaume Thierry, "How Shakespeare Tempests the Brain: Neuroimaging Insights," *Cortex* 49, no. 4 (2013): 913–19.

9. Attempts to come to terms with what constitutes the "digital humanities" are many. For a good introduction, see Matthew K. Gold, ed., *Debates in the Digital Humanities* (Minneapolis: Univ. of Minnesota Press, 2012). More focused examples of corpora-based approaches to literary analysis can be found on Ted Underwood's excellent blog, The Stone and the Shell, http://tedunderwood.com. The blog I maintain with Jonathan Hope, www. winedark sea.org, Wine Dark Sea, might provide further examples.

10. On the centrality of the philological tradition to humanities practice, particularly its emphasis on repeated comparisons across contexts of reference and use, see James Turner, *Philology: The Forgotten Origins of the Modern Humanities* (Princeton, NJ: Princeton Univ. Press, 2014).

11. See Michael Witmore, "Text: A Massively Addressable Object" in Gold, ed., *Debates*, 324–27, and at http://winedarksea.org/?p=926.

12. On the possible origins of this practice, see Paul Saenger, *Spaces Between Words: The Origins of Silent Reading* (Stanford, CA: Stanford Univ. Press, 1997). Of course, words were notionally distinguishable before they became visually distinct on the page.

13. The digital encoding of texts makes this practice even more precise, for example in the "univocal" address of this same text segment using the following URL: http://www. perseus.tufts.edu/hopper/text?doc=urn:cts:greekLit:tlg0059 .tlg024.perseus-grc1:70a. Alternatively footnotes, anchored often at the ends of sentences, take as their unit of address a segment of preceding text that deserves elaboration or citational support. On the history of the footnote, see Anthony Grafton, *The Footnote: A Curious History* (Cambridge, MA: Harvard Univ. Press, 1999).

14. Erasmus, *De recta Latini Graecíque sermonis pronuntiatione Des. Erasmi Roterodami dialogus; eiusdem Dialogus cui titulus, Ciceronianus, siue, De optimo genere dicendi* (Paris, 1528).

15. Making a list is just one way to specify features. There are also probabilistic and algorithmic mechanisms for identifying features or feature sets. A topic model, which infers distinct distributions of words (topics) that could have stochastically generated a set of documents, is a mathematical artifact that no human could generate spontaneously. For a useful nontechnical explanation of topic models, see Underwood's blog post, "Topic modeling made just simple enough," The Stone and the Shell, April 7, 2012, www.te-dunderwood.com /2012/04/07/topic-modeling-made-just-simple-enough/.

16. Even the decision to choose a topic model as a source of features involves interpretive judgments. One must choose, for example, how many topics to model for a given corpus.

17. For a recent treatment of the problem, see the essays in *Computability: Turing, Gödel, Church, and Beyond*, ed. B. Jack Copeland, Carl J. Posy, and Oron Shagrir (Cambridge, MA: MIT Press, 2013).

18. The corpus, counts, illustrations, and annotated Shakespeare texts referred to in this paper are available at http://winedarksea.org/?p=2619. In the corpus of Folger Editions of Shakespeare plays, I have removed speech prefixes, act and scene divisions, and stage directions so that we can focus on the spoken words that transact events onstage.

19. Those interested in counting things in their own texts can use a text tagger, Ubiqu+Ity, created by the Mellon-funded research project, "Visualizing English Print, 1530–1800." The tagger, which supports both custom tagging

(counting specified words) and a more general feature set known as DocuScope, can be accessed at http://vep.cs.wisc.edu/ubiq/.

20. If the features were distributed randomly, there would be little reason to compare the frequencies. Whether you view them as different in kind, texts and rainforests both supply features that do not distribute randomly.

21. The ability to correctly sort plays according to such terms is an example of what computer scientists call "domain knowledge," usually the preserve of the "domain expert." When they work on other people's problems, computer scientists are leveraging the domain expertise of their collaborators, since they are not, qua scientists, expected to have this kind of knowledge themselves.

22. Hope and I have adopted these genre assignments in our work on Shakespeare's plays. We find they are generally accepted among Shakespeareans. In doing so, we are following the "Catalogue" page of the First Folio of Shakespeare's plays, which represents the domain expertise of at least two people who knew Shakespeare, and nineteenth- and twentieth-century criticism on the "late romances" or "late plays." Our "metadata," in other words, express interpretive critical judgments. It is important to note that, while the diagrams explain the statistical patterns visually, the underlying process of establishing significance claims here is mathematical.

23. The picture is not necessary to such tests. I have chosen a diagram to illustrate these ideas because they give a geometric intuition into the multivariate (i.e., multifeature) space of comparison. For a tutorial on how such comparisons are automated through a procedure known as principal components analysis (PCA), see Witmore, "Finding 'Distances' Between Shakespeare's Plays," Wine Dark Sea, June 23, 2015, and July 6, 2015, http://winedarksea.org/?p =2225 and http://winedarksea.org/?p=2271.

24. Hope and I discovered that the most pronounced difference among Shakespeare's genres is that between histories and comedies, at least when they are described in terms of frequently iterated linguistic features. See Hope and Witmore, "The Hundredth Psalm to the Tune of 'Green Sleeves': Digital Approaches to Shakespeare's Language of Genre," *Shakespeare Quarterly* 61, no. 3 (2010): 357–90.

25. Shakespeare, *As You Like It*, ed. Barbara A. Mowat and Paul Werstine, available at http://www.folgerdigitaltexts.org/html/AYL.html.

26. Laura Mulvey, "Visual Pleasure and Narrative Cinema," in *Visual and Other Pleasures* (New York: Palgrave, 1989).

27. Subsequently, Mulvey's idea was explored quantitatively in a field called cinemetrics, which identifies and measures features in film and television. For an introduction to cinemetrics, see "Movie Measurement and Study Tool Database," Cinemetrics, http://www. cinemetrics.lv/index.php.

28. See Hope's post on the play and his work with actors at Shakespeare's Globe in London, "The Very Strange Language of *A Midsummer Night's Dream*," Wine Dark Sea, http://winedarksea.org/?p=1440.

29. According to JSTOR's Understanding Shakespeare website, the Shakespearean phrase "To be or not to be" is cited by more academic essays in the humanities than any other—at least within the journals they aggregate. See https://labs.jstor.org/shakespeare/hamlet.

30. On the representation of deep subjectivity, which also goes by the name "inwardness," during the early modern period, see Katharine Eisaman Maus, *Inwardness and Theater in the English Renaissance* (Chicago: Univ. of Chicago Press, 1995).

31. The fact that unsupervised statistical techniques such as principal components analysis (PCA) often unearth dynamics in these datasets that parallel human judgments is particularly useful for introducing technical people to the complexity of what humanists do.

32. Rens Bod, *A New History of the Humanities: The Search for Principles and Patterns from Antiquity to the Present* (Oxford: Oxford Univ. Press, 2015).

33. Earl Lewis, president of the Andrew W. Mellon Foundation, uses this phrase to describe what motivates our continued search for knowledge and understanding in the humanities.

34. Automated feature selection and unsupervised statistical techniques will, without a doubt, produce more statistically significant groupings and comparisons than we can name, and already have.

35. See Hope and Witmore, "The Hundredth Psalm," 374–82.

36. See Susan Snyder, *Shakespeare: A Wayward Journey* (Newark: Univ. of Delaware Press, 2002), 29–45, and *The Comic Matrix of Shakespeare's Tragedies* (Princeton, NJ: Princeton Univ. Press, 1979), 70–74. What quantitative techniques can add to Snyder's insights is a sense of how much and how often a given play veers into generic territory other than its own.

CHAPTER FOURTEEN

Anthropotheology

Latour Speaking Religiously

BARBARA HERRNSTEIN SMITH

> To talk about religion again. No one
> appointed him, nothing marked him out, if
> not the certainty that once we modify, as he
> has done (as he thinks he has done), the
> common version of the sciences, everything
> else can start to change—first and foremost,
> religion.   LATOUR, *Rejoicing: Or the Torments*
> *of Religious Speech*

BRUNO LATOUR IS A RELATIVELY recent taste in the Anglo-American academy. He has been publishing important work in the anthropology and/or sociology of science since the 1980s but, until the past five years or so, has been greeted largely with antagonism or indifference by science and humanities faculty alike. While Latour's work (or tendentiously selected passages from it) was a prime target of science warriors in the 1990s, people in the humanities have generally found his writings too remote from current concerns to seem interesting (he has had little to say, for example, about the politics of race or gender, at least explicitly) or too closely associated with the natural sciences to seem approachable. In recent years, however, invocations of ideas and approaches associated with Latour have become commonplace, along with citations of specific texts he has authored, especially those with irresistible titles.

For many readers, Latour is most closely identified with actor-network-theory (ANT), a set of radical concepts and sophisticated methods developed originally in the sociology of science. He is also well known, especially among people in the humanities, as a subtle analyst of modernity and, more generally, as a vigorous advocate of environmentalism. Less widely known are Latour's extensive writings on religion. These include, from the 1970s, a doctoral dissertation on biblical interpretation and a related study of the early twentieth-century writer, Charles Péguy; a long essay from 1996, *Petite réflexion sur le culte modern des dieux Faitiches*, later translated as "On the Cult of the Factish Gods"; an important lecture from 2002, "'Thou Shall Not Freeze-Frame,' or, How Not to Misunderstand the Science and Religion Debate"; and, also from 2002, a small but in many ways extraordinary book, *Jubiler ou Les tourments de la parole religieuse*, recently translated as *Rejoicing: Or the Torments of Religious Speech*. Religion, or religious "being" as a specific mode of existence, figures centrally in *An Inquiry into Modes of Existence (AIME)* and, along with Nature, is one of the major categories of analysis in Latour's 2013 Gifford lectures, "Facing Gaia: A New Inquiry into Natural Religion."[1] Indeed, the hope and effort to frame a proper and—in Latour's important term—"diplomatic" account of religion, and especially of its relation to science, have been central motivating forces in his work for at least the past two decades and, in some respects, from the beginning.[2]

Virtually any reader who undertakes the serious study of Latour's writings (as distinct from casual sampling or heresy-hunting) will find them engrossing, instructive, often exhilarating and always impressive. But "the humanities" make up a very mixed package of practices in the present Anglo-American academy, and people currently working in the fields so designated make up a very mixed multitude. The ways in which any of us take up Latour's work, to "recompose" that package or otherwise, will depend, of course, on our particular assessments of those practices and on our aims and angles more generally. Additionally, because attempting to do things "with Latour" will, sooner or later, involve encounters with his religious writings and with their particular concerns and perspectives, the ways we take up his work are also likely to depend on what he would call our "attachments." A detailed examination of Latour's writings on religion is beyond the scope of this article and its occasion in this book. What I hope to do here is suggest the interest of these writings for scholars in the humanities and also to indicate the ways in which they seem likely to create problems for such readers, including or perhaps especially for longtime admirers of his work.

I.

When readers fail to understand why I have continually changed fields, and when they do not see the overall logic of my research . . . their comments amuse me, for I know of no other author who has so stubbornly pursued the same research project for 25 years, day after day, while filling up the same files in response to the same sets of questions.   LATOUR, "Biography of an Inquiry"[3]

Alluding to his successive studies of science, art, politics, and law, Latour has described his general project as "the comparative study of the various ways in which the central institutions of our cultures produce truth" or, as he also calls those ways, "truth regimes."[4] The regime on which his earlier work focuses is that of the modern natural sciences. In empirical—archival and onsite—investigations conducted in the late 1970s and early '80s and in their theoretical elaborations as actor-network-theory, Latour has sought to demonstrate that what are commonly taken as scientific truths—facts, laws, discoveries, entities—are not, as commonly assumed, fixed, prior, and given by "nature" (itself radically reconceptualized by Latour) but, rather, the contingent products of dynamic networks of multiple, heterogeneous elements. The elements include both humans—scientists, technicians, bureaucrats, and sometimes farmers or fishermen—and nonhuman agents or actors, from sick cows and virulent microbes to pulleys and petri dishes. All these are moving in different, potentially conflicting directions, and some are stronger or weaker than others; but, in laboratories and other centers of calculation and control, some elements can be linked together to form associations that are effective in serving particular human ends. It is the pragmatically effective linking of such elements that secures what we call the truth of scientific facts (for example, the microbe theory of disease or the structure of DNA) and that sustains what we experience as the reality of the entities associated with those facts (for example, microbes or genes).[5]

This constructivist-pragmatist understanding of scientific truth and knowledge reflects an increasingly commanding tradition of research and theory that extends from Ludwik Fleck's *Genesis and Development of a Scientific Fact*, originally published in 1935, to the writings of a number of mid-twentieth-century historians, sociologists, and philosophers of science and, from there, to ongoing work in the field now known as science and technology studies (STS).[6] As formulated, elaborated, and promoted in writings by, among others, Latour, it has proved compelling to increasing numbers of humanities scholars, along with researchers and theorists in the social sciences, both as a set of conceptual and methodological resources for work in their own fields and also as a well-developed alternative to still-dominant positivist views. As research and teaching in the humanities continue to involve closer connections to the natural sciences, Latour's work in this tradition can be especially important and, in regard to the earnest or aggressive scientism sometimes displayed in these developments,[7] it can be especially instructive.

There has been no radical break between Latour's early and recent work on science and no reversal in the direction of his thought. Since his "coming out as a philosopher," however, he has supplemented and, in some crucial regards, sought to supersede ANT and empirical science studies more generally with an array of speculative methods and explicitly metaphysical projects.[8] He has also been increasingly explicit about what he evokes, especially in *Rejoicing*, as his particular task or responsibility: that is, to read aright the texts and inscriptions of the religion that, as he says, "matters" to him and to translate, transmit, and make effective its message for those he calls "Moderns."[9] Latour's thirty-year-long comparative investigation of truth-regimes was pursued in good measure in the service of that task. *AIME* can be seen as the consummation of the investigation and, with the Gifford lectures, as his most valiant venture to date as missionary to the Moderns.

II.

Is existence not among the perfections indispensable for respect, which
the idea of belief never allows us to preserve? Thus I had to come back to the
crack that runs between epistemological questions and ontological questions.

The new history of the sciences has allowed me to slip in between the two. LATOUR, "On the Cult of the Factish Gods"

Contrary to routine misunderstandings of constructivist accounts of scientific facts, to be *constructed*—made, built, fabricated, put together from heterogeneous elements—is not to be *unreal*. Abstract facts, like material artifacts, are assembled and composed, but both are "real" in the sense of being, at least provisionally, stable and consequential. The same can be said of gods and other religious beings: demons and divinities, spirits and fetishes.

As Latour tells the story in "On the Cult of the Factish Gods," Gold Coast natives, scorned by European traders and invaders, insisted that certain wooden dolls—dubbed *fêtiches* by the Portuguese—were gods. The natives, Latour observes, had "constructed" something "that went beyond them." But, he asks, is this not true as well of the facts constructed by Western scientists, for example, Louis Pasteur's "ferment of lactic acid," the existence of which emerges through laboratory instruments and tests?[10] Moderns, with all the apparatus of scientific rationality, no less than supposedly primitive people with their wooden divinities, invest things that they themselves have made with a power that goes beyond them. Facts and fetishes, demons and ferments: "All ask to exist," Latour writes. "None is caught in the choice . . . between construction and reality, but each requires particular forms of existence whose list of specifications must be carefully drawn up."[11]

Fetish-gods, like scientific facts, acquire their potency—or, as it may be called, their "truth" or "reality"—within a framework of specific ideas, habits, discourses, and material apparatus; but the potency of neither can survive outside those frameworks. Whether divinities or DNA molecules (and, as Latour extends the point in *AIME*, whether cats, mats, machines, political collectives, or characters in novels), the conditions of their continued existence—he calls them "felicity conditions"—are highly specific, not always in place, and always more or less fragile. In the case of religious icons, for example, they are breakable by the acts of impassioned iconoclasts or modern "critical thinkers."[12]

The imputation of an equivalent real existence to the facts of modern science and the divinities of putatively primitive religions—or, put differently, the acknowledgment of their equivalent ontological status—is an example of what Latour calls "symmetrical anthropology." He explains its method and aim in the essay: "By taking the most respected beings of a culture—our own—as examples, we can shed light on the most despised beings of another culture."[13] The most respected beings of our own culture are scientifically established facts and entities. The most despised beings are African fetish-gods, the demons afflicting the immigrant patients of a French ethnopsychiatrist and, not quite "of another culture," the Virgin as sighted at Lourdes. Latour's symmetrical anthropology can be seen as due scientific impartiality, as a generous exercise of the sympathetic imagination, or, perhaps, as practicing relativism with a vengeance. It can also be seen as a sophisticated elaboration of the rhetorical move known, especially in theological circles and in response to derisive iconoclasms, as *tu quoque*: "You, too! the supposedly enlightened ones: *you* do just what you scorn *us*, the supposedly benighted ones, for doing."

In describing the facts of modern science symmetrically with religious beings, Latour does not seek to demote the authority of the truth-regime of Western science. What he seeks to demote—indeed, to undo utterly— is a set of dichotomies and commonly skewed dualisms that have become central to modern Western thought: nature as divided from society, objects as divided from subjects, real as opposed to manmade or constructed, and existent as opposed to (merely) believed-in. But of course, and not incidentally, he thereby promotes the epistemic dignity of the experiences of those who fear demons or see visions of the Virgin, and the ontological dignity of those beings themselves.

In a classic constructivist treatment, our experience of the truth of scientific facts and the reality of visions of divinities would be understood in terms of more general, largely social-psychological dynamics. Thus Fleck, in *Genesis and Development of a Scientific Fact*, describes the complex processes involved in the formation and stabilization of what he calls "belief systems," with religious doctrines and scientific paradigms, along with political and other ideologies, as examples. In

Fleck's account, the coherence and stability of all such systems are preserved through the ongoing mutual adjustment of the perceptions, prior beliefs, background assumptions, and shared material practices of the interacting members of a social group or "thought collective." Fleck called the resulting shared sense of the truth of some fact or doctrine among the members of such a group a "harmony of illusions"—illusions not in the familiar and itself dubious sense that there was some otherwise verifiable set of objective facts that contradicted them, but insofar as that sense of truth was projected outward and regarded as an objective correspondence of idea and world.

Latour has repeatedly expressed admiration for Fleck's work, and the affinities of their respective accounts of facts and truth are evident.[14] The detailed historical-sociological narrative of the establishment of the microbe theory of disease in *The Pasteurization of France* closely parallels Fleck's narrative, in *Genesis and Development*, of the establishment of the Wassermann test for syphilis, including the way a key pathogen is coaxed into existence in the laboratory. Crucial to Fleck's accounts, however, is an analysis of the social-psychological dynamics involved whereas Latour rejects explanatory appeals to the psychological and, in *AIME*, banishes the term "belief."[15] Also, significantly, while both reject table-thumping empiricisms in favor of constructivist understandings of facts and truths, Latour invokes a rather obscurely defined "second empiricism" to ground *AIME*'s ontologies. These differences—as much matters of intellectual project and genre as of philosophical position—mark an important space between the tradition of science studies with which Latour's work has been associated and his recent writings, those on religion and more generally.

III.

There exists a form of original utterance that speaks of the present, of definitive presence, of completion, of the fulfillment of time . . . ; a form of speech whose sole characteristic is to constitute those it is addressed to as being close and saved; a kind of vehicle that differs absolutely from those we've evolved elsewhere to accede to the distant in order to control information about the world.   LATOUR, *Rejoicing*[16]

The perennially disputed relation between the truths of science and those of religion is addressed directly in Latour's essay, "'Thou Shall Not Freeze-Frame,' or, How Not to Misunderstand the Science and Religion Debate." Originally a talk for a lecture series titled "Science, Religion and the Human Experience," the essay offers a set of formulations regarding that relation that Latour develops in detail in *Rejoicing* and iterates in more recent writings. The essay also involves, contra iconoclasts of all persuasions, a crucially revised interpretation of the biblical commandment prohibiting images. Rhetorically reflexive throughout, the essay is, among other things, a mock (but not mocked) sermon. Latour writes: "Religion, at least in the tradition I am going to talk from, namely the Christian one, is a way of preaching, of predicating, of enunciating truth in a certain manner—this is why I have to mimic in writing the situation of an oration given from the pulpit" (TNF 28).

Latour begins with a strong contrast between "speaking religiously," evidently as in prayer or ritual utterance, and what he calls "double-click communication," that is, the idea or ideal of an unmediated transfer of information. The truth of a double-click message, if any such existed, would be its exact correspondence to an objectively determinable state of affairs. Religious speech acts, on the other hand, "transport" not information but persons. In religious speech as in love talk, what attests to the truth of an utterance is not its correspondence to some putatively objective reality but its renewal of speakers' and hearers' confidence in the reality of something vital: a sense of closeness; a promise of futurity (TNF 29–31). Here as elsewhere in Latour's writings on religion, claims are put forth largely through analogy, allusion, and intimation—which is not untypical, of course, of theological arguments or sermons.

Clearly, Latour observes, it would be improper, what he calls a "category mistake," to judge the truth of a religious speech act using double-click communication as a measure.[17] Just as it would be wrong to maintain that sentences such as "I love you" have no truth value just because they possess no informational content, it is wrong, in seeking to understand the angel Gabriel's salutation to the Virgin, to ask who Mary was, to ponder "whether or not she was really a Virgin," or to imagine that she might have been impregnated with "spermatic rays."

"Paradoxically," Latour writes, "by formatting questions in the procrustean bed of information transfer so as to get at 'exactly' what it meant, I would have *deformed* it, transmogrified it into an absurd belief, the sort of belief that weighs religion down and lets it slide toward the refuse heap of past obscurantism" (TNF 33). In *Rejoicing*, Latour describes—at length and with considerable scorn—religious scholars' efforts to explicate New Testament texts so as to make them more reasonable-sounding, more conformant to historical data or otherwise palatable to intellectual tastes corrupted, as he sees it, by Double Click (here and elsewhere personified and often associated ironically, or maybe not so ironically, with "the Evil One"). He continues in the essay: "The only way to understand stories such as that of the Annunciation is to *repeat* them, that is to utter again a Word which produces into the listener the same *effect*," one that "impregnates . . . with the same gift, the same present of renewed presence. Tonight, I am your Gabriel!" (TNF 33).

Seeking explicitly to evoke the power and effects of religious transmission, Latour turns from verbal to visual representation and comments on a set of strong images from Christian iconography. We do not, or should not, assess such images, he observes, by their fidelity to presumed true originals. Nor should we isolate or "freeze-frame" them from the flow of mediating representations that enable their truths to be realized (this being Latour's revision of the second commandment). He goes on to stress the comparably vital role of relays of inscriptions, images, and other representations in science (reports, charts, photographs, mathematical formulae, and so forth). "Truth," Latour writes, "is not to be found in correspondence—either between the word and the world in the case of science, or between the original and the copy in the case of religion—but in taking up again the task of *continuing* the flow, of elongating the cascade of mediations one step further" (TNF 46). The commonly supposed objective realities behind genes or the microbe theory of disease are like the mistakenly supposed "originals" of representations of the empty Sepulcher or of the arresting thorn-crowned face of Jesus in a trompe-l'oeil painting of the Veronica veil. In all these, what matters, what sustains the truth of the events and the reality of the figures in question, is the continuity of the practices of representation that mediate their existence.

Elaborating these points in the essay's concluding pages, Latour observes, in what operates as an important and continuing distinction, that, while the mediating chains of reference that secure the truths of science are counterparts to the flows of utterances and images that convey the truths of religion, the relays in each go "in two different directions" (TNF 46). In science, they bring what is far close (for example, through astronomical photographs, charts, and models), but religious texts and images bring us to what is near—our neighbor and our salvation.

Because Moderns have worshipped the false idol of Double Click, Latour maintains, they have misunderstood—indeed, reversed—how truth and reality are secured both in science and in religion. To correct what he calls this "comedy of errors," he offers a set of alternative characterizations of religious belief and scientific knowledge that are central to *AIME* and repeated, with variations, in his Gifford lectures. "Belief," he writes, "is not a quasi-knowledge question *plus* a leap of faith to reach even *further* away; knowledge is not a quasi-belief question that would be answerable by looking directly at things close at hand." Rather, a leap of religious faith "aims at jumping, dancing towards the present and the close, to redirect attention away from indifference and habituation." Conversely but comparably, knowledge in science "is not a direct grasp of the plain and the visible . . . but an extraordinarily daring, complex, and intricate confidence in chains of nested transformations of documents that, through many different types of proofs, lead toward new types of visions that force us to break away from the intuitions and prejudices of common sense" (TNF 45–46). The parallels and reversals in this set of comparisons are striking. Simultaneously vague and enthusiastic, they join an evocation of the most familiar and accessible experiences of religious faith to a celebration of the most heroic activities and exalted achievements of science while maintaining a sharp distinction between the two. They are nothing if not diplomatic.

IV.

In seeking to frame an account of the relations between science and religion that is both generally acceptable and also corrective of what he sees as past philosophical and theological errors, Latour has taken on a task that is immense and, as suggested in *Rejoicing*, variously—certainly rhetorically and perhaps, for Latour, conceptually as well—"tormented." Such an account must negotiate steep differences of view between Moderns, many of them invested in conventionally celebratory views of science and some of them scornfully antireligious, and Christian communicants, many of them invested in conventionally orthodox religious views and some of them resentfully anti-science. Thus, while secular-minded readers may welcome a theology that claims neither supernatural nor substantial status for its god(s) and that segregates religion from politics and morality, communicants might feel that something essential has been lost in the negotiations. Accordingly, Latour's accounts of religion vis-à-vis science operate with a good bit of euphemism, circumlocution, studied vagueness, and, it could be said, equivocation. For example, while Latour derides familiar theological allusions to realms "above" or "beyond" the natural or the material, the apparent heterodox force of such gestures is considerably defused by his equally strong efforts to undermine familiar understandings of "nature" and "matter." Similarly, while he seems to suggest that religion is immanence all the way down and all the way up, too, it is not surprising that fellow faithful sense, in his texts, assurances of something like orthodoxy.[18]

To speak religiously to Moderns, Latour has tied together a theoretically sophisticated account of scientific knowledge with a rhetorically deft Christian apologetics to forge a singular quasi-symmetrical anthropotheology. The writings that compose it are bold, inventive, and in many ways compelling. Structurally and stylistically, *Rejoicing, AIME,* and related essays are remarkable works of lyric philosophizing, recalling works by Kierkegaard and, in their strong personal voice, Nietzsche. Fellow theologians are likely to be most appreciative of the originality of their formulations and also most closely attuned to their distinctive idioms.[19] Other readers will find them a rich resource for ongoing, reprised, or newly conceived scholarly projects. Historians and theorists of Western modernity will profitably engage with Latour's theologically inflected takes on law, politics, and economics. Those in literary and visual studies will appreciate his suggestive accounts of the re-presencing effects of texts and images, religious and otherwise. And humanities scholars of all stripes will be delighted by passages of an order of wit and literateness—vernacular as well as erudite—not often encountered in the pages of theologians, not to mention social scientists. Readers and scholars in all these fields, however, are likely to be perplexed by various aspects of these writings and to find them, to various extents, intellectually or experientially alien.

V.

I am not going to speak of religion in general, as if there existed some universal domain, topic, or problem called "religion" that could allow one to compare divinities, rituals, and beliefs from Papua New Guinea to Mecca, from Easter Island to Vatican City. A person of faith has only one religion, as a child has only one mother.   LATOUR, "'Thou Shall Not Freeze-Frame'"[20]

In his writings on religion, Latour has been concerned with a relatively confined set of aspects of a vast and multifaceted subject.[21] The focus is on religious representation and utterance, which, in *Rejoicing* and related essays, are identified largely with Christian iconography, New Testament texts, and the verbal practices of Catholic communicants. In *AIME*, the religious mode of existence is explicitly restricted to the beings of Christianity while demons, ghosts, fetish-gods, and other exotic divinities are assigned to a separate, somewhat obscurely described mode labeled "metamorphosis." Beings of the latter kind are sustained not, as in religion-proper,[22] by flows of sacred texts and images, but by a process that Latour calls "psychogenesis"—associated with shamans, exorcism, psychotropic drugs, and psychoanalysis—and explains as "the exterior production of interiorities." Also, strikingly, no other major religious tradition is mentioned in *AIME*'s five-hundred-page-plus "Anthropology of the Moderns." Writing as a professed Catholic, Latour could not be expected to deal with other faiths in the same manner or detail as he deals with Christianity. Nevertheless, readers are likely to miss some acknowledgment of the existence of other religious traditions and also of their variety, both as observed and as experienced.[23]

Experience carries a great deal of weight in *AIME*. The inquiry's method, "a second empiricism," is, Latour explains, a developed or extreme version of William James's "radical empiricism": that is, the inclusion of nothing that is *not* in experience and the exclusion of nothing that *is*.[24] Moreover, the test of the truth of its accounts of Modern values is, he tells readers, the accord of those accounts with their own experience. With regard to religion, however, the appeals to experience are highly selective and readers may find them otherwise thorny.

Some of the difficulties can be seen in the following passages, in which Latour specifies the mode of existence of "religious beings"—that is, the beings of Christianity, also identified as "the beings sensitive to the Word"—and explains their categorical differences from what he calls "the beings of metamorphosis."

> Religious beings . . . are truly beings; there's really no reason to doubt this. They come from outside, they grip us, dwell in us, talk to us, invite us; we address them, pray to them, beseech them.
>
> By granting them their own ontological status, we can already advance quite far in our respect for experience. We shall no longer have to deny thousands of years of testimony; we shall no longer need to assert sanctimoniously that all the prophets, all the martyrs, all the exegetes, all the faithful have "deceived themselves" in "mistaking" for real beings what were "in fact nothing but" words or brain waves. . . . .
>
> It appears infinitely simpler, more economical, more elegant, too, to stick to the testimony of the saints, the mystics, the confessors, and the faithful, in order to direct our attention toward *that toward which* they direct theirs: beings come to them and demand that they be instituted by them. But these beings have the peculiar feature of *appearing* to those whose souls they overwhelm in saving them. . . .
>
> If we are to be empirical, then, these are the ones we must follow. . . .

Like the beings of metamorphosis, religious beings belong to a genre "susceptible to being turned on and off." With one difference: if they appear—and our cities and countrysides are still dotted with sanctuaries erected to harbor the emotions these apparitions have aroused— they *disappear* even more surely. Moreover, this intermittence has provided the basis for mockery, and has been taken as proof of their lack of being; the critical spirit has not held back in this regard. But the big advantage of an inquiry into modes of existence is that it can, on the contrary, *include* this feature in the specifications: one of the characteristics of religious beings is that *neither their appearance nor their disappearance can be controlled.* (*AIME* 308–309)

This seems to be saying that the existence of the beings of religion proper is (only) in the particular experiences of those who experience such beings and that the reality of their existence is secured by our agreeing—out of respect for those experiences—not to question that reality. It also seems to be saying that, in spite of evident similarities, the invisible beings proper to Christianity cannot exist in the same manner as the invisible beings of other religions because only the former conform to what Christianity teaches about such beings.[25] The advantage noted here ("the big advantage of an inquiry into modes of existence is that it can . . . *include* this feature in the specifications") is that the person conducting such an inquiry can specify as a singular feature of the ontology of the beings of his own religion—and, indeed, as a manifestation of their autonomous power (that is, to appear and disappear *uncontrollably*)—what might otherwise be taken as their compromised reality: that is, the nondemonstrability of their existence and the fitfulness of their presence even in the experience of the faithful.

There is, clearly, no arguing with the structure or elements of an ontological claim of this kind. Readers not party to the type of stipulative logic involved may feel there is something hocus-pocus about it or note the apparent self-affirming circularity. Latour, however, defends its rationality strenuously: "I hope the reader will do me justice on this point: not once in this inquiry have I required anyone to give up the most ordinary logic; I have only asked that, with the *same* ordinary reasoning, the same natural language, they follow *other* threads. [The beings of religion] are rational through and through. Like psyches. Like fictions. Like references" (*AIME* 307). And, in any case, one may find it hard not to be charmed by a universe emptied of "matter" and animated by invisible beings flitting among souls, sliding among the pages of old books, in company with Heathcliff and perhaps Athena, as real as quarks and as reasonable as cats or mats.

VI.

If I still dare speak, it's only because I think I can brush aside the shadow that the ways of science once cast over the ways of being produced by religion.   LATOUR, *Rejoicing*[26]

Labeled an "inquiry," *AIME* can be seen as Latour's final report on his thirty-year-long comparative investigation of truth-regimes. Aspects of the inquiry, however, clearly had foregone conclusions, a number of which operate as axioms or, in Latour's term, "pre[-]positions," that is, as proper attitudes taken or given in advance. The sharp distinction and mutual incommensurability of the modes of veridiction of science and religion appear to be axioms of this kind. As set forth in *AIME*, these features obtain across the board: all truth-regimes involve distinct modes of existence, which themselves involve distinct discursive tonalities, interpretive keys, and modes of veridiction. One of *AIME*'s central conclusions (or givens), however, is that Moderns have brought much unhappiness upon themselves, the rest of humanity, and the rest of creation through their confusion of the truth-regimes associated with science and religion *in particular* and through their failure to respect the differences between the respective interpretive keys and tonalities of each.

One may agree: there is something tone-deaf in seeking to establish the truth of the Annunciation the way one might that of a theory of biological evolution. One can also see the broad advantages of maintaining a clear distinction between the modes of veridiction associated with religions and the natural sciences: it protects visions of the Virgin from dismissal in terms of empirical facticity and evidence regarding Jupiter's moons from dismissal in terms of scriptural or ecclesiastical authority.

Indeed, a strict partition of "science" and "religion" has obvious benefits for both, as demonstrated by the recurrent efforts of advocates or defenders of each to establish one.[27] Nevertheless, in view of the close, extensive, and formative connections between the development of the modern Western sciences and the institutions of religion, one must question the extent to which their respective discursive tonalities or even truth-regimes can be distinguished, certainly historically and, in some regards, currently as well.[28] And, in view of the exceptionally heterogeneous and continuously shifting contents of the packages of ideas, practices, institutions, and communities that have been and could be assembled under each of these terms, "science" as well as "religion" (and the latter even if confined to Christianity), one must question the conceptual coherence and practical workability of any claim about the fundamental nature of either of them or of their relationship.

When Moderns "start talking about the 'conflict between Science and Religion,'" Latour writes, "they act as though it were a matter of opposing (or 'reconciling,' which is worse) two types of approach: one that would give us Matter, the 'here below,' the rational, the natural, and one that would offer us the spiritual, the beyond, the supernatural, the supreme values!" (*AIME* 322). The sort of opposition and/or reconciliation Latour describes here is familiar in the idea of "nonoverlapping magisteria," as proposed by biologist Stephen Jay Gould.[29] In Gould's division, authority over the realm of facts and accounts of the natural world is claimed for science while religion is granted authority over the realm of values along with instruction in moral conduct. Gould's apportionment of the epistemic and moral universe is endorsed by many scientists, who believe they have the best of the bargain, and is also accepted by many theologians, happy to be granted clear title to a piece of the territory.[30] Of course, partitions like Gould's perpetuate what Latour identifies as key problematic dualisms of Modern thought: facts and values, matter and spirit, nature and culture. But Latour has sought only to challenge the terms in which those partitions have been drawn, not their existence as such. Few contemporary theorists have been more alert to the problems of conceptual segregation than Latour or devoted as much energy to exposing the dubious divides of Western thought. To the extent, however, that *AIME* depicts "science" and "religion" as distinct and counterpoised monoliths, its revisionist ontology, even as it discards familiar dualisms or significantly redistributes their traditionally defining elements, goes some distance toward perpetuating one of the most dubious of them.

VII.

I've got better things to do than to portray the ups and downs of the children of last century: things like altering the arrow of progress[,] . . . giving another meaning to the long history of the West, doing away with modernization.   LATOUR, *Rejoicing*[31]

Latour does not claim to be a historian, but his work involves a good bit of historiography as well as important theorizing about historicity and temporality.[32] *The Pasteurization of France* is, among other things, a history of the emergence of modern theories of disease; *We Have Never Been Modern* is, of course, a thorough overturning of modernity's self-flattering autobiography; *Rejoicing* relates the successive efforts of Christian theologians to meet the successive challenges of rationalism, both classical and modern; and both *AIME* and the Gifford lectures involve significantly revised versions of major chapters of Western social, political, and intellectual history.

The fields and approaches that make up science studies, including actor-network-theory (ANT), are programmatically anti-whiggish. They reject familiar heroic-progressivist narratives of the history of science and comparable manifest destiny accounts of the history of technology. Moreover, they tell very different *kinds* of stories about both. ANT's defining method is the slow, careful tracing of the construction of contingent networks of multiple, heterogeneous, complexly interrelated elements. While ANT accounts register practical successes and failures, they do not score the ideas and artifacts whose construction they narrate as intrinsically grand or foolish, nor do they portray the human agents whose efforts they follow as blind or faithful to (the) truth.

In his role of missionary to the Moderns, Latour sets aside this commitment to symmetrical historiography. Seeking to "[alter] the arrow of progress," he flips it around to point backward. Where Latour's Moderns tell of a rise from darkness and superstition through Reason and Science, he tells of a fall from unity and faith through the embrace of those very (misunderstood) values. His tale is of a community assembled by a salvific message; of the entrance of malign forces offering knowledge and power; of the folly and fumbles of leaders; of a message obscured, a people left wandering, and a land in ruin; and of the chance, perhaps, of redemption and renewal.[33]

The tale is old and familiar. To be sure, that is no reason to dismiss it. Nevertheless, the idea of the Scientific Revolution and the European Enlightenment as catastrophes for humanity is likely to be resisted by many of Latour's academic readers, including—and in spite of their shared sense of the ills of modernity—a good number of us in the humanities. It is not that such readers endorse familiar celebratory accounts. Intellectual, literary, and social historians, along with political theorists, are more likely to regard both developments, along with the Protestant Reformation, the Industrial Revolution, and other chapters in standard histories of modernity, as very mixed bags with very complex and variously operating ingredients. It is, rather, that we have learned to be skeptical of myths of a Fall, whether into Technology, Commerce, Individualism, or Fragmentation, and also of moralized histories, whether triumphal or nostalgic.[34] Many of us are inclined to see not only the twentieth century but also the past two millennia and perhaps the entire history of humanity as a long series of, precisely, "ups and downs": of local gains and losses, dominances and defeats; of emergences and extinctions both large and small; but not of globally grand triumphs and/or great botches, in either order. And many of us find the idea of modernity, or "the secular age," or any age, as a "parenthesis" in human history—as if an interruption or aberration—very peculiar.[35] "But, of course!" Latour might exclaim. "That is because you are Moderns—or worse, Postmoderns!"

It is true: many of us are, to various extents, one or the other of these or both of them. Insofar as we are, even as we appreciate and appropriate Latour's work around the clock, we will be troubled and more or less alienated by an image of the West in which critical thought is cast as the enemy of the ways to truth, and the fools and knaves of intellectual history are named Galileo, Hume, Kant, Voltaire, Émile Durkheim, Sigmund Freud, and Jacques Derrida.

VIII.

Psychology is to the subject what epistemology is to the object. One must be countered as forcefully as the other in order for experience to be tracked. LATOUR, *AIME*

In spite of the homage that he pays to William James, Latour rejects the relevance of psychology to the understanding of experience, including religious experience.[36] The field or, rather, fields of psychology (there are, of course, many specializations and variants) have a lot to answer for in the way of simplistic accounts of, among other things, the nature, sources, and effects of religious beliefs and experiences. There are, however, quarters of these fields where the assumptions of classic epistemology are rejected as strenuously as Latour rejects them (and for many of the same reasons) and where questions of subjects, psyches, and persons are approached in ways that accord closely with his own elaborated views of them. The relevant approaches, called, variously, "nonrepresentational," "ecological," "embodied," or "enactive," also suggest ways to understand beliefs—religious, scientific, and other—that do justice to their complex phenomenological dynamics.[37]

Latour is comparably insistent that the religious mode of existence, and religion as such (or at least Christianity), cannot be approached by the social sciences more generally:

There is a risk, obviously, that [the] requirement to treat religion rationally will be mistaken for a return to the critical spirit, that is, to the good old "good sense" of the social sciences. But it should be clear by now that we can expect nothing at all from the "social explanation" of religion, which would amount to losing the thread of the salvation-bearers by breaking it and *replacing it with another*, while seeking to prove that "behind" religion there is, for example, "society," "carefully concealed" but "reversed" and "disguised." Such an "explanation" would amount to losing religion. There is nothing "behind" religion . . . since each mode is its own explanation, complete in its kind. (*AIME* 307)

But, of course, religion, religious experience, and the related operations of mediation that Latour describes in *AIME* in ontological terms can be and have been described otherwise, by no means always either reductively or critically. Ethnologically and historically informed accounts of religious ideas, practices, and institutions, Christian and other, along with subtle explorations of religious subjectivities, have been produced for more than a century by anthropologists, classicists, and other scholars of religion who have shown no interest in exposing anything "behind" the objects of their study or inclination to mock anything within them. If Latour makes little use of these accounts, it is not because he is unaware of them. It is because they are irrelevant to what he has taken to be his task. For Latour, to "speak well" of religion—that is, of Christianity—is to speak of it religiously, which means in its traditional scriptural, theological, and homiletic idioms *and not otherwise*. The propriety or cordiality required here is less a matter of language than of attitude and, indeed, of attachment. The attitudes and attachments of a person of faith occupying the role of communicant or theological apologist are crucially different from those of a scholar of religion comparing practices from Papua New Guinea to Vatican City (though they may, in fact, be the same person). *Emic* and *etic*, inside and outside, the experienced and the observed: the differences between them cannot be bridged; they can only be finessed.[38] This, the "hard problem" of the philosophical tradition, is also the hard problem of anthropotheological diplomacy.[39] As simultaneously anthropologist of and missionary to the Moderns, Latour has attempted to solve or negotiate it by forging an original idiom—a way of speaking—that joins compelling evocations of religious experience to passionate theorizing in the service of a prophetic summons to worldwide conversion. There is good reason to think the mission will fail. What has been constructed along the way, however, will reward our exploration for some time to come.

Notes

1. "On the Cult of the Factish Gods," in Bruno Latour, *On the Modern Cult of the Factish Gods*, trans. Catherine Porter and Heather MacLean (Durham, NC: Duke Univ. Press, 2010), 1–66; Latour, "'Thou Shall Not Freeze-Frame,' or, How Not to Misunderstand the Science and Religion Debate," in *Science, Religion and the Human Experience*, ed. James D. Proctor (New York: Oxford Univ. Press, 2005), 27–48 (hereafter cited as TNF); Latour, *Rejoicing: Or the Torments of Religious Speech*, trans. Julie Rose (Cambridge: Polity, 2013); Latour, *An Inquiry into Modes of Existence: An Anthropology of the Moderns*, trans. Porter (Cambridge, MA: Harvard Univ. Press, 2013) (hereafter cited as *AIME*). The Gifford lectures can be accessed at the University of Edinburgh website, http://www.ed.ac.uk/humanities-soc-sci/news-events/lectures/gifford -lectures/archive/series-2012-2013/bruno-latour.

2. See Latour, "Biography of an Inquiry: On a Book about Modes of Existence," *Social Studies of Science* 43, no. 2 (2013): 287–301. He writes: "In this article, I would like to . . . recount the chaotic emergence of a systematic argument whose persistence over more than 30 years is surprising even to me" (288). The argument concerns the relation of the problematic truth of religious texts to the privileged truths of the sciences. For Latour's early work and studies in theology, see Henning Schmidgen, *Bruno Latour in Pieces: An Intellectual Biography*, trans. Gloria Custance (New York: Fordham Univ. Press, 2015), 11–19.

3. Latour, "Biography of an Inquiry," 288.

4. Latour, *The Making of Law: An Ethnography of the Conseil d'État*, trans. Marina Brilman and Alain Pottage (Cambridge: Polity, 2009), ix.

5. The account summarized here is initially developed in Latour, *Laboratory Life: The Construction of Scientific Facts*, coauthored with Steven Woolgar, 2nd ed. (Princeton, NJ: Princeton Univ. Press, 1986); *Science in Action: How to Follow Scientists and Engineers Through Society* (Cambridge, MA: Harvard Univ. Press, 1987); *The Pasteurization of France*, trans. John Law and Alan Sheridan (Cambridge, MA: Harvard Univ. Press, 1988). It is further elaborated and elucidated in Latour, *Pandora's Hope: Essays on the Reality of Science Studies* (Cambridge, MA: Harvard Univ. Press, 1999), and *Reassembling the Social: An Introduction to Actor-Network-Theory* (Oxford: Oxford Univ. Press, 2005).

6. For a good account of the tradition, see Jan Golinsky, *Making Natural Knowledge: Constructivism and the History of Science* (Cambridge: Cambridge Univ. Press, 1998).

7. For examples and discussion, see Barbara Herrnstein Smith, "Scientizing the Humanities: Shifts, Collisions, Negotiations," *Common Knowledge* 22, no. 3 (2016): 353–72.

8. See Latour, "Coming Out as a Philosopher," *Social Studies of Science* 40, no. 4 (2010): 599–608.

9. Latour's usage of the term "Moderns"—and, in connection with it, either "we" or "they"—varies widely, and the specific reference of the term in his work

tends to be elusive. Most generally and neutrally, it seems to mean something like *(we) educated, post-Enlightenment more or less secularized Westerners.* Throughout his writings, however (most crucially and influentially in *We Have Never Been Modern*), and in tones ranging from affectionate irony to bitter sarcasm, Latour depicts the members of this group (or, in *AIME*'s ethnographic conceit, "tribe") as fundamentally benighted, self-ignorant, and arrogant: mistaken about the constitution of their world, mistaken about their own motives and values, and given to airs regarding those they regard as unenlightened. Since specific examples of individual Moderns (historical or contemporary) in Latour's work are few and far between, readers will be inclined to supply them from their own knowledge, experience, or imagination in accord with their own sense of Western or human history and their more general intellectual and/or cultural tastes and distastes.

10. Latour, "On the Cult of the Factish Gods," 16.

11. Latour, "On the Cult of the Factish Gods," 45.

12. See Latour and Peter Wiebel, eds., *Iconoclash: Beyond the Image Wars in Science, Religion, and Art* (Cambridge, MA: MIT Press, 2002). The recurrent quote marks are Latour's.

13. Latour, "On the Cult of the Factish Gods," 45.

14. See, for example, "Transmettre la syphilis, partager l'objectivité," Latour's postface to the French translation of *Genesis and Development, Genèse et développement d'un fait scientifique*, trans. Nathalie Jas (Paris: Les Belles Lettres, 2005), and *AIME* 91.

15. What Latour would banish is not the term "belief" as such (he acknowledges its innocuous usages) but its invidious or patronizing invocation, especially in relation to religious ideas. Thus his efforts "to slip in between" what he calls "epistemological questions and ontological questions" are related to his need, in establishing the respect-worthiness of divinities, to escape the choice between a dubious claim of objective existence for such beings and an unwanted ascription of their existence to ("mere") subjective belief. See the epigraph to this section, *AIME* 13, and the entry on BELIEF in the online glossary, http://modesofexistence.org/inquiry/?lang=en#b[chapter]=#3&b[subheading]=#41&a = SEARCH&c[leading]=TEXT&c[slave]=VOC&s=0&q=belief.

16. Latour, *Rejoicing*, 118.

17. Latour stresses that it is also a mistake, though of a different kind, to appeal to a putative correspondence-to-reality to explain the efficacies of the natural sciences.

18. See, for example, Tim Howles's rejoinder to Jan Golinski's appreciative but distanced reading of "'Thou Shall Not Freeze-Frame'" (Golinski, "Science and Religion in Postmodern Perspective: The Case of Bruno Latour," in *Science and Religion: New Historical Perspectives*, ed. Thomas Dixon, Geoffrey Cantor, and Stephen Pumfrey [Cambridge: Cambridge Univ. Press, 2010], 50–68). Howles writes: "Religious people, Golinski thinks, 'will still want to insist on the

ontological reality of the things they believe in and will not be happy to have their religion reduced to the manipulation of signs that lack any reference to the real world.' . . . However, in the light of this chapter [i.e., chap. 11 in *AIME*], I suggest we can put Golinksi's claim to bed as unfounded. Latour does not lead us into the realm of apophatic theology and the beings of [REL] are not to be taken as merely Feuerbachian projections. There is ballast to Latour's theology." (AIME Research Group site, http://aimegroup.wordpress.com/2014/07/08 /chapter-11-welcoming-the-beings-sensitive-to-the-word/#more-145.)

19. See, for example, the appreciative account of these writings by theologian Adam S. Miller, *Speculative Grace: Bruno Latour and Object-Oriented Theology* (New York: Fordham Univ. Press, 2013). Miller's style, like Latour's, is highly allusive and, in Miller's case, also exceedingly gnomic.

20. Latour, "'Thou Shall Not Freeze-Frame,'" 28.

21. The scope of the concept "religion" and the meanings of the term are, of course, extensively contested. On the concept, see Benson Saler, *Conceptualizing Religion: Immanent Anthropologists, Transcendent Natives, and Unbounded Categories* (Leiden, The Netherlands: Brill, 1993); Daniel Dubuisson, *The Western Construction of Religion: Myths, Knowledge, and Ideology*, trans. William Sayers (Baltimore: Johns Hopkins Univ. Press, 2003). On the term, see Jonathan Z. Smith, "Religion, Religions, Religious," in *Critical Terms for Religious Study*, ed. Mark C. Taylor (Chicago: Univ. of Chicago Press, 1998), 269–84.

22. My term "religion-proper" here and below refers to the mode that Latour names "religion" and identifies, usually indirectly but always recognizably, with Christianity.

23. Latour's references to non-Catholics or non-Christians tend to join them together in terms that are, at best, vague: for example, "those outside" versus "those inside" the Church or, sometimes, "the indifferent" versus "the faithful."

24. In a note on EMPIRICISM in the online *AIME*, Latour writes: "Rereading James allows us to take radical empiricism as a watchword, but the phrase 'radical empiricism' takes on a more developed sense in *AIME*. . . . *AIME*'s radicalism is even more extreme." http://modesofexistence.org/inquiry/?lang =en#a=SEARCH&s=0&q=Empiricism.

25. "There is a constant risk," Latour writes, "of interpolating, confusing the two, failing to respect the contrasts. To care for is not to save. To initiate the circulation of psychogenics is not at all the same thing as letting oneself be overwhelmed by angels" (*AIME* 304).

26. Latour, *Rejoicing*, 122.

27. For samples of such efforts, see Paul Kurz, ed., *Science and Religion: Are they Compatible?* (Amherst, NY: Prometheus, 2003).

28. On the historical connections, see John Hedley Brook, *Science and Religion: Some Historical Perspectives* (New York: Cambridge Univ. Press, 1991); Peter J. Bowler and Iwan Rhys Morus, *Making Modern Science: A Historical Survey* (Chicago: Univ. of Chicago Press, 2005) 341–66; Peter Harrison, *The Territories of Religion and Science* (Chicago: Univ. of Chicago Press, 2015). For

continuing connections, see David F. Noble, *A World Without Women: The Christian Clerical Culture of Western Science* (New York: Knopf, 1992).

29. See Stephen Jay Gould, *Rocks of Ages: Science and Religion in the Fullness of Life* (New York: Ballantine, 1999).

30. On the problems with Gould's partition, see B. H. Smith, "Science and Religion, Lives and Rocks," *New York Times*, January 25, 2010, http:// opinionator.blogs.nytimes. com/2010/01/25/science-and-religion-lives-and -rocks/?_r=0. On the limited success of theological attempts at reconciliation, see B. H. Smith, *Natural Reflections: Human Cognition at the Nexus of Science and Religion* (New Haven, CT: Yale Univ. Press, 2009), 95–120.

31. Latour, *Rejoicing*, 66.

32. See, for example, the chapter titled "The Historicity of Things: Where Were Microbes before Pasteur?" in *Pandora's Hope*, 145–73, and, with notably different concerns and emphases, Latour, "Charles Péguy: Time, Space, and *le Monde Moderne*," *New Literary History* 46, no. 1 (2015): 41–62.

33. Elements of the tale recur throughout Latour's writings and lectures. For recent instances, see "Charles Péguy," 50, and "Inside the 'Planetary Boundaries': Gaia's Estate," the Gifford lectures, lecture 6, http://www.ed.ac.uk /humanities-soc-sci/news-events/lectures/gifford-lectures/archive/series-2012 -2013/bruno-latour/lecture-six.

34. Representations of modernity as lapse, loss, and degeneration are a staple of traditional religious moralism and recur in more sophisticated forms in current so-called postsecularist thought. See, for example, Alasdair Macintyre, *After Virtue: A Study in Moral Theory* (Notre Dame, IN: Univ. of Notre Dame Press, 1981); Charles Taylor, *A Secular Age* (Cambridge, MA: Harvard Univ. Press, 2007); Brad S. Gregory, *The Unintended Reformation: How A Religious Revolution Secularized Society* (Cambridge, MA: Harvard Univ. Press, 2012); Thomas Pfau, *Minding the Modern: Human Agency, Intellectual Traditions, and Responsible Knowledge* (Notre Dame IN: Univ. of Notre Dame Press, 2013).

35. For the idea of a "modern parenthesis," see Latour, "Why Has Critique Run out of Steam? From Matters of Fact to Matters of Concern," *Critical Inquiry* 30, no. 2 (2004): 234.

36. See the epigraph to this section and the entry on PSYCHOLOGY in the glossary to the online *AIME*, from which the epigraph is drawn: http:// modesofexistence.org/inqu iry/?lang=en#b[chapter]=#17&b[subheading]=#289 &a=SET+VOC+LEADER&c[leading]=VOC&c[slave]=TEXT&i[id]=#vo cab-421&i[column]=VOC&s=0&q=Psychology.

37. For important works exemplifying these approaches, see Francisco J. Varela, Evan Thompson, and Eleanor Rosch, *The Embodied Mind: Cognitive Science and Human Experience* (Cambridge, MA: MIT Press, 1991); Esther Thelen and Linda B. Smith, *A Dynamic Systems Approach to the Development of Cognition and Action* (Cambridge, MA: MIT Press, 1994); Edwin Hutchins, *Cognition in the Wild* (Cambridge, MA: MIT Press, 1995); Rafael E. Núñez and

Walter J. Freeman, eds., *Reclaiming Cognition: The Primacy of Action, Intention, and Emotion* (Thorverton, UK: Imprint Academic, 1999); Alva Noë, *Action in Perception* (Cambridge, MA: MIT Press, 2004); Louise Barrett, *Beyond the Brain: How Body and Environment Shape Animal and Human Minds* (Princeton, NJ: Princeton Univ. Press, 2011). On their implications for current controversies in epistemology, see B. H. Smith, *Belief and Resistance: Dynamics of Contemporary Intellectual Controversy* (Cambridge, MA: Harvard Univ. Press, 1997), 37–51, 125–52. For their relevance to the understanding of beliefs, religious and other, see B. H. Smith, *Natural Reflections*, 5–19.

38. The alternatives here, also commonly seen as the (merely) "subjective" and the (putatively) "objective," are what Latour has sought, in his writings on religion and more generally, "to slip in between" (see note 15 above)—or, precisely, to finesse.

39. It is, for the same reasons, a hard problem for interdisciplinary studies involving both humanities and natural-science fields. For discussion, see B. H. Smith, "Scientizing the Humanities."

CHAPTER FIFTEEN

Politics Is a "Mode of Existence"

Why Political Theorists Should Leave Hobbes for Montesquieu

GERARD DE VRIES

PARAPHRASING VON CLAUSEWITZ'S WORDS that war is the continuation of politics with other means, in *The Pasteurization of France,* Latour wrote, "Science is not politics. It is politics by other means."[1] Often condensed as "science is politics by other means," this statement has been taken for a call to critically analyze the gender and other ideological biases of—what passes for—objective scientific knowledge. Latour never endorsed the agendas of critical theory and the social constructivist sociology of scientific knowledge, however. He followed another trail.

Latour's life project is aimed at empirically redescribing modern science, technology, politics, and institutions like law and religion as what one may call various "ways of worldmaking." Nelson Goodman, an analytic philosopher, used that phrase as the title of a book in which he offered his reflections on the different versions of the world scientists and artists make and remake.[2] In contrast, rather than studying the *versions* that are produced to represent or articulate the world symbolically, Latour's interest is the making of worlds *themselves*. Latour's business is ontology, not epistemology.

Latour set out to develop a richer, fairer vocabulary to account for who we, "the Moderns," are and what we value. He changed our views of science. In this chapter, I'll argue that his approach opens up another conception of politics as well, one that diverges from the by now conventional understanding of politics that Max Weber—taking Hobbes's concept of the state for granted—referred to as "the leadership, or the influencing of the leadership, of a political association, hence today, of a state."[3]

Weber's conception of politics fills the columns of the politics sections of our newspapers, as well as academic treatises. To describe what goes on in politics, it suggests one has to ask questions like "who leads, that is, who has gained the power to rule legitimately?"; "whose ideology, or interest, dominates?"; or even, in Lenin's brutal words, "кто кого?" (who whom?—who does what to whom for whose benefit?)

To get at an alternative Latourian conception of politics and to see why it's worth our attention, some background is necessary—and some help from a friend, Montesquieu.

Science Is Not Politics; It Provides Politics with New Forms of Power

Latour developed his philosophy hand in glove with his empirical work.

The backbone of Latour's philosophy is a *relationist* ontology, that is, a worldview according to which continuity of existence is not guaranteed by pre-given, hidden essences, but by a being's relations with other beings. In his philosophy, the existential question which any existent faces is not "to be or not to be," but "to be or *no longer* to be." To sustain, to maintain itself, and to continue its existence, an existent has to be related, to be connected, to other existents. Collectives of entities related by (for the time being) stable connections make up reality, the world. In Latour's philosophy, what other philosophers took for essences are the characteristics and competences, the role, place, and meaning of entities that eventually make up a collective with *stable* relations, one that continues to exist.[4]

The empirical component of Latour's project is deviously simple. Turning away from previous approaches in philosophy of science which had searched for epistemological justifications of scientific claims, Latour set out to redescribe "science in action" by approaching scientists in the same way as one attends to builders or cooks, that is, to people who use their skills and instruments to reorganize existing relations between humans and nonhuman entities to create something new— something which is both constructed *and* real, that is, strong enough to resist trials.[5] The generic term he introduced for the work, the movements by which new connections are instituted is *translation*. The term captures the idea that by instituting new relations, entities get both a new place among other existents and a new meaning. By focusing on the translations that are performed, Latour redescribed what science *is*: a "way of worldmaking" in which facts are produced that are both constructed and real.

So, in *The Pasteurization of France*, Latour followed Pasteur's moves, the chain of translations Pasteur had to realize *in* the world—a world which comprises both humans and nonhumans, both of which he had to relate to and interact with. Due to the work—the translations— Pasteur performed in his laboratory, formerly unknown, invisible, and dangerous enemies, whose existence was doubtful, were turned into visible, identifiable and manipulable entities, living organisms, microbes, which could be defeated by vaccines and by taking precautionary measures. By using his laboratory as a fulcrum, Pasteur succeeded in shifting the balance between humans, animals and deadly diseases. His work enabled humans to change the scales, to get leverage, and become stronger. Pasteur taught farmers what they had to do and what to avoid to substantially reduce the risk of their cattle dying of anthrax; hygienists and public officials learned what to take on to advance public health more efficaciously; the colonization of large parts of Africa was enabled by vaccinating French soldiers, missionaries, and colonists against parasites that previously had limited the extent of empires.[6] Pasteur's science provided *new means*, a fresh form of power, to pursue economic, medical and political interests in ways that farmers, hygienists and politicians could not have foreseen.

If we want to call Pasteur's science "politics by other means," *that* is his politics, not the ideas, the ideology, and the interests he may have had and shared with other members of the French bourgeoisie. Pasteur brought about a revolution in veterinary practice, in public health, and in society. For that he is honored with statues in French cities. The names of most other members of the nineteenth-century French bourgeoisie are forgotten.

However, Pasteur's success presents a puzzle as well. If science provides politics with fresh, better, more powerful means in order to pave the way to a well-ordered, healthy, and prosperous nation and to colonize large parts of Africa, the question arises why we would still need the more traditional forms of politics. Are those people right who—with Plato—loath politics for its endless disputes and its lack of rationality, and long for experts to take over, or who—being fed up with experts as well—put their trust in some former CEO, a Great Dealmaker, who knows—or at least claims to know—what works? Now that in science and in the business world new, more efficacious, means to move society forward have been developed, should the French perhaps start removing the statues which honor their statesmen from their cities, to store them in some museum depot, next to all other remains of times past?

Before we find ourselves living under autocratic or technocratic rule, we should try to understand what politics *is* and why we should cherish it.

Unscrewing Hobbes's Leviathan

Modern political theory started when Hobbes declared his intention to undertake "a more serious search into the rights of states and the duties of subjects."[7] Hobbes's declaration marks the end of one distinct phase in the history of political theory as well as the beginning of another and more familiar one, Quentin Skinner has argued. "It announces the end of an era in which the concept of public power had been treated in far more personal and charismatic terms. It points to a simpler and altogether more abstract vision, one that has remained with us ever since," namely the concept of the state as an omnipotent yet impersonal power. By now, "the idea that the confrontation between individuals and states furnishes the central topic of political theory has come to be almost universally accepted."[8]

Hobbes conceived the state as a union in which, by a hypothetical covenant, the many had become One, a single agent, an artificial man, the Leviathan, "that *Mortall God*, to which wee owe under the *Immortall God*, our peace and defence."[9] An organic metaphor, the body politic—which Hobbes used as a synonym for the state—accounts for the idea that multiple parts can be integrated to become the Form and

Matter of an artificial person, the sovereign, to whom acts can be attributed. The idea is famously captured by the frontispiece of *Leviathan*. The multitude makes up the body politic; its soul, and so its animating force, is the sovereign who keeps watch over city and countryside and who has been authorized by the many to rule.

In a paper which marks the birth of Actor-Network-Theory (ANT), Callon and Latour "unscrewed" Hobbes's Leviathan.[10] They point out that Hobbes's argument hides a problem: *how* does a multitude of individuals, of micro-actors, become the State, a macro-actor? Hobbes's suggestion that a hypothetical covenant lifted the multitude out of the state of nature—when there was a war of all against all and "the life of man, solitary, poore, nasty, brutish, and short"[11]—is no longer acceptable. History, anthropology, and ethology have proven this explanation impossible, Callon and Latour state. So how can we understand a multitude to act as one person?

The answer Callon and Latour suggest comes in three steps. First, like Hobbes, they conceive all actors—individuals and macro-actors like the state alike—as being isomorphic. The best way to understand this is to consider all actors as *networks*, they state. When a network is connected to another network, the latter is enlarged, but it neither turns into something else, nor jumps to another level; it remains a network. Once *all* actors are viewed as networks, the idea that there are micro- and macro-actors that occupy different levels evaporates; everything is and remains on the same plane. Secondly, Callon and Latour suggest that we view the construction of the body politic (which Hobbes accounted for in terms of a hypothetical covenant) as "merely a specific instance of a more general phenomenon, that of translation," that is, the process of enlarging networks by whatever means: "negotiations, intrigues, calculations, acts of persuasion and violence, thanks to which an actor or force takes, or causes to be conferred on itself, authority to speak or act on behalf of another actor or force."[12] Thirdly, they observe that to continue its existence as an enlarged network, the associations that are established have to last longer than the interactions that formed them, which is made possible by enlisting more *durable* materials—that is, by extending a network by including nonhumans, rather than only human actors. Without including nonhumans, the enlarged network would soon fall apart.

This, Callon and Latour claim, solves what they call "Hobbes's paradox," the puzzle of how a multitude can act as one person, the problem that—from Parsons to Giddens—has haunted twentieth-century sociologists who, to account for social order, were faced with the problem how to integrate micro- and macro-sociology theoretically and methodologically.

The sociologists of the twentieth century, Latour wrote later,

> have been looking, somewhat desperately, for social links sturdy enough to tie all of us together or for moral laws that would be inflexible enough to make us behave properly. When adding up social ties, all does not balance. Soft humans and weak moralities are all sociologists can get. The society they try to recompose with bodies and norms constantly crumbles. Something is missing, something that should be strongly social and highly moral. Where can they find it? Everywhere, but they too often refuse to see it.[13]

What the sociologists refused to see, Latour argues, is that society would not last a minute if it were held together by social relations between humans alone. To account for its *continuity*, one will have to take relations with nonhumans on board as well, like the material conditions and facilities people share and the mundane artifacts, tools, materials, and technologies they use. Nonhumans provide the solidity and stability which allow groups of people to exist longer and to extend further than human interactions and institutional rules and norms alone. They solve the riddle of why—despite the weakness of the human will, our feeble memory, and unsteady norms and morals—social order is remarkably robust.

Hence, instead of conceiving the social as a domain of reality made of what sociologists conceive as the proper object of their discipline (institutions and subjectively meaningful human interactions), in ANT one conceives the social rather as a *type of connection* between things that are not themselves social.[14] Attention is shifted to the *translations* which create the connections between a wide array of heterogeneous—

human and nonhuman—entities that together establish, stabilize, and maintain some collective, an "actor-network," to which we may subsequently attribute action—the kind of collective we call in conventional terms a company, a research project, a court of law, the Navy, or the market for complex financial products.

Address the Normative Issue by Way of Introducing "Modes of Existence"

Does ANT solve Hobbes's problem? Partly. It shows a way to understand how a multitude can turn into One, a macro-actor, the state. But Hobbes wrote *Leviathan* to address a bigger problem; he set out to undertake a search for the rights of states and the duties of its subjects to account for the *authority* of the sovereign. For Hobbes, "[t]he matters in question are not of *Fact*, but of *Right*."[15]

Having dealt at length with Hobbes, Callon and Latour concede that in their article they "are not interested in political science."[16] The case study they use to illustrate the kind of analysis they propose—how Renault succeeded in subverting the plans of EDF (the French electricity provider) to introduce electric vehicles in the 1970s—is not from the domain we usually call "politics." It gives a view of how power is played out and how society is made out of heterogeneous elements, but not of the rights of the French state and the duties of its citizens. Renault may have succeeded in defeating EDF, but that doesn't mean that it has been granted political authority. Having subsumed Hobbes's covenant under "the general phenomenon of translation," Callon and Latour miss out on what is *specific* about the kind of translations Hobbes restricted himself to, the kind of translations known as "political representation."[17]

By subsuming everything under the generic terms *translation* and *the social*, we lose sight of the specificity of what goes on in science, politics, law, and the business world. To be sure, there is a lot of interaction between these domains. But in spite of this, nobody will take the headquarters of a political party for a court of law or for the boardroom of a company, nor would we mistake a court of law for a laboratory—even though what is discussed, or is in the papers which are consulted, may originate from one of the other buildings. The collectives which are set up by "the general phenomenon of translation" come in many forms and they may perform quite different functions. How to distinguish them?

Politics, law, the economy, and science are different *institutions*, sociologists of former times used to say; by emphasizing that all actor-networks are established by connecting and translating heterogeneous elements, ANT has so far failed to account for what used to be called the different functions of institutions and the different values they serve.

Latour found the entry to address this issue in the simple observation that once an actor-network is *set up*, it may allow something, a specific being, to *circulate* along the trajectory of associations that make up the network.[18] His work in science studies suggested that for science this specific being is *reference*, a constancy that passes through translations and that allows scientists to step back in the trajectory of translations to retrieve information, if needed.[19] In his empirical study of the French High Court for administrative law, the *Conseil d'État*, he argued that in law something different circulates, namely what French jurists call a *moyen*, a legal ground or reason.[20] Latour discovered that his study of ways of worldmaking had to incorporate ways of truthmaking, of veridiction, as well—with truth being understood not as *adaequatio rei et intellectus* but as in accordance with the *internal* normativity of a specific *mode of existence* (the term Latour introduced to account for what sociologists used to call institutions) which are provided by *conditions of felicity*.[21]

In *An Inquiry into Modes of Existence*, he fully fleshed out the idea and showed how it enables him to *contrast* the various modes of existence that make up the world of the Moderns by registering the worth and form of eloquence of each of them, their specific form of truthmaking. It saved him from having to *compare* science, law, politics and religion by either taking science as a standard of truth and rationality that none of the others can meet, or by relativizing all of them by calling them "cultures" and thus losing what science—and each of the other institutions—makes unique.

All of this worked beautifully—except for politics. When Latour tried in *An Inquiry into Modes of Existence* to account for the mode of existence of politics, the argument is still performed in the long shadow cast by Plato. Of course, having shown at length in his work in science studies that rationality is not what makes science stand out, there is no point in arguing that politics is less rational than science.[22] But the experience of people who are *disappointed* by politics still framed the inquiry. To account for this experience, Latour turned again to Hobbes. He identified politics with the phenomenon of political representation. Politics, he writes in *An Inquiry into Modes of Existence*, starts with

> a multitude that does not know what it wants but that is suffering and complaining; then, by a dizzying translation/betrayal, . . . a version of its pain and grievances [is invented] from whole cloth; . . . a unified version that will be repeated by certain voices, which in turn—the return trip is as least as astonishing as the trip out—will bring it back to the multitude in the form of requirements imposed, orders given, laws passed; requirements, orders, and laws that are now exchanged, translated, transposed, transformed, opposed by the multitude in such diverse ways that they produce a new commotion: complaints defining new grievances, reviving and spelling out new indignation, new consent, new opinions.[23]

And then everything starts over again.

What makes political reasoning stand out is that it never goes in a straight line, but moves in what Latour calls "the Circle," "pass[ing] from one situation to another and then com[ing] back and start[ing] everything, *everything*, all over again in a different form"[24]—a continuous movement, impossible to trace, which allows people to group and regroup and which "has been celebrated under the name of autonomy."[25] While conceding that political moves are "object oriented"[26] (without issues, there is no politics), Latour replaced Hobbes's hypothetical covenant by the even mysterious process of "the Circle."

So, although he framed the argument in the wrong terms, Plato had a point. People who expect democratic politicians to represent their interests and wishes in a straightforward way will be deeply disappointed. But why should they, even for a minute, obey the orders and the laws which "astonishingly" come out of this bizarre process of "dizzying" translations and betrayal—a process that then starts over again, so that the laws and orders they are supposed to obey may be replaced in the next round of "the Circle"? Is that what one calls "autonomy"?

Latour's analysis of politics leaves us little room: to cherish democracy, he suggests, one has to lower one's expectations about political representation and eventually about politics as such; the alternative is to put one's trust in someone who promises to talk straight, or in technocrats who know more efficacious means to move society.

To save democratic politics from strong men and technocracy, we need a better account of politics than "the Circle."

Latour may be excused. To explore science and law as modes of existence, he could rely on his extensive empirical work; for politics he could not. To get a better idea of what politics *is*, we will need the help from someone who did the necessary empirical work. For that we have to turn to Montesquieu, rather than Hobbes. We need to read *De l'esprit des lois*, Montesquieu's comparative, empirical study of three kinds of government, namely monarchies, republics, and despotic states, published in 1748. As we will see, guided by a relationist ontology quite similar to the one Latour introduced two and a half centuries later, Montesquieu redescribed what government and politics are.

Montesquieu's *De l'Esprit des Lois*

Comprising several hundred pages, Montesquieu's masterwork is divided into thirty-one books. Each book contains chapters which discuss the laws, customs, climate, terrain, the life of peoples, their wealth, commerce, religion, and their manners and morals. An amazing number of details is discussed, covering the societies of ancient Greece and Rome, various European countries, Islamic states, and the Far East.

Book I, "On laws in general," however, differs in style from all subsequent books. In philosophical, rather than empirical terms, it sets the stage for what will follow. It opens with the following lines: "Laws, taken in the broadest meaning, are the necessary relations deriving from the nature of things; and in this sense, all beings have their laws: the divinity has its laws, the material world has its laws, the intelligences superior to man [i.e., angels] have their laws, the beasts have their laws, man has his laws."[27]

By speaking univocally about God's laws, the laws of nature and man's laws, Montesquieu makes clear that he will treat religious and moral rules, laws of nature, and political and civil laws on a par. A few lines later, it becomes clear that for his general concept of law, he takes his lead from natural philosophy, the physics of his time.

Rejecting Aristotelian physics, seventeenth-century natural philosophers had started to describe the physical world in terms of laws of nature. But instead of interpreting these laws in *epistemological* terms as formulating known relations between observed phenomena, Montesquieu uses natural philosophy's concept of law to introduce a *relationist ontology*. Observing that "the [physical] world, formed by the motion of matter and devoid by intelligence, still continues to exist," he argues that "its motions must have invariable laws; and [that] if one could imagine another world than this, it would have consistent rules or it would be destroyed."[28] Without its laws, without relations between moving bodies, the natural world would not *continue* to exist; everything would fall apart and pass away.

An important consequence is immediately drawn: "[i]t would be absurd that the creator, without these rules [i.e. the laws of nature], could govern the world, since the world would not *continue* to exist without them."[29] Even God, the creator, has to acknowledge this. So, the creator can be relegated to the wings; to understand the world that exists and continues to exist, we have to focus on laws, on relations, not on who or what has created or authored them.

Montesquieu may have sidelined the creator, but what about humanity? Human beings write their own laws, Montesquieu concedes. As intelligent beings, they have reason, they can guide themselves. However, they are also limited beings, prone to lapsing into ignorance and error. While the laws of nature apply with invariable necessity, this is not the case for human laws; they impose a lesser degree of necessity than the laws of nature. So, people have to be stimulated in one way or another to follow their own rules and to be called back to their duties. With another subtle move, Montesquieu has pushed not only the creator but also the authors of human laws to the wings. We are invited to focus exclusively on relations, on laws, and on the ways by which people are incited to follow them.

This move sets the stage for hundreds of pages in which Montesquieu analyses and compares three kinds of government in empirical terms. Their names are the traditional ones: despotism, monarchy, and republic—the last comprising both democratic and aristocratic republics.[30] But whereas the tradition of political thought defined each of them by pointing to the *source* (the authorship) of power—like the number of rulers, their virtues, their use of brute force, the way succession is regulated or—in *Leviathan*—the way the many have authored the sovereign's rule by some hypothetical covenant—Montesquieu distinguishes them by putting the spotlight on how in each of them power is *exercised*.

A short discussion provides the argument for this conceptual shift. In both despotism and monarchy, a single person rules; but there is a significant difference between the two. In despotism, "one alone, without law and without rule draws everything along by his will and caprices," whereas in a monarchy, "one alone governs according to fixed or established laws," which "necessarily assume mediate channels through which power flows."[31] So, the same source of rule can be exercised in such different ways that the types of government must be said to be different. Hence, Montesquieu defines the *nature* of a government—"that which makes it what it *is*"—by the way power is *exercised*, that is, whether or not by way of law and mediate channels. This definition not only suffices to distinguish monarchy and despotism, but also to specify the nature of a democratic republic—the kind of government in which the people as a body have sovereign power and in which the people are, in certain respects, the monarch, and in other respects subjects. As they can be the monarch only through their votes, Montesquieu argues that the laws that establish the right to vote and regulate how, by whom, for whom, and on what issues votes should be cast are fundamental in this kind of government.[32] So, once again, laws take center stage.

However, it is not sufficient for a government to be formed and established. A government should be able to act and to maintain itself as well. As human laws apply with a lesser degree of necessity than the laws of nature, for a government to be able to act, to rule, to exercise its power, something else is also needed. Montesquieu suggests that *passions* provide the impetus that sets governments in motion and allows them to maintain themselves. He refers to the kind of passion which is needed as a government's *principle*. Which principle allows a government to act and to maintain itself depends on the nature of the government, he argues, to conclude that the principles of republics, monarchies, and despotic states are, respectively, virtue, honor and fear. Only when a government's nature and its principle match will a government be able to act and maintain itself: without the nobles pursuing honor, a monarchy will fall apart; a republic will be corrupted if its people lose virtue, that is, their love of democracy, of equality, and the preference of the public's interest over one's own; and when the people no longer fear his ever-raised arm, the despot is lost. The principles play a pivotal role in *sustaining* government. History shows that "[t]he corruption of each government almost always begins with that of its principles," he argues.[33]

At this point, Montesquieu has defined the various kinds of governments in generic terms by identifying the nature and principle of each of them. But governments do not fall out of the blue, nor do they operate in a vacuum. They are established *in* a specific region, they govern people who share a region's geographical and climatic conditions and who are already related to each other by their way of life, their trade, customs, religion, and their morals and manners. Establishing a government means instituting new relations *in addition to* the ones that are already present, namely relations that determine how a ruler's power will be exercised. As this power will have to be exercised *in* the world, the way power is exercised will have to *relate* to the mundane—physical and social—conditions of the society that has to be governed. In other words, to *set up* a government, a region's—physical and social—conditions have to be *translated* into the way power is exercised, that is (with the exception of despotism) into laws.

Once a government is established, to be able to act and maintain itself, passions (appropriate to the kind of government which is established) have to be present as well. With the exception of fear, they will not emerge spontaneously. Hence, to sustain itself, a non-despotic government has to incite and support the passions which will allow it to *continue* to exist. To achieve this aim, appropriate civil and criminal laws will have to be introduced.

This argument sets the course and the full scope of the empirical inquiry Montesquieu undertakes in *De l'esprit des lois*. For each kind of government, he discusses at length not only how its laws must relate to its nature and principle, but also how they should be

> related to the physical aspects of the country; to the climate, be it freezing, torrid, or temperate; to the properties of the terrain, its location and extent; to the way of life of the people, be they plowmen, hunters, or herdmen; they should relate to the degree of liberty that the constitution can sustain, to the religion of the inhabitants, their inclinations, their wealth, their number, their commerce, their mores and manners; finally, the laws are related to one another, to their origin, to the purpose of the legislator, and to the order of things on which they are established.[34]

This whole complex of relations is laid out and discussed in great—and sometimes confusing—detail for each kind of government.

First Despotism. Without law, by exercising brute force, despots rule over timid, brow-beaten people. Living in fear of the despot's fury, "men's portion, like beasts', is instinct, obedience, and chastisement. It is useless to counter with natural feelings, respect for a father, tenderness for one's children and women, laws of honor, or the state of one's health; one has received the order and that is enough."[35] Despotism "leaps to view, so to speak; it is uniform throughout; as only passions [i.e., fear] are needed to establish it, everyone is good enough for that."[36] For Montesquieu, despotism is the default condition of humanity, it is what one gets if nothing better is available.

In contrast to despotism, both monarchies and republics are complex, fragile, political constructions. They do not come into being just because human beings love liberty and hate violence—people who have to live under the yoke of a despot will also share those feelings.[37] To establish a monarchy or a republic, the mundane—physical and social—relations that make up the society have to be translated—step by step—into laws that channel the way power is exercised; and for a government to be able to act and maintain itself, *additional* laws have to be introduced that incite and support the appropriate passions. But once all of this is in place, the ruler's exercise of power will be guided by laws and will be reviewed in mediating institutions, which brings reflection into the ruler's business: before his orders are executed, they will be reviewed and if necessary amended, or even annulled. Therefore, Montesquieu concludes, monarchies and republics provide *moderate* government; their subjects and citizens are granted security of their person and property. To provide for *liberty* as well, he argues, additional constitutional provisions are needed, namely separation of the legislative, executive and juridical powers.

Politics as a Mode of Existence

Montesquieu's argument that liberty requires a separation of powers was embraced by the US Founding Fathers and subsequently by constitutionalists in many other countries. The theory of government that led to this conclusion was discarded, however. As Skinner, quoted above, points out, in political theory Hobbes sets the tune. To be sure, political theory has evolved since Hobbes. More complex social contract theories have replaced his hypothetical covenant and nineteenth-century philosophy and twentieth-century sociology have taught political theorists that individuals cluster in groups or classes which—to pursue their interests, and to get recognition and respect—compete for what Weber called the "leadership or the influencing of the leadership of the state." Nevertheless, the Hobbesian framework suggests that, basically, political analysts have to address only two issues: who have gained power, an empirical question; and, secondly, what is the source of their legitimacy to rule, a normative one.

If that is all there is in political analysis, why would anybody still take the trouble to pore over the hundreds of pages of Montesquieu's book— that "labyrinth with no route [through it], and no method [in it]," as Voltaire let a spokesperson scoff in *L'A B C*, his book of political dialogues?[38] To be reminded of the importance of the separation of powers? For that, these days, occasionally reading the *New York Times* will do. For its empirical insights, then? Quite unlikely. For empirical evidence, Montesquieu had to rely on ancient authors, untrustworthy travel reports, and his own ad hoc observations in several European countries. It led him to argue that democratic republics could be established only in small (city-)states; that the climate of the tropics is incompatible with moderate government; and that although slavery is despicable, its spread in large regions of the world was inevitable. History has shown him to be wrong on these and many other points. For normative guidance, then, perhaps? On this score, too, we seem to have got the wrong man. In his Preface, Montesquieu expressly wrote that he did "not write to censure that which is established in any country whatsoever," and that in his book "[e]ach nation will find the reasons for its maxims."

However, Montesquieu has a stronger position than it may seem when his work is perceived through the lens of Hobbesian-style political theory. What distinguishes Montesquieu from Hobbes is not a lack of interest in the rights of states and the duties of individuals, but the way in which they are addressed. While Hobbes wrote that "[t]he matters in question are not of *Fact*, but of *Right*," Montesquieu argues that rights come into being only where the exercise of power is channeled by laws and mediate institutions. Neither Providence, nor reason, nor normative theories, nor our good will provide for them. But once a moderate government is in place, and power is exercised according to law, rights are both constructed and real. (If in doubt about the latter, compare the life of a citizen with that of a stateless person, or an illegal immigrant who has been denied citizen's rights.)

Looking back, we may observe that Montesquieu performed a shift in political theory similar to the empirical turn Latour, two and a half centuries later, performed in science studies. Instead of seeking a justification for ready-made *products* (for science: the claim that scientific knowledge expresses—or approximates—truth; for politics: the claim that a sovereign's rule is legitimate), attention is shifted to the *process* (to science in action; respectively to the exercise of power). And this shift enables to account for the way in which something (scientific facts; respectively rights and duties) is produced that is both constructed *and* real.

Once the empirical turn Montesquieu performed in political theory is appreciated, it becomes clear that Voltaire was mistaken. *De l'esprit des lois* is anything but a labyrinth. In fact, by using the terms Latour introduced in *An Inquiry into Modes of Existence* to specify a mode of existence—hiatus, trajectory, conditions of felicity, being to institute, and alteration[39]—the way Montesquieu reconceptualized government and politics can be recapitulated in eight steps:

1. Suppose despotism—rule based on fear and violence—to be the default condition of humanity. However, as a matter of empirical fact, apart from despotism other kinds of government exist, namely monarchies and republics. So, the question becomes: what does it take to set up and maintain each of these two different kinds of government?

2. Governments do not fall out of the blue and they don't exercise power in a vacuum. Establishing a government means *adding* new relations to the many mundane—physical and social—relations that are already present in a region, namely relations that determine how a ruler's power is to be exercised.

3. The way power is exercised defines the *nature* of a government—that which makes it what it is: a monarchy, a republic, or a despotic state. In despotism, one alone rules without law; in monarchies and republics, power is exercised by way of law and mediate channels. As a government has to act and maintain itself *in* the world, to set up a monarchy or republic, its laws have to relate to (that is, *translate*) the physical and social conditions of society.

4. In monarchies and republics, laws and mediate channels bring reflection into the rulers' business. They form a *trajectory* along which orders are communicated, reviewed, and—if needed—amended or annulled. As a consequence, monarchies and republics offer *moderate* government.

5. However, as humans are limited beings, it is not guaranteed that they will abide by their government's rule. In other words, the exercise of power may be confronted with *hiatuses*. To overcome them, and to allow a government to act and maintain itself, specific *conditions of felicity* have to apply (which Montesquieu refers to as the *principle* of

government, the specific kind of passions which should guide those involved in governing).

6. Which conditions of felicity are required depend on the nature of the government: a monarchy's principle is honor, a republic's is virtue. In contrast to fear of a despot's fury, these passions do not emerge spontaneously; education and civil and criminal laws will have to be introduced to animate and support them.

7. Once all of this is in place, a *new being*—legitimacy—is instituted which can circulate on the *trajectory* along which power is exercised. As the government's orders and decisions (those of the monarch, republican legislators, executives, and magistrates, etc.) will be reviewed and, if necessary, amended or annulled, the exercise of power will be considered as legitimate.

8. As laws and mediate channels bring reflection into the business of government, in countries with moderate government people do not have to live in fear of a despot's will and caprices and ever-raised arm. Hence, the *alteration* a moderate government allows is security of one's person and property, and—provided additional constitutional provisions (namely separation of powers) are put in place—liberty as well.

Taken together, these steps explicate politics as what Latour calls a specific *mode of existence*. Once in place, this mode replaces the default condition of humanity—exercise of power based on fear and violence—with a political process in which a society's social and physical conditions are translated into laws, power is exercised according to law, and rights and duties are distributed and allocated. As a mode of existence, politics is not naturally given, nor provided for by Providence, reason, or good will. To establish it requires "a masterpiece of legislation that chance rarely produces and prudence is rarely allowed to produce."[40]

Conclusion

Obviously, both societies and the way they are governed have radically changed since *De l'esprit des lois* was published. Apart from law, new forms of exercising power have emerged.[41] Representative democracy was introduced, with political parties competing for power in elections.

In welfare states, the government's concerns far exceed the "peace and common defence" Hobbes's multitude expected from their sovereign. But apart from expanding their relations with humans, governments have to deal increasingly with the nonhuman world as well: to care for public health, essential technological infrastructure, the environment, and even the global climate. The techniques and principles to establish democratic legitimacy have evolved too: while for most of the twentieth century elections and bureaucratic rationality served that function, since the 1980s the Western world has entered a "new age of legitimacy," with increased emphasis on image and communication.[42]

Given these developments, it would therefore be rather pointless to discuss what Skinner singled out as the central topic of political theory—the confrontation of states and individuals—without taking into consideration the mundane, human and nonhuman, world *in* and *on* which governments have to act and to maintain themselves. As Montesquieu's theory addresses this problematic, in spite of having been published almost three centuries ago, his conception of politics is remarkably up to date—not only for pointing out the still important separation of powers, but also for being realistic—in the sense of situating politics in an explicit, down-to-earth fashion.

To seriously discuss contemporary politics, we had better leave Hobbes for Montesquieu, to subsequently conceive his theory in Latour's terms as explicating politics as a specific mode of existence.

Firstly, it will provide an antidote to the naïve optimism that a moderate government can be established anywhere by introducing a few mechanisms for collecting and expressing the will of the people. In 2005, at the *Making Things Public* exhibition Latour curated in Karlsruhe's Center for Art and Media (ZKM), Sloterdijk ironically made the point in a video installation in which he impersonated a businessman eager to sell an inflatable, pneumatic parliament that could be dropped from an airplane to offer instant democracy anywhere, even in a desert.[43] As the so-called coalition of the willing was to discover in Iraq, things don't work out quite that way.

Secondly, it emphasizes that even once it is established, the continuity of a democratic government's existence cannot be taken for granted. While Carlyle famously declared, "invent the printing press and democracy is inevitable,"[44] today, we find ourselves discussing whether democracy is sustainable in the age of digital technologies and climate change.[45] As Montesquieu's theory emphasizes, democratic legitimacy is a delicate being. It requires both a large body of laws and institutions that channel the exercise of power *plus* citizens' compliance with democracy's specific conditions of felicity (which Montesquieu identified as their "virtue")—a point often missed in current discussions about the crises of democracy that are framed by opposing the will of the people and democratic rule of law.[46]

Finally, in spite of the fact that in many domains, science, politics, and government are mixed up, "science is *not* politics by other means." They are different modes of existence. Science makes it possible to reach remote entities and may deliver the means to control them; the being this mode of existence institutes is *reference*. In contrast, politics allows power to be exercised in a moderate way rather than by brute force, which provides security for one's person and liberty; the being it institutes is legitimacy. The *contrast* matters: where it is not respected, one or both of them will be corrupted: a "politicized" science will corrupt the search for truth; where government is left to science one ends up with technocracy.

In *De l'esprit des lois*, we encounter a coherent, Latourian conception of politics, a richer vocabulary than Hobbes offered, one that may help us to discuss what will be required to maintain democracy under the threat of climate change[47] and in the digital age.[48] It shifts the key political question from the one the Hobbesian framework suggests—how to authorize a government that speaks for us and that respects our interests and identities—to a both more profound and simple one: can we *continue* to live together on the one earth we inhabit, without fear, that is, can we *sustain* life and liberty?

Notes

1. Bruno Latour, *The Pasteurization of France* (Cambridge, MA, & London: Harvard Univ. Press, 1988), 229 (§ 4.6.2.1). The original French text (*Les Microbes*, Paris: Ed. A.M. Métailié, 1984, 257) follows Von Clausewitz even more closely: "La science, c'est la politique continue par d'*autres* moyens."

2. Nelson Goodman, *Ways of Worldmaking* (Indianapolis & Cambridge: Hacket Publishing Company, 1978).

3. *Max* Weber, *Gesammelte Politische Schriften* (Tübingen: J. C. B. Mohr, 1988), 506.

4. Apart from Latour, philosophers who have defended relationist ontologies include Nietzsche (cf. Alexander Nehamas, *Nietzsche: Life as Literature* [Cambridge, MA, & London: Harvard Univ. Press, 1985]), chap. 3; Alfred North Whitehead, *Science and the Modern World* (New York: Free Press, 1967 [1925]), and—as will be argued—Montesquieu. Darwin's definition of a species as a population of individuals separated from other populations by a reproductive gap is based on relationism as well, cf. Ernst Mayr, *The Growth of Biological Thought* (Cambridge, MA, & London: The Belknap Press of Harvard Univ. Press, 1982), 272; cf. also Gerard de Vries, *Bruno Latour* (Cambridge: Polity Press, 2016), 159–60. The ontology is counterintuitive for those whose intuitions have been shaped by mainstream Western philosophy. But the experience with the world wide web may help us to grasp it, Bruno Latour et al. argue in "The Whole is Always Smaller Than Its Parts. A Digital test of Gabriel Tarde's Monads," *British Journal of Sociology* 63, no. 4 (2012), 591–615.

5. Latour, *Pasteurization*, 158 (§ 1.1.5).

6. Latour, *Pasteurization*, 140 ff.

7. Thomas Hobbes, Preface of *De Cive* cited in Quentin Skinner, "The State" in *Political Innovation and Conceptual Change,* ed. Therence Ball, James Farr, Russell L. Hanson (Cambridge: Cambridge Univ. Press, 1989), 90.

8. Skinner, "The State," 90.

9. Thomas Hobbes, *Leviathan,* ed. C.B. Macpherson (Harmondsworth: Penguin Books, 1968 [1651]), 227.

10. Michel Callon and Bruno Latour, "Unscrewing the Big Leviathan: How Actors Macro-Structure Reality and How Sociologists Help Them To Do So" in *Advances in Social Theory and Methodology—Towards an Integration of Micro- and Macro-Sociologies,* ed. Karin Knorr-Cetina and Aron V. Cicourel (London: Routledge and Kegan Paul, 1981), 277–303.

11. Hobbes, *Leviathan,* 186.

12. Callon and Latour, "Unscrewing the Big Leviathan," 279.

13. Bruno Latour, "Where Are the Missing Masses? The Sociology of a Few Mundane Artifacts" in *Shaping Technology/Building Society—Studies in Sociotechnical Change* ed. Wiebe E. Bijker and John Law (Cambridge, MA: MIT Press), 225–58, 227.

14. Bruno Latour, *Reassembling the Social—An Introduction to Actor-Network-Theory* (Oxford: Oxford Univ. Press, 2005), 5.

15. Hobbes, *Leviathan,* 727.

16. Callon and Latour, "Unscrewing the Big Leviathan," 296.

17. Cf. Hannah F. Pitkin, *The Concept of Representation* (Berkeley and Los Angeles: Univ. of California Press, 1967).

18. Bruno Latour, *An Inquiry into Modes of Existence: An Anthropology of the Moderns* (Cambridge, MA: Harvard Univ. Press, 2013), 32

19. Bruno Latour, *Pandora's Hope—Essays on the Reality of Science Studies* (Cambridge, MA: Harvard Univ. Press, 1999), chap. 2.

20. Bruno Latour, *The Making of Law* (Cambridge: Polity Press, 2010).

21. Cf. John L. Austin, *How to Do Things with Words* (Oxford: Oxford Univ. Press, 1976 [1962]) and De Vries, *Bruno Latour*, 107 ff and 172.

22. Latour, *Pandora's Hope*, chap. 8.

23. Latour, *Inquiry*, 341.

24. Latour, *Inquiry*, 338.

25. Latour, *Inquiry*, 345.

26. Latour, *Inquiry*, 337.

27. Montesquieu, *The Spirit of the Laws*. tr. and ed. A. M. Cohler, B. S. Miller, and H. S. Stone (Cambridge: Cambridge Univ. Press, 1989 [1748], cited, as usual, by book and chapter), I, 1.

28. Montesquieu, *Spirit of Laws*, I, 1.

29. Montesquieu, *Spirit of Laws*, I, 1 (italics added).

30. As aristocratic republics are halfway between monarchies and democratic republics, in the following I'll discard them and will understand under a republic a democracy, popular government.

31. Montesquieu, *Spirit of Laws*, II, 1; II, 4.

32. Montesquieu, *Spirit of Laws*, II, 2.

33. Montesquieu, *Spirit of Laws*, VIII, 1.

34. Montesquieu, *Spirit of Laws*, I, 3.

35. Montesquieu, *Spirit of Laws*, III, 10.

36. Montesquieu, *Spirit of Laws*, V, 14.

37. Instead of despotism, contemporary anthropologists point to tribalism as the default condition of mankind, cf. Robin Fox, *The Tribal Imagination* (Cambridge, MA: Harvard Univ. Press, 2011). But they agree with Montesquieu on the crucial point that (in Fox's words, *Tribal Imagination*, 60) "the institutions we . . . prize are not the product of a freedom-loving human nature but the result of many centuries of hard effort to overcome human nature."

38. Voltaire, "The A, B, C or Dialogues between A B C, translated from the English by Mr. Huet," in Voltaire (1994 [1768]) *Political Writings,* ed. D. Williams (Cambridge: Cambridge Univ. Press), 89–90. The secondary literature shows that the perplexity has persisted to the present day. Cf. for example Cohler's introduction to the 1989 Cambridge University Press translation of *De l'esprit des lois*: "The best analogy for the book is the complex mosaic and embroidery of the eighteenth century, or even rococo painting. Although there is no over-arching, organizing image, there are similarities and contrasts that send the viewer, or reader, across the painting, or through the book" (Montesquieu, *Spirit of Laws*, Introduction, xxi).

39. Latour, *Inquiry*, 488–89, pivot table, and De Vries, *Bruno Latour*, 163.

40. Montesquieu, *Spirit of Laws*, V, 14.

41. Michel Foucault, *Discipline and Punish—The Birth of the Prison* (Harmondsworth: Penguin Books, 1979).

42. Pierre Rosanvallon, *Democratic Legitimacy—Impartiality, Reflexivity, Proximity* trans. A. Goldhammer (Princeton, NJ, and Oxford: Princeton Univ. Press, 2011).

43. P. Sloterdijk, and G. Mueller von der Hagen (2005), "Instant Democracy: The Pneumatic Parliament," in: B. Latour, and P. Weibel, eds., *Making Things Public—Atmosphere of Democracy* (Cambridge, MA: MIT Press, 2005), 814–25. The video, shown in 2005 at the "Making Things Public" exhibition, curated by Latour in Karlruhe's ZKM, is available a.o. in the Centraal Museum, Utrecht, the Netherlands.

44. Cited in John Dewey, *The Public and its Problems* (Athens: Swallow Press/Ohio Univ. Press, 1954 [1927]), 110.

45. Cf. among many others David Runciman, *How Democracy Ends* (London: Profile Books, 2018).

46. Cf. e.g., Yasha Mounk, *The People vs. Democracy* (Cambridge, MA: Harvard Univ. Press, 2018) for framing the problems democracies currently face by setting the power of the people against the rule of law.

47. Cf. Bruno Latour, *Où atterrir?* (Paris: La Découverte, 2017).

48. Cf. a.o. Gerard de Vries, "Do digital technologies put democracy in jeopardy?" in *Life and the Law in the Era of Data-Driven Agency*, ed. Mireille Hildebrandt and Keith O'Hara (Cheltenham and Northampton, MA, USA: Edward Elgar Publishing, 2020) and Paul Nemitz, "Constitutional Democracy and Technology in the age of Artificial Intelligence," *Royal Society Philosophical Transactions A (2018)*, doi:10.1098/RSTA.2018.0089.

CHAPTER SIXTEEN

Art as Fiction

Can Latour's Ontology of Art Be Ratified by Art Lovers?
(An Exercise in Anthropological Diplomacy)

PATRICE MANIGLIER

A NEW DIPLOMAT HAS BEEN APPOINTED in the city. His name is Bruno Latour. But why would we need a new diplomat? Because, he claims, we are at war. This war is not physical and it is not between countries: it is metaphysical and it is between domains of action. Each one of these domains of action claims to have a privileged access to Reality as such. Typically, sciences would claim to describe the world, while politics or religion would only be about rules of conduct. But reciprocally, religion rejects science as an untrustworthy witness to what truly exists. In consequence, everything is progressively being swept up in a terrible escalation of unreality, to the point where it seems we end up living only among ghosts.

Our understanding of the humanities is largely dependent on those metaphysical wars. C. P. Snow's famous two-cultures split can be construed as a conflict between two claims for existence. Indeed, the idea that only science provides access to reality removes arts, literature, and the humanities in general to the realm of entertainment or personal expression. Many of us, however, still feel that the humanities are capable of a specific kind of truth, and that they give access to a particular aspect of reality. Latour's *An Inquiry into Modes of Existence* has no other aim than helping to support this intuition.

However, the way Latour characterizes art in the book might seem unacceptable to those who have the keenest interest in it, because he argues that art shares in the mode of existence of *fictions*. This characterization is likely to trigger two kinds of concerns on the part of those I will call the "militants of art," that is to say, those concerned with vigorously defending and protecting art's mode of existence, since it might seem both too narrow and too broad. Too narrow, in the sense that it seems to restrict art to *figurative* art, thus leaving aside the entire realm of modern and contemporary art. Too broad, because it conflates art with things like TV shows, ordinary sentences, and all kinds of images. Is the notion of fiction thus totally unfit to capture the ontology of artworks and their specific way of establishing existence?

Those concerns rely, I contend, on mistaken interpretations of the Latourian concept of fiction. In the following pages, I will try to defend this concept and to show how it helps us to arrive at a better understanding of the very form of truth that is at work in art, as a crucial part of the humanities. In the first section, I will offer a short summary of Latour's diplomatic method as I understand it. We will see that it definitely requires the ratification of those "militants" seeking to defend the existence of the various beings of the world. In the second section, I will examine the first reason the militants of art are likely to reject Latour's characterization, namely that it would be incompatible with modern and contemporary art. In the third and last section, I will examine the second source of resistance, its perceived tendency to dissolve art's proper form of truth by conflating it with other practices in a manner that art lovers would consider disrespectful.

However, I want to emphasize from the onset that the concerns raised by the militants of art are not *objections* as such, and that my aim is not to reply to possible objections—unless you want to call objections any official protest made by a voter against a representative for not correctly representing her, for betraying her mandate, for doing a wrong to her by dispossessing her of her voice. The problem is not whether it is true that art is a kind of "fiction" or not, but whether this claim can ideally be made acceptable to those who devote a significant part of their life to art. The ultimate criterion of validity of Latour's text is diplomatic and not theoretical. This shift is so new that we really have to work at it. This is why the following pages can be seen as an attempt to illustrate Latour's conception of "anthropology," redefined as a form of comparative metaphysics. Our question is thus simply whether the proposal made in *An Inquiry into Modes of Existence* can be ratified by those whom it is supposed to represent—that is, ultimately, by *all of us*.[1]

I. The Rules of Latour's Diplomatic Method

The *Inquiry* poses a simple enough question: how many different kinds of reality are we, "the Moderns," ready to accept, and what are they? It takes the form of an investigation conducted by an imaginary anthropologist who visits Western cultures for the first time and lets herself be surprised by what sounds natural to us. The most surprising thing is that the Moderns believe that they constitute a radical exception because, so they claim, they have an exclusive and direct access to reality. Consequently, they deny that many aspects of their life, like science, technology, the economy, etc., are "cultural"; they are mere expressions of reality as such. This is the only claim Latour's anthropologist rejects. But she cannot use the notion of "culture" (or social construction) for the same reason: because she would thus maintain, if only on a phantasmatic mode, some idealized version of "reality," and would hence reiterate the division that she is supposed to challenge. She will therefore try to study not constructions of reality, but different realities, corresponding to the different domains in which the Moderns have tried to find some kind of truth: sciences, arts, religions, morality, etc. The question thus becomes: how many forms of reality, or, to use Latour's vocabulary, "modes of existence," have been installed by the Moderns?[2]

The answer to this question is constrained in certain ways, which define the protocols of the *Inquiry*. We can extrapolate from Latour's book an implicit set of rules, which constitute the rules of the new method he introduces in this most recent work, a method centered on the notion of *diplomacy* and based on the idea that anthropology must be redefined as a form of metaphysical inquiry.

First, it must be noted that such an investigation does not emerge simply from a sense of intellectual curiosity, as if, like Sartre's Roquentin, I woke up one morning amazed or horrified that there was something rather than nothing and exclaimed: "My God! What could *being* be?" Rather, the question emerges from metaphysical controversies, from conflicts of reality, such as science versus religion, or art versus business. Metaphysics, or rather this particular sort of metaphysics, is not invented in the theorist's armchair; it comes from the heart of the battle. It presupposes that there are adversaries or "militants," who work so that certain beings do not perish. They are dedicated to these beings, they look after them, they promote them, and they put themselves at their service. In the case of art, for example, the category of militants would include critics, curators, reviewers, art historians, and artists themselves. And a "being" such as an artwork is first of all something *that one is attached to*, or, to pick up a Heideggerian term, *something one cares about*, which is to say, *something that is the object of some activism*.

This necessarily means that metaphysics always comes second. The metaphysician doesn't set up the existents, but is content to describe the mode of existence of beings proposed by others. Let's call this first methodological requirement the *principle of positivity*. It explains why Latour's argument in his most recent book takes the form of an "inquiry." The report one comes up with about a mode of existence must be based on the effective practices of the militants who defend this particular mode of existence, allowing one to see just what they are attached to. Reporting on the being of "law," for example, means attending carefully to what the practitioners of law say and do. *Positive metaphysics* (which we could oppose to *dogmatic metaphysics*) is thus characterized by the fact that it makes itself conditional on positions that it declares itself incapable of imposing. The task is not to say *what is*, but to say that being can be understood in several ways and that this variety of meaning is what allows conflicts to be smoothed over. Therefore, Latour's inquiry will not allow, methodologically, any a priori decision about being. The being that is of interest is not the being that the metaphysician establishes through the sovereignty of his or her logical or illogical mind; it is the being that must be saved in conflict situations: being as a *diplomatic problem*. Consequently, metaphysics must follow the militants, situating itself as close as possible to the points that they say they cannot give up.

This first principle leads to a second. There are no grounds for describing a mode of existence unless misunderstandings as to the *specificity* of its type of reality would jeopardize mindful interaction with the beings that are typical of this mode. Any *denial of reality*—for example, the refusal to admit the existence of some god ("this is just an idol!") or some legal qualification ("there is no such thing as persons")—is a signal and a call for metaphysical intervention. Not that one necessarily has to accept all the militant's claims for reality as indisputable. There may be realities that can be sacrificed (for example, because their existence cannot be affirmed without denying the existence of all others, or, what comes to the same thing, because they are radically incomparable). But any denial of reality necessitates by rights a metaphysical inquiry. The example of a militant of art, or law, or religion holding fast to one point, asserting that yielding this point would mean allowing the beings he is devoted to to perish and therefore giving up his own practice—this is the rightful condition for a metaphysical enterprise. Without such points of *resistance*, without these cries of protest, there are simply no grounds to ask questions about being. This not only means that metaphysics comes second, and that it presupposes the practices of militants, but also that metaphysics is always *in the service of* a threatened reality. It is necessarily *engaged*, taken up, in dealing with the resistance of a being to what we must call its *reduction*. In the case of such a reduction, a *wrong* is inflicted on certain beings, and it is a matter of delivering justice to these misunderstood beings. Art, for example, has frequently been subject to such forms of reduction from the side of economics or politics.

However, this doesn't mean that the metaphysician should become, purely and simply, a militant on behalf of any given truth. The difference between metaphysician and militant is that the former does not enter directly into the conflict. She tries to move around the field of the different meanings of "being," so as to shift from an opposition to a difference. While the militant holds onto an existence, the metaphysician differentiates between ways of being. The former says, like Galileo, "And yet, this is." The latter adds, "Yes, but in *this* way." She has won if she can say to two militants who are each defending their modes of existence: "Look, you are both negating each other because you don't see that the assertion of one does not imply the negation of the other, because you mean it in several different ways." She *gives value* to the singularity of one mode of existence vis-à-vis another, by showing that these beings are not subject to the protocols that one wants to impose on them. We can call this the principle of *irreduction*, because indeed it is a question of showing the irreducibility of certain beings to others.[3]

The third rule is, in a way, symmetrical with the preceding one. In the same way that an existent one cares for should be saved in its ontological singularity, so this should happen in such a way that it doesn't, in turn, crush other modes of existence. Therefore, Latour insists, the metaphysical description must never lose sight of the need to reconcile with other beings the existence of the being that one is concentrating on not giving up. This requirement imposes particular constraints on a description, which explains why activists might not immediately recognize themselves in the account given by the metaphysician of the beings they are looking after. Latour urges us to be ready to fend off any possible reduction, but this will not, or rather should never, turn the being that is described into a being of absolute privilege or radical exceptionality. To take seriously the distinctiveness of art, for example, does not mean buying into existing accounts of art's radical autonomy or its separation from the social world. This constraint could be called the incompleteness principle, or the *counterhegemonic principle*.

Finally, the description should lend itself, at least by rights, to being signed off by the militants. Since the metaphysician is only a representative, she is necessarily limited by her mandate. And because she has an additional diplomatic aim, she has to verify that her description protects beings in *all* the ontological controversies that engage the militants that represent them. So, once a compromise is reached and a description finalized, she has to verify with the relevant militants that the essential has been effectively saved, allowing for the continued existence of the beings they care about. For example: will art historians, art critics, curators, and artists be willing to sign off on the descriptions that this essay proposes?

Yet this final requirement should not be interpreted as the need for an empirical consensus, which would, of course, be illusory. The militants don't necessarily understand our metaphysical operation; they may not be interested in it, or they may even refuse to accept the diplomatic protocol (for example, the counterhegemonic principle). The question is nevertheless one of rights: the metaphysical text must be *correctable* by militants of good will. It is not addressed to specialists who could be divided amongst themselves; it is addressed to all those who are *interested* in the beings it describes. This means, first of all, that a text should not be written in such a technical or professional language that it preselects a very specific audience. It also means that the metaphysician should continue to remain vigilant, listening for the *alerts* the militants put out: the wrongs that they see as not having been redressed. She should verify that any reluctance militants might have in accepting the proposed compromise (i.e., the provisional report entitled *An Inquiry into Modes of Existence*) can by rights be dissipated: in other words, that it is based on a poor comprehension of the metaphysical text or even a refusal of the principle of diplomatic philosophy. Factual misunderstandings can therefore be used by the metaphysician in the correction of her description. She must be able to *respond* to the reluctance of activists defending a particular mode of being.

In summary, I have four methodological principles to guide this metaphysical inquiry.

A *principle of positivity*: metaphysics does not decide what is, but describes the manner of being of entities located elsewhere.

A *principle of irreduction*: metaphysics doesn't describe random modes of being, but only those which seem to be the object of a denial or a wrong, in order to show that these entities can coexist with others on condition that they respect each other's format and reality.

A *counterhegemonic principle*: each mode of existence must be described in such a way that it doesn't crush another.

A *responsibility principle*: compromises reached must be submitted for ratification, and we must be prepared to improve the metaphysical description.

With these four principles in mind, we can now go back to what Latour says about the ontology of works of art, while engaging the militants representing this mode of existence. We will see if Latour's proposition allows the beings of art to be saved, at the cost of some possible sacrifices that should, however, not be *fatal*, in other words that should not threaten the existence of that which is dear to the militants; they would only require rearrangements.

We previously identified two concerns about Latour's inquiry that could be raised by those art militants of good faith who are necessary for the proceedings of the diplomatic method. We will treat these concerns in turn.

II. Is Art Fiction? The Question of Modern and Contemporary Art

Take the first of these concerns. Does calling works of art fictions mean that all works of art bring imaginary entities into existence, whether in a story (Homer's *Odyssey* brings Odysseus, an imaginary personage, into existence), or in an image (Gustave Moreau's *L'Apparition* brings into being an imaginary scene where Salomé points her finger at the appearance of St. John the Baptist's head)? This would be to arbitrarily reduce the field of art to a very small part of it, modeled on stories. We cannot confuse fiction with narration. Odysseus and Don Quixote are fictional beings, but Manet's asparagus is one to the same degree, even if it relates to the register of description rather than narration. A still life is just as much a fiction as a history painting is. An experience of feelings of love upon reading Alphonse de Lamartine's *Le Lac*, and the story of the entry of the Trojan horse into the city about to be sacked, are also fictional beings. Let's call nonnarrative fictional beings *representative fictions*, by taking up Aristotle's old distinction between *mimesis* (representation) and *muthos* (narration, *fabula*).

So it is clear that Latour is not proposing to reduce all artists to *storytellers*. But does he not oblige us to think of all art as representative, "mimetic"? The modernist avant-garde has succeeded in convincing us that art can, and perhaps should, be nonfigurative. The idea that there has to be a representative dimension for a thing to have artistic characteristics seems to be regressive as far as the history of twentieth- century art is concerned. The search for *literalness* in artistic expression has been the main business of artistic modernity. To make a work that relates to nothing but itself, that works as a tautology ("I am exactly what you see, *and nothing else*, look no further"), that breaks down the spectator's tendency to get distracted and obliges her to concentrate on what is there, and there alone—all this was one of the great passions of the twentieth century. One could also think of minimalist sculpture, conceptual art, even Brechtian theater.[4] Can we say that Kazimir Malevich's "White on White" is a fiction (in the sense that it makes something imaginary exist)? Can this be said of Piet Mondrian's paintings? Or of Pierre Huyghe's dog?[5] Is she not a real dog, even though she has one pink leg? And Eduardo Kac's rabbit: is it not a real fluorescent rabbit, not a rabbit painted with a fluorescent color, but a rabbit born fluorescent because of genetic engineering? In what sense can it be said "fictional"?

These questions allow us to get to the nub quite quickly. By fiction we usually mean a reality composed of two asymmetrical halves. On the one we would have a material half, a book, for example, or a canvas covered with pigment, sound waves passing through a human mouth; and on the other immaterial beings, pure mental "representations." No one would be inspired to go and tip a little vinaigrette on Manet's asparagus, because we know it is not asparagus but an image of asparagus, and all one would eat would be glue and paint. But Latour points out that fiction is a much more complicated thing. What characterizes fiction is *the inseparability of the material and the figure*. One cannot separate the asparagus that appears in Manet's painting from Manet's painting; one cannot separate Odysseus from Homer's epic, etc. No doubt Odysseus and Don Quixote are removed from the specific material supports that were there when they first appeared: they move into foreign languages; they reside in our memories; they begin to live elsewhere: in children's illustrated books; in the apocryphal series of adventures of a sad-faced knight, which Cervantes was confronted by, and which he turned around to reintegrate in his fiction throughout the second volume of the authentic Don Quixote; in scholarly rewritings (Joyce) as well as in an anime television series (*Ulysse 31* was a series created in the eighties by a French-Japanese team). So one essential property of fictions is that of being able to be picked up again and moved to another world. And yet one necessarily comes back to the materials to which these beings are hooked up. They are never *totally* detached from them. That is what Latour means in this important passage:

This happens every time a little cluster of words makes a character *stand out*; every time someone *also* makes a sound from skin stretched over a drum; every time a figure is *in addition* extracted from a line drawn on canvas; every time a gesture on stage engenders a character *as a bonus*; every time a lump of clay gives rise *by addition* to the rough form of a statue. But it is a vacillating presence. If we attach ourselves to the raw material alone, the figure disappears, the sound becomes noise, the statue becomes clay, the painting is no more than a scribble, the words are reduced to flyspecks. The sense has disappeared, or rather *this particular sense*, that of fiction, has disappeared. But—and this is its essential feature—the figure can never actually *detach* itself, either, from the raw material. It always remains held there. Since the dawn of time, no one has ever managed to *summarize* a work without making it vanish at once. Summarize *La Recherche du temps perdu*? Simplify Rembrandt's *Night Watch*? Shorten *Les Troyens*? And why? To discover "what they express" *apart from* and *alongside* their "expression"? Impossible, unless we imagine Ideas embodying themselves in things. This impossibility is the work itself.[6]

What characterizes the general system of fictions is the inseparability of matter and form. This is not simply due to the ontological law requiring that any imaginary (or mental or incorporeal) content must be supported by something material to be said to *exist*. For the opposite is also true: if one wants to separate the vision of the asparagus from its interpretation as asparagus, we have nothing left, not even the articulation of stuck-on pigments. What then would the "painting" consist of? Why not include the frame, and the weft of the cloth and even the dust that is sometimes found on it? In such a scenario, as soon as one dusts off the painting, it will no longer be "the same." What reasons do we have here to speak of one and the same object, except insofar as it is the support for a representation? In the same way, if you don't hear the way sounds are organized in a melody, you simply no longer hear the *same* sounds. The "figure" thus gives as much being to the "material" as it gets from it.

So if one concedes that the fictional being does not simply designate the "mental" or "imaginary" part of the work, but its unstable totality, the whole movement of both breaking away and inseparability, then one understands that a Mondrian painting or an Yves Klein monochrome can be just as much "fictions" in their being as Flaubert's *Madame Bovary* or Leonardo's *Mona Lisa*, even if they "represent" nothing. The latter two can no more be separated from their "materials" (yet without being reduced to them) than the first two can be separated from a kind of figuration of themselves. They exist in addition to their own material reality. In fact, isn't this really the whole point of contemporary art? The people who walk on Carl Andre's *144 Tin Square* (1967) are witness to the fact that the problem is not one of *understanding* such a work, but of *perceiving* it. In order to perceive it, some understanding is necessary, and inversely, understanding it means first of all paying attention to it, becoming sensitive to it. The same applies to Andy Warhol's *Brillo Boxes*. If they are works of art, is it not the case that they have been, dare I say, "fictionalized real," as in the expression, "frozen alive"? Modern art's characteristic claim to "literalness" is in fact an attempt to renew the resources of fiction.

This argument allows us to better understand an important aspect of the aesthetics of contemporary art. This aesthetics is often reproached for taking pleasure in works that can only be appreciated if one reads the "instructions for use," so to speak, as if only the *interpretation* of works made their value, as if these works were not content in themselves with their aesthetic power, but only received it through discourses about them, from the outside, in the manner of garden stakes for plants that can't stand up by themselves.[7] One must first reply that this is also largely true of works prior to the Duchamp revolution, not only because classical painting assumed knowledge outside of pure visual experience (that of *historia*, for example, and even the aesthetic conventions and critical debates linking artistic practice to particular theories), but also quite simply because, if one didn't see *a smile* in the Mona Lisa (if one could only see stuck-on colors), one would not appreciate it as much, since it is the back-and-forth play of matter and figure that makes it so marvelous. This brings us to the important point: there is no difference between perceiving the emergence of this mysterious smile (the mystery does not hang on the ambiguity of its meaning as much as on the uncertainty of its reality: is it really a smile, almost a smile, or the dream of a smile?) coming through the superimposed layers of varnish applied almost five centuries ago, and reflecting on the meaning of the first editorial of *Art & Language*, or glossing Robert Rauschenberg's gesture when he erased Willem de Kooning's drawing. In all these cases, one has to move from the work to *something else* (an image, an idea, an emotion) and come back from the "something else" to the work. Sometimes this "something else" is a perceived figure; at other times it is a commentary. But the commentary is a sort of "figuration" too, if one means by that a slight breaking away from the here and now of the work but not utterly detached from it.

So, even in the case of the de Kooning drawing erased by Rauschenberg, it is not a matter of just anyone's drawing (but an artist whom Rauschenberg admired, his erasure gesture being both an homage and a sacrifice; and also a representative of abstract expressionism, the aesthetic philosophy of which is thus "captured" by Rauschenberg in the sense that it becomes an instrument of figuration through the "commentary" he is making in erasing the drawing); and not just any drawing (but a drawing chosen by de Kooning himself because of the richness and intricacy of its lines, its resistance to erasure); and it is not erased in any random way (but with an eraser, meticulously, over several weeks), etc. So even in this work where there is "nothing to see," as Rauschenberg himself attests, the singularity of its here and its now (which constitutes its "material") keeps *relaunching* the "comprehension" of the work, that is, keeps allowing us to become more sensitive to it. The more we perceive it, the more we understand it, and conversely. Hence when one arrives in front of this slightly crumpled little rectangle flattened in its plain frame at the San Francisco Museum of Modern Art, one *perceives* something that one would not notice were one not primed with everything we just mentioned, that is, if one had not *learned* it. But this is no different from the person who knows nothing about the theological problems of the Incarnation being less sensitive to the extraordinary mode of appearance of Leonardo's *St. John the Baptist*. De Kooning's drawing erased by Rauschenberg is therefore a fiction, even if it doesn't represent anything. It is a fiction simply because it is possible *not to see anything in it*, as Daniel Arasse said about works requiring more classical regimes of *sensitivity*.[8]

So the concept of "fiction" is a long way from being inapplicable to contemporary art, and it allows us to characterize what Jacques Rancière called the "contemporary regime of art." This regime is distinctive in that it has flattened out and equalized all the modalities of fiction, whether it is a question of figuration in the strict sense (the emergence of an interpretable figure: "this is a smile," "this is the battle of Waterloo," etc.; or the identification of an *historia*), of states of mind (I feel melancholic in reading this poem, or looking at this painting by Caspar David Friedrich), of interpretation (under the guidance, or not, of the knowledge of iconographic programs[9]), or of theory (which necessarily accompanies works in *Art & Language*, for example). All of this equally relates to the same mechanism that prolongs the work beyond its "material." In the same way that it is not absurd to say that Lamartine's "Le Lac" *figures* love, even though it does not depict it in an allegorical or descriptive mode, so one could also say that Warhol's *Brillo Boxes* figure the ontological ambiguity of the commodity and the art object, even if this is done through a series of gestures that appeal to the spectator's *reflections*. One might well be amazed by the worlding (*cosmogenesis*) of Jean-Baptiste-Siméon Chardin's peaches, as the Goncourt brothers were,[10] and thus of the *passage* from the dried colored paste on the canvas to the recognizable, almost palpable object, or one could be passionate about the effects that a work has on art theory (hence the way that Donald Judd's sculptures "comment" on the traditional constraints of sculpture by transgressing them). In every case there is play in the *passage* from one level to another. It doesn't matter whether we are dealing with the passage from *matter* to *percept* rather than the passage from *gesture* to *sense*, or from a narrative *topos* to a *visual interpretation*. This, then, is why contemporary works of art are fictions.

We are now in a position to provide the most general definition of the concept of fiction. The problem of fiction is not that of the existence of imaginary beings, but of the perception of a certain category of objects, which have the peculiar characteristic of being perceptible only on condition of being extended along a particular plane. Thus a rock concert is simply noise for someone who doesn't know the connections that this being demands (and that are conveyed through the social and bodily gestures of the concert, as described by Antoine Hennion[11]); in the same way, a virtuoso chord in a Gustav Mahler symphony would slip by completely unnoticed by those who can't perceive such pertinent nuances in "classical" music (and this perception can be accompanied, for instance, by the need to go to the concert, or to lie on one's couch selecting from one's playlist).[12] This risk, uncertainty, or "hiatus" is what characterizes "fictions" in Latour's sense. In order to extract the fragile supplement that makes fictions what they are, very diverse conditions are needed. Hennion demonstrates along these lines how some people need to watch the countryside go by from a train window in order to fully hear music, or at least music that they had previously not attended to; others, instead, have a need to dance. Nevertheless, the main idea is to apprehend a fragile, uncertain being, which only exists under the precise condition of being taken up, welcomed, commented, designated, and reactivated.

A fictional object is defined by the condition of being extended into something else: the work of art exists because someone speaks about it, in exactly the same way that it only exists through the goose bumps I have when I see it. There is a great variety of extensions, but without them the work does not *exist*. We can distinguish three types. First, states of mind; second, critical reactions (on the one hand, ordinary conversations on coming out of the cinema or an exhibition, for instance; on the other, texts by professional writers of criticism or theory—the work only exists by being spoken about, even if it does not exist just because it is spoken about). Thirdly, there are other works, as one of the most essential prolongations of a work of art. A fiction necessarily gives rise to other fictions.

One nevertheless has to specify that this does not mean that a work of art is only an effect of discourse or representation. It just means that it has to pass through something else, that it must be extended via something else. But Latour insists on saying that it is the work of art that solicits us, and not we who give it being in some sovereign fashion. It solicits us in two ways: from the position of its creator who makes the work come about, and from the position of the receiver, who takes care to continue the work. Creator and receiver together contribute to the existence of the work, without being ex nihilo inventors of a reality that could be reduced to an effect of representation. Here Latour is alluding to two well-known phenomena: on the one hand inspiration, and on the other delight. In both cases, we are passive in the face of the work of fiction. We see that it is calling us to create it, and we put ourselves at its service in order for it to exist, even going so far as to invent the necessary subjectivity it needs, as André Gide did, in words cited by Maurice Blanchot, saying that he had the personality that was needed to write novels.[13] On the other hand, delight is the state in which the reader or the spectator forgets herself, and delight can go quite a long way in this self-forgetting in order to make the work exist, or rather to valorize as much as possible that which she has perceived. So the problem with fictions, in fact, is that it is easy to miss them. They are fragile. It is easy not to see how inspired a jazz improvisation is. It is easy not to see how inspired Latour's proposition is. So it has to be highlighted (which is what I am trying to do here for Latour's work). These are operations that are at the heart of the *making-be* (faire-être) of these objects. So we can sum up Latour's thesis by saying: fictions exist as they pass through us, but that does not mean that we invent them; they are more like viruses that can reproduce only as parasites to host organisms, who, by spitting, coughing, and sweating, encourage the propagation of these entities.

Here we find three moments that are actually characteristic of all "existence" according to Latour, each mode of existence being defined by the distinctive manner in which it goes through these moments. First, it has relations with heterogeneous things (for example, Leonardo's hand and Arasse's books, but also the walls of the museums, the memories one retains, the other Leonardo works, and so on). A being that stops, that does not move about or "pass," disappears, falls into the void. Secondly, this passage is nonetheless finalized by the *instauration* of certain beings: one needs the relation between the strings of the electric guitar, my neighbors swinging their hips, the amplifiers, and even the beer, the night, etc., so that "the music" can exist, but these relations are prescribed for the instauration of this being, "music." It is for this reason, for example, that we would not include among the regime of relations the fact that I am in a relation to a paper factory via the ticket I bought to get into the concert. In effect, this relation is not relevant for the instauration of this particular being (but it is from another point of view, that of attachment, as Latour calls it). The fictional mode of existence is therefore defined by way of *contrast* between the material and the figure, through this "shifting" out to another level (as Latour calls it), without which there would simply be no being. And thirdly, this contrast is never guaranteed once and for all; it always has to be renewed, begun again. So, three forms of alteration: being through *others* (music is caught up in a network); being through *contrast* (music is not sound); being through *retakes*, iterability. To make something be, one has necessarily to *instaure* a new being (contrast) by mobilizing several different beings (network) through a number of takes (prolongations).

III. Is Art a Fiction Like Others?

Let us now turn to the other point of resistance art militants might have in regard to this requalification of their beings as fictional beings: does one not lose the specificity of art in the fictional regime? In fact, as we have just seen, the Latourian notion of fiction is a very broad one. It includes not only all images, but also the simplest signs and the most commonplace phrases. Wouldn't one's grasp of the being of works of art necessarily be hindered if there were no difference between ordinary language and literature, between conversation and verbal art, between mechanical drawing and fine drawing, between a death mask and a Rodin sculpture, between a strip from a photo booth and a Richard Avedon print, between the self-invention of character by Kim Kardashian versus that of Salvador Dali, Warhol, or Joseph Beuys, etc.? Even if one were to accept works of art as fictions, one would not have come to terms with their essential ontology, because this originality relates precisely to the way in which they tear themselves away from the ordinary fictional regime.

This hesitancy should be treated with a lot of respect. It is not a matter of knowing whether there is a difference between art and other fictions or not (obviously there is), but of knowing if this difference is of degree or in their very nature; in other words, if it deserves the creation of another "mode of existence" or not. Latour fuses two major lines of traditional metaphysical thought: that of dealing with fictional beings on one side, and conversely that of dealing with works of art. He does not reduce the latter to the former, but redefines them both into a new "mode of existence." This fusion corresponds to the set of specifications that is found in the *Inquiry*: not the maximization of the differences among genres of being, but their capacity to be diplomatically compatible at the heart of a system. Yet is this not committing such a grave "wrong" to these particular beings, these "works of art," that it deserves to be rescinded?

"Fiction" is what the headlines in the daily paper and a Mallarmé poem have in common. Both require a *figure* to be released, otherwise they would not even exist. Suppose you are confronted for the first time with a headline written in Hindi, a language with which you happen to be utterly unfamiliar. It is not enough to say that you don't understand what is written; you don't even know *what there is to understand*, since you don't know how to break the text into words or letters. To learn how to read is to train our perception to see patterns, and not only to become capable of assigning a sound to a visual shape. A foreign language is not a mere concatenation of perceptions (sounds for oral language or shapes for writing) to which we cannot ascribe any interpretation. It is a confused perception. In other words, the most simple linguistic experience implies a *figuration* problem, just like the most subtle artwork does. They are all fictional beings in Latour's sense.

Mallarmé was famously insistent on the ontological difference between a poem and the "words of the tribe." But does one need to push the difference to the point of tearing the poem's mode of existence away from fiction, just to satisfy their irreducibility? Can one not be content to note that there is a difference of degree: the poem is a second-degree fiction, since it is made from first-degree fictions, i.e., forms of language. One could borrow Bertrand Russell's theory of types. If language forms are type 1 fictions, then poetic forms are type 2 fictions: they are "metafictions," "superfictions," fictions of fictions. This might actually provide the first "criterion" for recognizing the specificity of art among fictions. Let's call it "stratification." The more a fiction is stratified, the greater the number of disarticulated layers it needs. In order to understand a phrase like "ship sinking off the coast of Normandy," you need fewer strata of shifting (*strates de débrayage*) than you do to understand "À la nue accablante tu," Mallarmé's famously untranslatable poem that describes something that is more or less a shipwreck. Art is a *fictional supplement*, an extra fiction. It is made by shifting shiftings. Therefore it is intrinsically relative. It transforms what it touches into "material," and it significantly "materializes" other fictions. This is the reason why art rises above itself, each work needing to "go further" than its predecessors, playing on "intertextuality," etc. This derivativeness (*secondarité*) of art has been theorized by one of the major aesthetic currents in modernity, "formalism." Viktor Shklovsky suggested that art was the ensemble of procedures through which an effect was produced based on an already fictional material.[14] Art *recharges* fiction.

More generally it seems that what we call "art" corresponds to a higher bid (*surenchère*) in fiction, a matter of extracting fiction from where it is least likely to be found. Art is a tendency to *virtuosity* in fiction. There are all sorts of virtuosities, but the essentially artistic one is that turned toward the most uncertain fictions. This gives us a second, quite traditional, criterion: art is the practice through which "the Moderns" (as Latour calls them) gave themselves the task of not only isolating this mode of existence in itself, but also of making it the object of a specific and passionate quest. Language is a fiction, but it is undoubtedly useful for many things. Concentrating on dealing with what is fictional in it is a way of beginning to make an artistic medium out of it. In the same way, drawing is an ancient medium, as witnessed by cave paintings, but it was at the same time used to instaure beings other than fictions: religious beings, for example, or those to do with "metamorphosis" that Latour also discusses. Making fiction itself an exclusive aim, and therefore bidding at a higher level, is perhaps the gesture that constitutes what we mean by art. There is a technical word that in the aesthetic tradition designates the property of fiction taking itself as the goal of fiction: *autotelic*, that which is its own aim. But we could put it more simply: *free*. Art searches for the liberty *of fiction*. What we call art seems therefore to be a space of intensification, even a chain reaction, of fictional effects.

So artistic fictions are both more stratified and freer. The higher bid that is characteristic of fiction nevertheless takes a particular form. It is characteristic of fiction to be subject to uncertain perception, because the contrast between figure and material is always in play. The higher bid of fiction will therefore pivot on obtaining the maximum of figurative contrast with the minimum of material difference. This criterion could be called *finesse*. It is the third criterion, and, once again, it is a matter of degree rather than nature. There is the same kind of difference between the Kardashian sisters and Dali as there is between a bad Disney cartoon and a good one. One is easy, elicits simple figures out of its material, and *instaures* a quickly grasped contrast; the other, in an opposite way, runs very close to a certain "flatness," is easy to miss, etc., but is a delight for those who know how to get a fix on it.

Artistic fictions are more stratified, freer, finer, but do they have other differences compared to other fictions? Yes, they can be said to be more or less "long." The most powerful fictions are those that "send" us the furthest, those that best make us "forget" the world we are in, not because we have a need to forget it (because that would be for "psychological" reasons relating more, according to Latour, to the mode of existence of "metamorphoses"), but because they seem to have more consistency than all the other beings that surround us. The more powerful a fiction is, the more it is capable of replacing our world by another. This is the meaning of Orhan Pamuk's work of fiction, *The New Life*, in which a man reads a book that changes his life once and for all. This is a "fiction," of course, a perfect fiction of fiction. But we are familiar with the everyday expression of this criterion. We know that a fiction, to exist, has to be extended via other materials, has to become the object of "commentaries," which are nothing but ways of bringing out the contrast with the figure, or delivering up other works, etc. The longer I have to write to really bring out the feeling of the figure of which I consider myself to be the recipient, the more intense the fiction is. The proof that a Leonardo painting is a very intense fiction is that, in order to reconstitute what I have seen, to valorize and safeguard the being that has visited me on this occasion, I have to write a very long book. Thus, artistic fictions intentionally extend the trajectories of fiction. The material/figure contrast is here achieved by way of very long and sinuous trajectories of instauration.

This is not all. Not only are artistic fictions more stratified, freer, finer, and longer than the others, but they must also be more "complex." By this I mean two things: on the one hand, fiction "sends" us in a greater or smaller number of *different* figurative directions; on the other, it can make lines of figuration interact with one another in a *more or less intricate* way. A fiction will be all the "richer" for opening up more "readings," which in turn differ from each other, and also by allowing these readings to refine or reinforce each other. *Moby-Dick* is a work likely to be the object of numerous readings. These readings do not cancel each other out. They curiously integrate themselves, so that we end up with a quite particular perceptual experience, which is *Moby-Dick* itself. *Moby-Dick* is charged with all this complexity, and its fictional character is as strong as this complexity is broad and intricate.

These are only, in the end, different ways in which fictions can be *more fictional*. Basically, works of art are beings that are more fictional than others, or, more precisely, beings that incorporate an *intentional differential of superfiction (surfiction)*, in the "formalist" sense. We can therefore see that none of these criteria requires an *exit* from the mode of existence that is fiction. It seems to me that there is but one essential criterion, which is perhaps that of nature rather than degree. And this is a difference between art and ordinary fictions that has not yet been discussed. Works of art, no doubt, "detach" us, "release" us, "send" us, and allow us, as in the banal expression, to "escape," but at the same time they oblige us to come back very close to ourselves. Art militants might well protest at this point that there is an important property of these beings they are defending, which is not reducible to a difference of degree. *Delight*, they might end up saying, *is not an escape*. Artistic fiction is not content with sending us off into another world; it organizes relations with the worlds from which it is detached (and from which it detaches us) that are much more complex. If you fail to recognize it, you reduce art to a simple practice of *entertainment*. Now, the *delight* characteristic of artistic fiction, that which makes it different from the entertainment characteristic of the ordinary regime of fiction, is that it is made up of a distraction that is at the same time a greater attention brought to bear on that from which it is detaching us. This is perhaps the formula for art: to distract us from our situation, but in order to lead us back via extremely difficult-to-follow paths; sending us far away, but bringing us back closer to ourselves (or rather closer to the situation where we find ourselves on an n-1 level, compared to where we were at the start). There is thus a *circle of art* (as there is for politics, according to Latour), and this circle would allow artistic fictions to escape the mode of existence of other ordinary fictions.

It would do so, that is, if it were true that this property was not established for all fictions. Perhaps all the art militant has done here is put her finger on an aspect of Latour's report that must be rectified. In general, all fictions would go *in both directions*: not only from the material to the figure, but also from the figure to the material. One can in fact show that this hypothesis is correct. Thus, if one takes a very "ordinary" fiction, natural language, it would be false to think that the "signifier" is the material, while the "signified" is the "form." In fact, the material is as much semantic as phonetic. The "figuration" process goes both ways. It might be that language is not specific in this respect and that all fictions are animated by this double movement, like a sort of reciprocal shifting. Developing this point would take too much space.[15] Suffice it to say that in this case, there would be a return movement in all fictions, a double release, a coming and going characteristic of all "signs" in general. Works of art would be characterized only by the intensification and acceleration (ideally pushed to the infinite) of this coming and going. Incidentally, this corresponds to the "pendulum" image that Paul Valéry used to define the play of art.[16] We might then have to rewrite Latour's diplomatic account of fictions in order to accommodate works of art. But this is good news: it shows that the diplomatic process has not been in vain: referring back to the militants has been fruitful.

IV. Conclusion

We can be satisfied. The test of diplomatic compromise that the *Inquiry* has undergone, its discussion "in the agora," as Latour demanded, with the "militants" of a mode of existence (idealized militants, no doubt, but that cannot be a valid objection), this manner of taking seriously the proposal that is *An Inquiry into Modes of Existence*, has not only enabled a precise, and one hopes interesting, coming to terms with a "modern" phenomenon that is particularly dear to us ("art"), but it has also allowed, if not the development, then at least the preliminaries for the correction of the conceptual apparatus of the *Inquiry* by refining the concept of fiction. We became, as Latour hoped, "co-researchers."

Of course, this is not the end of the matter. A lot remains to be done if we want the diplomatic procedure to be complete. We would need to check that this recharacterization of "art" helps to settle the main ontological conflicts in which art and fictions in general are engaged; we might also ultimately have to draw an image of the "common house" (as Latour calls his own systematic attempt) in which all the modes of existence are composed. But that is an essential feature of the diplomatic attitude: it has no end. The inquiry will go on.

Notes

I wish to thank warmly Stephen Muecke for his generous and outstanding work of translation.

1. This article is the outcome of a series of workshops organized for one part by Paris West University Nanterre La Défense and the École supérieure des beaux-arts de Montpellier Agglomération (ESBAMA), and for another part as one of the diplomatic encounters organized by Bruno Latour in the course of the AIME project (for more details, see modesofexistence.org). I wish to thank my partners in crime for this work, Laetitia Delafontaine, Grégory Niel, Juan Luis Gastaldi, and Michel Martin, as well as Christian Gaussen and, of course, Bruno Latour. I am also grateful to the students of Montpellier Art School and other artists, art critiques, art historians, and even art dealers, who have agreed to discuss Latour's proposal about art as fiction on those occasions. They are the "militants of art" I have in mind.

2. I have offered an introduction to Latour's *An Inquiry into Modes of Existence* (in particular, by resituating it in the context of what is known as the "ontological turn" in anthropology), in Patrice Maniglier, "A Metaphysical Turn? Bruno Latour's *An Inquiry into Modes of Existence*," *Radical Philosophy*, no. 187 (2014): 37–44.

3. Of course, this is also an allusion to Latour's first metaphysical text, *Irreductions*, published in *The Pasteurization of France* (Cambridge, MA: Harvard Univ. Press, 1993).

4. Let us recall that Michael Fried characterized minimal art as a "literalist." See "Art and Objecthood." I have myself tried to explore in new directions the notion of literalness in contemporary art in "Du conceptuel dans l'art et dans la philosophie en particulier," in *Fresh Théorie II*, ed. Mack Alizart and Christophe Kihm (Paris: Léo Scheer, 2006), 504 sq.

5. Emma Lavigne, *Pierre Huyghe* (Paris: Centre Georges Pompidou Service Commercial, 2013), 187–97.

6. Latour, *Inquiry*, 244.

7. This might be the reason for Gilles Deleuze's rather vicious attack against conceptual art in *What is Philosophy?* See Deleuze and Félix Guattari, *Qu'est-ce que la philosophie?* (Paris: Editions de Minuit, 1991), 187; Deleuze couldn't find there the *consistency* at the level of *composition* that, according to him, characterized art.

8. Daniel Arasse, *On n'y voit rien: Descriptions* (Paris: Gallimard, 2000). The book has been translated in English under a slightly different title: *Take a Closer Look*, trans. Alyson Waters (Princeton, NJ: Princeton Univ. Press, 2013). In French it meant something like: "One cannot see anything here."

9. Iconographic programs are sets of precise instructions given by the commissioner of a painting to the artist in order to specify what the painting is meant to represent. Those programs were very common in the history of Western painting and show that the "reading" of visual art was codified. Erwin Panofsky and his school were very influential in putting those programs at the center of art history. See, for instance, Panofsky, *Meaning in the Visual Arts: Papers in and on Art History* (New York: Doubleday, 1955).

10. "C'est là le miracle des choses que peint Chardin: modelées dans la masse et l'entour de leurs contours, dessinées avec leur lumière, faites pour ainsi dire de l'âme de leur couleur, elles semblent se détacher de la toile et s'animer, par je ne sais quelle merveilleuse opération d'optique entre la toile et le spectateur, dans l'espace." Edmond and Jules de Goncourt, "Chardin," *Gazette des beaux-arts* (July 1863), republished in Goncourt, *L'Art du XVIIIe siècle* (Paris: Rapilly, 1873).

11. See Antoine Hennion, *La Passion musicale: une sociologie de la méditation* (Paris: Métailié, 1993), 317.

12. Hennion, "Une sociologie des attachements," *Sociétés* 3, no. 85 (2004): 20.

13. Maurice Blanchot, *L'Espace littéraire* (Gallimard, 1955; Paris: Folio, 2007), 105.

14. On this notion of "formalism," see my article, "Du mode d'existence des objets littéraires: enjeux philosophiques du formalisme," *Les Temps Modernes* 5, no. 676 (2013): 48–80.

15. I have exposed the details of this argument in "The Embassy of Signs: An Essay in Diplomatic Metaphysics," in *Reset Modernity!*, ed. Latour (Cambridge, MA: MIT Press, 2016).

16. Paul Valéry, "Poésie et pensée abstraite," *Œuvres I* (Paris: Gallimard, 1993), 1332.

CHAPTER SEVENTEEN

Actor-Network Aesthetics

The Conceptual Rhymes of Bruno Latour and Contemporary Art

FRANCIS HALSALL

A ROOM WHERE THE LIGHT GOES on and off; a renovated row of houses in Chicago; a series of brightly colored box structures measuring 300×150×160 cm; a large metal spiral covered in fabric in front of a large photograph of donkeys in a graveyard; a partially cultivated wasteland populated by plants, bugs, and an emaciated dog with a dyed pink leg. The subjects and objects of contemporary art can easily read like a "Latour Litany," the term coined by Ian Bogost and Graham Harman to describe the lists that frequently appear in Bruno Latour's work. Today, anything, it would seem, can be art. That is, none of the following describes accurately the contemporary condition of artistic practice: stylistic development that can be grouped according to an epoch or "-ism"; individual art objects that can be prioritized over the systems of their production; or understanding art according to specific mediums.[1]

This condition of eclecticism is the basis for this paper, which begins from the assumption that clear distinctions cannot be drawn between contemporary artistic and theoretical practices. My argument is, ultimately, that Latour operates like a contemporary artist. In arguing this point, I borrow Bruce Clarke's concept of conceptual rhyming, which he uses to describe an equivalence between the narratives told in sociology and in fiction. As Clarke says: "The stories Latour tells about sociotechnological quasi-objects present a range of transformative interactions that rhyme conceptually with narrative fictions of metamorphic changes."[2] I use the term here because it nicely captures an aesthetic dimension to this equivalence between art and actor-network-theory.

If contemporary art can take anything as its medium and subject matter, then the means of production available to the contemporary practitioner are similarly multiple and open-ended. Art is now, as Hal Foster puts it, the "paradigm of no paradigm."[3] This eclecticism of medium, style, or modes of production is, I argue, mirrored in the approaches by which Latour redescribes the world: as something in which everything is ready at hand to be assembled and reassembled according to his analytic methods and to whatever mediums or platforms are available. Deploying aesthetic strategies such as metaphor, allusion, jokes, fictions, and so on, Latour's modes of analysis mirror the methods and style of contemporary artistic practice.

I begin by giving an account of the conditions of art in terms of three categories, all of which are rendered ambiguous through contemporary practice: style/epoch, object, and medium. My aim is to demonstrate both the diverse nature of contemporary practice and, hence, its conceptual rhyming with Latour's practice. This equivalence returns in the conclusion, where I argue that Latour's strategies are the same as those of a contemporary artist; that is, to mobilize the various platforms ready to hand and *perform* theory. To be clear, my claim is not that terms from Latour's theory can be used to better understand what's going on in contemporary art. Instead, my claim is that there is an equivalence between the way that Latour works and contemporary art practices. In short, Latour works like an artist.

Next an account is given of Latour's notable engagement with the art world and with artists.[4] He finds in art a useful mode of inquiry that complements his own engagement with sociological and anthropological analysis. This turn is important because not only does it provide access to resources (such as people, money, spaces, and vocabularies), but also because in doing so it can address vital concerns such as climate change and the condition of the Anthropocene, where the imbrication of culture and nature that Latour attributes to modernity seems likely to produce ecological catastrophe.

Subsequently, I offer something akin to another Latour Litany by identifying several of his key concepts that relate directly to contemporary artistic practice. These are: the collapsing of the modern nature/culture distinction; actor-network-theory (ANT); network; quasi-objects; flat ontology; mediators/translators; and irreductions. This is, admittedly, a partial list. It is not intended to be a thorough description of the richness of Latour's thought. And I do not claim that any of the examples of art I discuss are reducible to a single theoretical paradigm. I do, however, hope to have given both a sense of what is happening in art right now, and to demonstrate its conceptually rhyming with Latour's practice.

Such a partial listing and overview of Latour's position are problematic because he is constantly developing his position (from anthropological history of science, through an "actor-network" account of the social construction of facts, on to metadiscursive reflection on method), even to the extent of reevaluating his previous claims, such as with the AIME project, which revisits his claims in *We Have Never Been Modern* and offers a very different account of modernity. But in this respect, I conclude, Latour is also behaving like an artist. Consider, for example, Bruce Nauman who, since the 1960s, has steadfastly refused to stick to one medium, working with films, video pieces, performance, neon sculptures, and land art interventions, and even staging a sort of retrospective of his work in audio form, in *Raw Materials*. Latour, like Nauman, also moves through different styles and mediums. Both artists use their practice to stage a series of experiments in both form and content in order to reflect upon their strategies, mediums, and outputs.

(Post) Contemporary Art, Object, and Medium

As my opening examples illustrate, nowadays anything, more or less, can be art. For Arthur Danto, this has been the case since the historical turning point of pop art, after which "there is no special way works of art have to be."[5] This is the condition of contemporary art. After the End of Art there is: an end to the historical development that culminated with high modernism; a dissolution of distinctions between mediums in the spirit of artistic pluralism; and an increasing self-reflexivity on the part of artistic practice, whereby its aesthetic content becomes indistinguishable from its theoretical content. Here I characterize this condition according to three modalities related to: style/epoch, object, and medium. In doing so, I offer a constellation of different tendencies to give an account of what contemporary practice is like. In all three cases these modalities overlap; the cubist epoch, for example (named by Douglas Cooper as the period from 1907–25[6]), was characterized by a *style* enacted in the *medium* of paint and presented through the *objects* of paintings.

First, there is what Peter Osborne names the postcontemporary, that is, the uncoupling of contemporary practice from a particular style inhering within a particular time. This phrase describes a situation rather than a style and "the historical-ontological condition for the production of contemporary art in general."[7] Unlike previous periods that can be both stylistically and historically delineated, such as the baroque or impressionism, postcontemporary art is rather identified through certain characteristics, including its conceptuality, its aesthetic dimension, and "a radically distributive—that is, irreducibly relational—unity of the individual artwork across the totality of its multiple material instantiations, at any particular time."[8]

Second, there is the condition that Lucy Lippard calls "the dematerialization of the art object."[9] She uses this phrase to refer to art after modernism, when an interest in singular objects was replaced by work that explored its relationship with its various systemic environments through strategies including conceptualism, performance, earthworks, and systems art. While Lippard's frame of reference was the late 1960s, her definition presents a still-current challenge to the art object on the part of contemporary practices that continue to foreground the relationships between the object and its various environments. This move of aesthetic and theoretical attention from discrete art objects to the institutional systems of support, display, and distribution remains a key characteristic of art today, evident in two ways. On the one hand, it is apparent in the still-dominant legacy of Duchamp's readymades as a historically self-conscious acknowledgement of the relationship between art and the institutional frames of art history, the museum, and the market. This acknowledgment applies to both contemporary versions of conceptualist immaterialism (such as Tino Seghal's staging of conversations in gallery spaces); and to so-called new materialisms (such as Martin Creed's use of everyday objects in his work), in which the distinction between individual art objects and their environments is rendered ambiguous. On the other hand, in the so-called "social turn" in poststudio arts practice (referring to relational practices and socially engaged or community-based art), art is acknowledged as inextricable from social networks of distribution and display.[10]

Third, Rosalind E. Krauss's account of art after modernism presents it as being in a "postmedium condition." The technology of photography challenged the status of the specific aesthetic object of art, both as a unique object and as a mode of image-making. This challenge plays out in a lack of faith concerning medium specificity and in a reinvention of the conceptual coupling between artistic production and its mediums: "And if photography has a role to play at this juncture, which is to say at this moment of postconceptual, 'postmedium' production, Benjamin may have already signalled to us that this is due to its very passage from mass use to obsolescence."[11] Due to technological imperatives, Krauss thus argued, postmedium art required a jettisoning of a certain modernist understanding of medium specificity. She advocated instead a definition of medium as the "technical support" for the work of art. This expanded definition would not reduce medium to "the specific material support for a traditional aesthetic genre."

More recently, responding to the new technological imperatives of the twenty-first century, David Joselit and Lane Relyea have revisited the problem of medium in contemporary art practice vis-à-vis technology. Joselit identifies new screen technologies (iPads, phones) as allowing images to circulate freely between different means of distribution and display and, hence, as creating the conditions under which contemporary art is now determined by *format*, defined as "a heterogeneous and often provisional structure that channels content."[12] Formats are opposed to *objects* that are characterized by "discernible limits and relative stability [that] lend themselves to singular meanings," and these formats "regulate image currencies (image power) by modulating their force, speed, and clarity."[13] "*After* art," Joselit explains, "comes the logic of networks where links can cross space, time, genre, and scale in surprising and multiple ways."[14] Due to the flattening effects of the network as medium, moreover, the distinctions between artists, curators, writers, and so on is becoming increasingly hard to discern. This blurring can be seen in the way that curating is becoming indistinguishable from art-making. It is further evident in recent curatorial strategies where historical exhibitions now compile supporting material such as documents, objects, images and so on alongside the primary works when constructing a historical survey or wider-themed display.[15] Or consider how a pop star like Katy Perry can claim to have "personally curated" a cheap jewelry range, *Prism*, for Claire's Accessories.[16] Likewise, for Relyea, contemporary art is framed by "the rise to dominance of network structures and behaviors and their enabling manifestations: the database, the platform, the project, and the free agent or do-it-yourselfer."[17] One consequence is the blurring of the lines between artist, cultural agent, and entrepreneur. This theme returns in the conclusion, where I propose that Latour is also such a figure: that is, someone who blurs the lines not only between academic disciplines but also between artistic, cultural, and entrepreneurial activities.

Latour and Art

Latour has a deep connection with the contemporary art world. He is a rare example of a theorist who has also collaborated with artists and worked as a curator, and whose work is highly visible in art discourse (as well as in design and architecture which are not discussed here).[18] His work appears in the widely read online art journal *e-flux*;[19] he was awarded the Nam June Paik Art Center Prize (2010); and he has been involved with three large-scale exhibitions at the Center for Art and Media in Karlsruhe (ZKM): *Iconoclash: Beyond the Image Wars in Science Religion and Art* (2002), *Making Things Public: The Atmospheres of Democracy* (2005), and *Reset Modernity!* (2016). Latour also established an experimental research program (with Valérie Pihet) in art and politics called Sciences Po École des Arts Politiques (SPEAP)[20] that is conceived in the spirit of exploring similarities between modes of inquiry in different disciplines:

The autonomy of everybody is disappearing, the autonomy of scientists as well as artists. So it is the idea of autonomy which is disappearing. But on the contrary the skills and competences that are necessary for an artist are becoming much more, so to speak, *respectable*, and this is because art is connected now as much to research as well as science. So there is a loss of autonomy in the sense that we are not interested in something which is just artistic, but we would be very much interested in something which would use skills in design and art in order to reopen possibilities which had been closed down by the divide between art and science.[21]

This engagement in art is as much a tactical and pragmatic move as it is an epistemological one insofar as Latour admits that it allows access not only to new strategies for interrogation and new questions to be asked of the world, but also to new resources by which to address these questions. Speaking of his own SPEAP initiative, he says: "We find that in people applying to the SPEAP program there are many more artists interested in talking with social scientists and philosophers than there are philosophers who are aware that it's in the art world where resources are often possible."[22] He refers to this mode of working as a "useful kind of wishful thinking"[23] that draws an equivalence between the inquiries of science and art: "If politics is the art of the possible . . . then we need political art to open this up, this 'possible' or to multiply this possibility."[24] Both art and science have the potential to achieve the same ends; they are just different strategies for producing narratives around shared social and historical realities. Art and science are merely different ways of narrating the world: "There's a narrativity, and an urgency also, shared by people who are completely different in their approach."[25]

By being equivalent to other modes of inquiry, art is *useful* for Latour.[26] It can be used to secure resources that are not only practical (such as funding, or access to gallery spaces and audiences) but also epistemic. For example, in his remarks "on the usefulness of art history to make sense of scientific practice," Latour argues that art, like science, religion, and other modes of social mediation, is underwritten by a tacit faith in transcendence. Art, then, can be used to offer a critique of this situation, by giving the opportunity to reflect on it: "Once the aestheticians and their ahistorical Beauty have been pushed aside, it is slightly easier to recompose the quality of a Rembrandt, out of a motley crowd of small mediators, than, say, the second law of thermodynamics. It remains fair to say that Beauty is more easily seen as a construction than is Truth. . . . In art it remains slightly easier than in science to be constructivist and realist at the same time."[27] His argument is that the inherent constructivism of art is more overt, and hence more obviously observable, than in science. The observations of science become naturalized as facts.

On the other hand, we appreciate art precisely because it doesn't deal with facts or objectivity. Art thrives on its distinction from the world it represents, and this distance forms part of its subject matter. It is because we know we are experiencing art that we don't run onto the stage to prevent Hamlet's death or try and sit on one of Joseph Kosuth's chairs. There is also, for Latour, a pressing urgency for recognizing the need to be constructivist and realist at the same time. This is the condition of the Anthropocene and of potentially catastrophic climate change. The Anthropocene is both a human and a natural occurrence. It is generated by human activity but registers on a meteorological and geological level. Art, in other words, is necessary right now because it can mobilize resources (intellectual as well as practical) in the light of unavoidable ecological issues faced by the modern dissolution of the distinction between the domains of nature and culture.

Nature/Culture

This distinction, Latour claims, begins with modernity, which established it in both discursive and ontological terms. It did so through its valorization of science as the guarantor of facts in contrast to culture as the producer of values. Of this rift, he invokes the metaphor of the Gordian knot of nature/culture that, he argues, was severed by modernity.[28] Latour claims to be attempting to retie this knot by "criss-crossing, as often as we have to, the divide that separates exact knowledge and the exercise of power—let us say nature and culture."[29] In perhaps his most notorious example, his discussion of Louis Pasteur in *The Pasteurization of France*, Latour proposes that the natural facts of microbes are as much a product of culture as they are of science. Before Pasteur's revolution microbes did not exist, and yet after it they had existed all along.[30]

However, it is not enough for the microbe to be recognized merely by an individual experimenter, or even by a scientific community. In order for it to be established as a fact, it needed to be socially acknowledged as such: "It is not enough to say simply to the Académie, 'Here's a new agent.' It must be said throughout France, in the court as well as in town and country, 'Ah, so that was what was happening *under* the vague name of anthrax!' Then, and only then, bypassing the laboratory becomes impossible. To discover is not to lift the veil. It is to construct, to relate, and then to 'place under.'"[31] Crucially, once the existence of microbes is established socially, then these microbes are established as transhistorical facts. Latour remains adamant that his position is not one of social constructivism and relativism. Rather, it insists that "Nature and Society are no longer explanatory terms but rather something that requires a conjoined explanation," leading to the inextricable interleaving of facts and values to which my conclusion returns.[32]

This situation, Latour argues, has now become more obvious and vitally important in the era of the Anthropocene, wherein the inseparability of culture and nature becomes an unavoidable condition of contemporary existence:

> Because of the very logic of the Anthropocene, you are inserted into the phenomena you study in a way that is unexpected and still unfathomed. The idea of a science that emerges from the dispassionate study of external phenomena is now much more difficult to sustain. The very distinction between the social and natural sciences breaks down because the argument that you are not supposed to be involved in what you study can no longer be maintained. A Chemist working on CO_2 is fully integrated into a feedback mechanism, whatever he or she does, in a way that resembles an economist involved in policy, or a sociologist involved in statistics.[33]

Hence, as already mentioned, the role for cultural practices to represent and, potentially, to address this potentially catastrophic situation.

One example is Pierre Huyghe's contribution to Documenta 13 (2012). Huyghe claimed that the medium for his work *Untilled* was "live things and inanimate things, made and not made."[34] Specifically the installation occurred on a plot of land in the Karlsaue park in Kassel, Germany. It included a compost heap and the replanting of many plants, including marijuana and poisonous fruits. A centerpiece of sorts was provided by a lifelike stone sculpture of a nude woman with a live beehive for a head. Alongside the compost, sculpture, dirt, bugs, plants, and curious humans, the installation was also populated by a white Podenco Canario dog with a pink leg. When I visited the installation, there were tortoises wandering around and ladybirds copulating in the foliage. Where art began and nature stopped was impossible to discern. *Untilled* embodied, and hence gave an aesthetic form to, the collapsed conceptual distinction between nature and culture that Latour also wishes to observe. In doing so, it affords the opportunity to reflect, *through direct embodied and aesthetic experience*, upon a key Latourian claim: that of the inextricable interlacing of nature and culture, facts and values, in complex matrices of actors and networks.

Actor-Network Aesthetics

The overarching framework for Latour's claims for the inseparability of facts and values is actor-network-theory (ANT). ANT considers the inextricable relations between social, cultural, historical, and natural entities, or actors. Its basis is that there is no form or reality external to these relations. In ANT there is no simple divide between nature and culture, facts and society, or objects and situations. As Latour puts it in talking about microbes: "There are not only 'social' relations, relations between man and man. Society is not made up just of men, for everywhere microbes intervene and act."[35] As a result, all entities are considered to be actors inseparable from their relations to other actors; as Harman observes, Latour's main thesis is that "an actor *is* its relations."[36]

When related to contemporary art, this thesis casts its objects not only as products of a particular set of historically and socially mediated relations, but as those which take these relations as their subject matter. Take, for example, the curator Nicolas Bourriaud's celebrated and contested defense of a *relational aesthetics* as a dominant paradigm in contemporary art discourse. Relational art, he argues, takes as its theoretical horizon "the realm of human interactions and its social context rather than the assertion of an independent and private symbolic space."[37] *Relational aesthetics* was, initially, focused on a small group of artists who had largely come to prominence in Europe in the 1990s, and it is now normally associated with a particular style of collaborative and participatory practices. However, in an expanded sense it can be argued that *all* art (like all other Latourian social facts and values) is a product of relations between various actors. Art is positioned within these networks and can also, therefore, in a mode of self-reflexivity, take these relations as its subject matter and place an "emphasis on a parallel engineering, on open forms based on the affirmation of the trans-individual."[38] In doing so, it can re-present conditions that are often hidden and naturalized and offer them up for observation. Hence Bourriaud proposes that art can participate in social networks in terms that resonate with Latour's valorization of the usefulness of art.

A pertinent example (not used by Bourriaud) is Theaster Gates's *Dorchester Projects* (2009), described as a "2 year design-build project . . . Acquisition of an Abandoned 2 story property for reuse as a Library, Slide Archive and Soul Food Kitchen."[39] Gates acquired several abandoned properties on Chicago's South Side, a neglected area that is historically associated with social deprivation and a predominantly black population. He moved into a house in a former shop on South Dorchester Avenue and then renovated the buildings to create what he calls "a vibrant cultural locus . . . and [to] reactivate the home as a site of community interaction and uplift." This project included further redevelopment, grant applications, and development of cultural and social activity. As one commentator observed, critically, Gates was using a system of real estate as his primary medium: "Put cynically, Gates's 'ecological system' involves the Rebuild Foundation acting as a kind of feel-good money laundering facility for the commercial art world and corporate developers, and this is what enables his status as a 'populariser.'"[40] In doing so, Gates was working with both social objects, such as real estate and grant applications, and social relations, such as existing communities and the subsequent emergent interrelations between people, objects, and environments. In treating objects and relations as his medium, he drew an equivalence between them that Latour could recognize as the equivalence between objects and relations within a network. Gates thus operates as an actor within a network of relations that are, amongst others, legal, economic, architectural, racial, and so on. And in doing so, he brings their conditions into view.

Network

Latour uses the concept of network alongside that of actor to allow himself methodological flexibility—that is, to move between the fields of science, technology, and society and to trace the complex interdependencies of these fields in an attempt to "retie the Gordian knot" that had united nature and culture before modernity. Contemporary art exemplifies this same condition of interdependency. It is inherently dependent upon a network of display and distribution in order to exist *as art*. Writing on contemporary painting, Joselit claims that "the most important problem to be addressed on canvas since Warhol" is *"How does painting belong to a network?"*[41] The same problem is posed by the painter Martin Kippenberg's statement: "Simply to hang a painting on the wall and say that it's art is dreadful. The whole network is important! . . . When you say art, then everything possible belongs to it. In a gallery that is also the floor, the architecture, the color of the walls."[42] This statement raises questions of how painting relates to the challenge of mechanical reproduction and how this contingent status can be made explicit.[43] Joselit observes: "Certainly, painting has always belonged to networks of distribution and exhibition, but Kippenberger claims something more: that, by the early 1990s, an individual painting should explicitly *visualize* such networks."[44]

In the specific instance of painting, its reliance upon a network of relations was part of the logic and aesthetic of modernism from the middle of the nineteenth century. Consider Picasso's quoting of El Greco in *Les Demoiselles d'Avignon* (1907) or Manet's appropriation of motifs from Raphael, Titian, and Giorgione in *Le Déjeuner sur l'herbe* (1863) and *Olympia* (1863). The latter two were identified by Michel

Foucault as the first "museum" paintings because they acknowledged "the new and substantial relationship of painting to itself, as a manifestation of the existence of museums and the particular reality and interdependence that paintings acquire in museums."[45] Such gestures demonstrate a self-conscious awareness on the part of the artist that the medium of painting concerns a historically configured network of protocols and references that painters draw on as their "technical support" (in Krauss's terms).

This self-reflexivity is seen already in Courbet's *The Painter's Studio* (1854–1855), identified in its title as a "real allegory" of the technical and historical conditions of the medium of painting. It offers representations both of the studio as a site of artistic production, and hence an integral part of the medium of painting, and of the dominant genres of painting such as landscape, nude, portrait, genre, and history painting. Michael Fried argues that the painting is an "allegory of its own production," in which Courbet depicts himself as "already immersed" in his medium and "physically enclosed, one might say subsumed, within the painting he is making, wherever the ultimate limits of that painting are taken to lie."[46] Such strategies continue in contemporary artworks such as Sarah Morris's *Midtown*, which consists of both a 16mm, nine-minute film from 1998 taking New York as its subject matter and a series of 1999 paintings of brightly colored, abstract grids that simultaneously allude to: the grid of New York (the home of postwar American modernism); the modernist architecture of the city and galleries in which the art is exhibited; and the grid as a signal structure in modernist painting.[47] Conceived in terms of networks, Morris's practice overtly reflects upon its positioning within a network of spatial and social relations. Hence, it is both a continuation of this modernist legacy and exemplary of the working practice that I'm claiming for Latour: one that identifies relations of actors in networks and takes these relations as its primary subject matter.

Quasi-Objects

The actors and networks described above are populated by what Latour, following Michel Serres, calls quasi-objects. These entities emerge from human and nonhuman activities. They are contingent upon the networks within which they subsist, while also being irreducible to those net-works. They rely upon their relations for their identity and in order to function, yet they also transcend those relations. These quasi-objects are "strange new hybrids" of things and situations.[48] Serres describes his own theory of the quasi-object through the example of a ball used in a game like football. Such a ball is "not an ordinary object, for it is what it is only if a subject holds it. Over there, on the ground, it is nothing; it is stupid; it has no meaning, no function, and no value. Ball isn't played alone. . . . Skill with the ball is recognized in the player who follows the ball and serves it instead of making it follow him and using it. The ball is the subject of circulation; the players are only the stations and relays. The ball can be transformed into the witness of relays."[49] Considered in these terms, Liam Gillick makes quasi-objects. His practice is voraciously wide-ranging. It involves writing, designing, teaching, architectural in-terventions, acting, and so on to produce strange new hybrids of things, ideas, and situations. In this way, Gillick, like Latour, provides strategies for revealing their conditions and the operations of social networks.

Gillick makes objects that look like art. They are the tokens that give him access to a number of different systems of distribution and display, circulating in the economic and institutional networks that form the support structure for contemporary art, such as the market, galleries, biennales, art fairs, and so on. He has also engaged with design compa-nies (collaborating with Pringle in producing products under the brand "liamgillickforpringleofscotland" [2011]); education (as a professor at Columbia and Bard); art discourse (as a prolific writer[50]), and even cin-ema (recently starring as a successful contemporary artist in *Exhibition* [Joanna Hogg, 2014]). However, rather than being entirely complicit with such networks, Gillick presents a discursive engagement with them, in order to "confront a socio-economic system that bases its growth and collapse upon 'projections.' "[51] Gillick's objects declare their status as quasi-objects, however, by performing their social and cultural construc-tion and drawing their contingency into view by making this contin-gency their subject matter. Conversely his objects are also, in their con-ceptual and stylistic rhyming with the legacy of modernist autonomy, declarative of their autonomy from social systems.

Consider, for example, Gillick's *Complete Bin Development* (Kerlin Gallery, Dublin, 2013). In formal terms, the work is a series of abstract, brightly colored boxes that recall the minimalist forms of Judd. They are sculptural frames standing three meters high and paneled with brightly colored, transparent Perspex. The objects stylistically allude to networks in so far as, according to the descriptions for their fabrication, any number of structures can be used as long as they are separated by at least 1.50 m. Hence the work is conceived of as a series of open frameworks, allowing for multiple iterations according to different exhibition conditions (indoors or outdoors, for example). Furthermore, the work also refers specifically to standard processes of mechanical engineering. In particular, Gillick is drawing on his extensive research into processes of automobile production, such as the Kalmar plant in Sweden, which produced cars for Volvo between 1974 and 1994. As he states, here one can find a model of how social systems in general operate in a post-Fordist context.[52]

Gillick engages in an aesthetics of networks in challenging the autonomy of both the art object and the process of authorship. In doing so, his objects function much like Serres's ball; they are quasi-objects that, while inanimate, move between different networks of circulation and control. These networks include the galleries, art fairs, and exhibitions where the objects are shown; economic systems in which they appear as commodities; and the discursive systems of contemporary criticism and art history wherein they circulate as discursive objects. Yet these objects are also, crucially, not entirely reducible to such networks. They disrupt them through strategies of contradiction and obfuscation.

Flat Ontology

The concatenation of the inhuman and human in Latour is the condition of what he calls a "variable-ontology world . . . the result of the interdefinition of the actors."[53] These "tacit metaphysical" assumptions are referred to by Harman as "flat ontology."[54] For Latour, Harman claims, "Every entity is both enslaved in a cultural/functional or perspectival system of meaning *and* has an undeniable reality to which human life is held hostage."[55] A flat ontology doesn't recognize any ontic hierarchy between elements, agents, and things, whether natural, artificial, imagined, or physical. This same perspective is often employed by contemporary artists, who work with a flat ontology insofar as anything is available as their medium. Everything is up for grabs: ideas, words, situations, or stuff.

Martin Creed works according to the principles of such a flat ontology. From one perspective his work exemplifies a certain paradigm of conceptual art. His artworks have included simple stacks of plywood, crumpled balls of paper, and a piece of Blu-Tac with an impression of a finger. He has made short films of people making themselves vomit and also produces music in the form of a pared-back, low-fi pop/punk that typically uses a very few basic chords, repetition, and simple lyrics, such as *Thinking / Not Thinking* (2011) which repeats the phrase "I was thinking and then I was not thinking" over a two-chord riff (a chord for each state) for about two minutes. His most notorious piece, *Work No. 227: The lights going on and off* (2000), is a room in which, as the title suggests, a light goes on and off. Such gestures resonate with the tradition of the readymade and the Duchampian legacy of an anti-art appropriation of everyday objects that was continued through the anti-aesthetic turn attributed to conceptual art. Thierry de Duve observes that this legacy necessitated a shift in judgment from one of taste—"Is this beautiful?"—to an ontological one—"Is this art?"[56] This tradition collapses the gap between art and life (as Robert Rauschenberg famously put it) by emphasizing the nonaesthetic elements of the art work through prioritizing the conceptual elements.

Creed's work, however, can also be considered in exactly the opposite way: as anti-readymades. That is, far from being anti-aesthetic, Creed's objects provide a means of rethinking our encounters with all objects in the world in aesthetic terms. In other words, he encourages us to imagine that all things in the world are equal both ontically and aesthetically. In his work, a flat ontology and a flat aesthetics begin to rhyme conceptually.

Mediators/Translators

When actors work in relation to one another, they are engaged in acts of what Latour calls "translation." In contradistinction to "purification," translation is the process though which "everything may be made to be the measure of everything else."[57] Through this process entities are abstracted in relation to one another according to the operations through which each entity engages with the world, as a means of "linking one thing with another."[58] Translation, then, "creates mixtures between entirely new types of beings, hybrids of nature and culture."[59] These mixtures occur through the action of mediators that transform but also distort the content of the elements they mediate according to their own operations. So when salt is extracted as brine from the mines in Bad Reichenhall, refined through evaporation, packaged, sold, and then sprinkled on Semmelknödel, the salt is being translated according to networks of extraction, refining, distribution, and consumption. The salt mine and gift shop are mediators. The raw salt excavated in the mine is not a representation of the process of extraction, just as the salt in the gift shop is not a representation of the natural brine it was made from. Instead, the salt (as a quasi-object) has been translated by the refining network that serves as its mediator.

An analogous process can be seen in the work of Phil Collins, who embeds his practice in social, cultural, and mediums systems to explore their operations as social mediators. This work has included the installation of a television production company (Shady Lane Productions) in Tate Britain, produced for his nomination for the Turner Prize in 2006; and *the world won't listen*, a project in which he recorded Smiths fans in Indonesia, Columbia, and Turkey singing the songs from The Smiths' greatest hits. Meanwhile, in *How to Make a Refugee* (1999), Collins filmed the journalistic photo shoot of a Kosovan boy and his family at Czagrane on the Kosovan border. The film is only eleven minutes long, but it makes for extremely uncomfortable viewing. The shaky hand-held camera that Collins uses emphasizes the voyeuristic position of camera crews and viewer. We can hear an awkward conversation going on between the news crew and the family through a translator. It is difficult to not feel complicit in the representation that is taking place here, while simultaneously being conscious of its artificiality. Alex Farquharson describes what is going on:

> The photographer asks the Kosovar boy to remove his shirt. He's 15, he says, but his chest is still that of a child. A bullet wound circles his navel, and his leg is in plaster, right up to the groin. "Should he put his baseball cap back on?" the translator asks the photographer. "Yeah, hat on," comes the reply. After a minute or two a vase of flowers comes between the boy's torso and us. The maneuver, in itself modest, seems full of significance. We sense, for the first time, the presence of the video camera through which we view the scene, and with it the ethical and emotional distance that separates its operator from the stills photographer and the feature writer, who are out of the frame but whose speech we overhear. In retreating behind the bouquet, it seems as if the camera is being directed by our own feelings of discomfort at having been implicated in this choreographed exploitation of another's misfortune. Only a moment ago he was being used as a cipher for war and its victims, but now, with his head and shoulders protruding from the colorful flora, the boy—whose name is Besher—is no longer a social type but a fulcrum of individualized ideals: youth, health, beauty, happiness, sensitivity, etc. He could be the privileged, poetic subject of a society painter circa 1890, rather than the victim of a brutal civil war. While this transfiguration takes place, the bouquet scrolls through various functions: it metaphorically dresses Besher's wounds; it screens his modesty; it hides our shame; it pays tribute to his beauty and bravery; it consoles his pain. [60]

The title further emphasizes what is happening here: a refugee is being made. And it is the network of news reporting and broadcasting as much as the complex situation on the ground that is responsible for the construction of the reality that we experience in western Europe. Hence the role of broadcast mediums as mediators in these situations is foregrounded through Collins's actions.

Irreductions

My final Latourian term, irreductions, is the term coined in the closing section of *The Pasteurization of France*.[61] It captures well what artists do and have always done. They make things that resist being reducible to other things. Artists recomplicate the world.

Despite Latour's flat ontology and the complex inseparability of actors/network and nature/culture, the quasi-objects and mediators in Latour's networks are not reducible to their relations. His networks are also comprised of irreductions. The salt that is mined, packaged, sold, and consumed is mediated and translated by a number of networks, but it is neither created nor exhausted by those networks. This is precisely what allows it to be transferred between networks; if there were no irreduction, there would be nothing to move between different networks. "Nothing is, by itself, either reducible or irreducible to anything else. I will call this the 'principle of irreducibility,' but it is a prince that does not govern since that would be a self-contradiction."[62]

In Isabel Nolan's art, we can see an emblem for this irreducibility and are presented with an instance of art embodying the conditions that Latour claims for the world in general. In the last room of the exhibition *The Weakened Eye of Day* (Irish Museum of Modern Art, 2014), a large sculptural form fills most of the space. This form is a spiral shape made of round metal the width of fat rope, covered in gray fabric that has been hand-stitched onto it. Longer than it is high, it is slightly bigger than human size. There doesn't seem to be anything represented here; the shape doesn't even resemble anything in particular. The title, *The Weakening Eye of Day* (2014), gives little away, although a sense of winding down, collapse, or dissipation of energy is suggested by the ever-decreasing yet inconsistent coils. At a push, an analogy might be made to a mathematical model from chaos theory of a system mapped in phase-space that is stuttering, randomly, to a halt.[63] It's a puzzling, awkward object, and the experience of encountering it is similarly puzzling and awkward. It gives the impression of trying to resist any attempt to grasp it, whether literally, conceptually, or aesthetically. I would even struggle to call it beautiful, although it does have a type of

FIGURE 1. Isabel Nolan, *The Weakening Eye of Day*, 2014. Installation at the Irish Museum of Modern Art, Dublin, Ireland, 2014. Courtesy of the artist and Kerlin Gallery, Dublin.

gauche grace. Behind the sculpture on the last wall of the exhibition is a large-scale photograph of two donkeys in a graveyard, one of which stares stupidly, stubbornly, and inscrutably over the heads of the audience. The donkeys and the sculpture, while separate pieces, are analogues for each other. They each have their own agency. They are both irreducible to their environment or to each other. And in them we have a metaphor not only for art but also for Latour's world.

Latour as Artist

My concluding example of an artist is Latour himself. Instead of finding examples of art that rhyme conceptually with Latour's claims, here I flip the emphasis of my discussion and argue that he, in his research and writings, also acts like an artist—that is, as a practitioner engaged in "reassembling the social."[64] In terms of medium/subject, the consequence of Latour's actor-network aesthetics is that he doesn't observe any disciplinary distinctions between different discursive objects. In other words, he is not bound by established anthropological or sociological categories, mediums, or subjects. Like a contemporary artist, anything is available for him to work with, hence the characterization of his work as a flat ontology.

Latour's modes of production are also artistic. He operates with the same procedures as contemporary artists in the sense that Relyea proposes; ones that use whatever platforms they have available to them. Describing the move from medium to platform in contemporary practice, Relyea says: "[The word comes from] discourses on engineering, design, and business management . . . to describe increasingly decentralized forms of coordination, whether in computer systems, product design, or interactions between different industrial manufacturers, suppliers, and work teams. According to them, a platform denotes a basic, underlying architecture or system, a common workbench that, while itself stable and enduring, is open and flexible enough to allow for a high variety of interfaces, a range of inputs and outputs."[65] In such practices it becomes increasingly difficult to distinguish between artist, designer, entrepreneur, and celebrity as they all operate as cultural practitioners mobilizing those networks that are at hand. This condition accurately describes Latour's ongoing AIME (*An Inquiry into Modes of Existence*), a "collaborative research platform" and project involving research seminars, an interactive website, a database of "crossing," an exhibition, and a book.[66] In these terms Latour operates through his many curatorial interventions, public talks, dialogues, and other activities, like the other artists introduced above. He is also a cultural agent working with those platforms and economies that are available to him.

Finally, stylistically, Latour's writing is as much a creative and aesthetic endeavor as it is an analytic and epistemological one. His work employs creative techniques and literary devices such as jokes, dialogues, metaphors, and personification. He describes in *Reassembling the Social* how ANT can learn from literary techniques used in fiction. "It is only through some continuous familiarity with literature that ANT sociologists might become less wooden, less rigid, less stiff in their definition of what sort of agencies populate the world. Their language may begin to gain as much inventiveness as that of the actors they try to follow—also because actors, too, read a lot of novels and watch a lot of TV!"[67] His writing frequently presents extended and evocative descriptions of "mundane objects" such as this: "Early this morning, I was in a bad mood and decided to break a law and start my car without buckling my seat belt. My car usually does not want to start before I buckle the belt. It first flashes a red light 'FASTEN YOUR SEAT BELT!', then an alarm sounds; it is so high pitched, so relentless, so repetitive, that I cannot stand it. After ten seconds I swear and put on the belt."[68] In such strategies Latour interleaves literary techniques with sociological observation to produce narratives in which the factual and fiction begin to conceptually rhyme. This demonstrates the importance of aesthetic strategies in his methodology as a means to find new ways to redescribe the world.

Latour also creates fictional characters like the imaginary anthropologist who is a key protagonist in *We Have Never Been Modern*. Consider, too, his creation of a fictional "technologist" Jim Johnson, Columbus Ohio School of Mines, as the author of the published text, "Mixing Humans and Nonhumans Together: The Sociology of a Door-Closer."[69] Here we see Latour as a literal actor within a discursive network, behaving almost as a performance artist (like Marvin Gaye Chetwynd né Alalia Chetwynd), creating and inhabiting a persona in order to make work.

In short, Latour tells stories about society.[70] *Aramis*, for example, is in one sense an account of a moment in the history of technology. The book recounts an unrealized plan for a Personal Rapid Transit (PRT) system for Paris, a system of transportation involving a complex network of individual cars operating as trains. It was eventually abandoned as unpractical and too expensive. Yet, rather than being presented as a social or technological history, *Aramis* is written as a murder mystery that opens with the question "Who killed Aramis?" On the opening pages we read:

> "It's truly a novel, that story about Aramis . . ."
>
> "No, it's a novel that's true, a report, a novel, a novel-report." 'What, a fake love story?'
>
> "No, a real technology story.'
>
> "**Nonsense!** Love in technology?!"[71]

That Latour would choose to present his anthropological and sociological observations in the form of creative writing is not merely a stylistic tic or cute trick. Instead it conveys a key element of his entire thought: the collapsing of boundaries between "Matters of Fact" and "Matters of Concern," or between epistemic and aesthetic judgments. Latour poses the question: "What is the *Style* of Matters of Concern?" To which he responds in part with a statement—"Surely you would agree that there should be a philosophy that allows ecological relations to be added to those of science creation and to the grasp of poetry"[72]—and with another question: "Can we, too, open the window and follow the poet who directs us to carefully follow the behavior of the bird?"[73]

The main thrust of my argument has been that in contemporary artistic practices, historical and discursive distinctions between artistic and philosophical methods begin to blur. This development has two consequences. First, it establishes an equivalence, which I'm calling here *conceptual rhyming*, between Latour's practices and those of contemporary artists. In other words, Latour operates just like a contemporary artist, that is, by using aesthetic strategies to rethink the world, its structures and relations. In these terms his claim in *AIME* that he's been "doggedly pursuing" for twenty-five years the question "is there another system of coordinates that can replace the one we have lost, now that the modernist parenthesis is closing?" resonates with the spirit of invention that characterizes artistic endeavor.[74] This spirit of invention allows the freedom to both think and work without disciplinary constraints.

Second, however, it raises a question about how these practices might be judged. If, as is claimed, Latour's work is like contemporary art, then its epistemic and ontological claims might be assessed accordingly. This entails recognizing aesthetics as a sphere of validity that is not subordinate to either scientific or judicial/ethical claims. On the one hand, this opens up the possibility of a use for aesthetics in metaphysics: that an aesthetics as first philosophy is possible.[75] On the other hand, this second conclusion forces a question about Latour's poetic speculations. If they are aesthetic acts performed in the face of the world, then they will not be verifiable in the terms of other modes of philosophical, sociological, epistemological, or empirical discourses. This does not mean that his claims are without meaning, but rather that the way in which their meaning is ascertained is not through conceptual or empirical verification, but through aesthetic judgements of taste. In other words, the question posed of all his work is not is it right or wrong, or even is it logically consistent, but rather: is it any good?

Notes

I'm sincerely grateful for comments from Rita Felski, Stephen Muecke, Declan Long, Paul Ennis, Anthony Warnick, and Johanna Gosse.

1. I've written elsewhere on how this condition can also be brought under the sign of *systems aesthetics*, whereby system emerges as a medium in art after modernism, and have used Jack Burnham and Niklas Luhmann to do so. Subsequently in this paper there is some slippage between system and network as metaphors for organization. Latour has been skeptical of the use of system in place of his own use of network and has made several critical comments on Luhmann, who had a different object and commitments and was not addressing metaphysical questions. See Francis Halsall, *Systems of Art: Art, History and Systems Theory* (Oxford: Peter Lang, 2008).

2. Bruce Clarke, "The Metamorphoses of the Quasi-Object: Narrative, Network, and System in Bruno Latour and *The Island of Dr. Moreau*," *Revista Canaria de Estudios Ingleses* 50 (2005): 39.

3. Hal Foster, *Design and Crime: and Other Diatribes* (London: Verso, 2010), 128.

4. That Latour is becoming an increasingly familiar figure in the art world might be related to the status of his commentator Graham Harman, who placed seventy-fifth in the 2015 *Art Review* "Power List" of 100 most important people in the art world: http://artreview.com/power_100/graham_harman/. This might, subsequently, lead to the ontological dimensions of Latour's work being emphasized in art discourse.

5. Arthur C. Danto, *After the End of Art: Contemporary Art and the Pale of History* (Princeton, NJ: Princeton Univ. Press, 1998), 47.

6. Douglas Cooper, *The Cubist Epoch* (London: Phaidon Press, 1971).

7. Peter Osborne, *Anywhere or Not at All: Philosophy of Contemporary Art* (London: Verso 2013), 51.

8. Osborne, *Anywhere or Not at All*, 48.

9. Lucy R. Lippard, *Six Years: The Dematerialization of the Art Object from 1966 to 1972* (Berkeley and Los Angeles: Univ. of California Press, 1997).

10. Claire Bishop, "The Social Turn: Collaboration and Its Discontents," in *Rediscovering Aesthetics*, ed. Halsall, Julian Jansen, and Tony O'Connor (Stanford, CA: Stanford Univ. Press, 2009), 238–55.

11. Rosalind E. Krauss, "Reinventing the Medium," *Critical Inquiry* 25, no. 2 (1999): 296.

12. David Joselit, *After Art* (Princeton, NJ: Princeton Univ. Press, 2013), 43.

13. Joselit, *After Art*, 53.

14. Joselit, *After Art*, 88–89

15. See, for example, the 2015 Barbara Hepworth show at Tate, the reopened Rijksmuseum, or the curatorial approach of Carolyn Christov-Bakargiev at Documenta 13 (2012), which included much nonart.

16. https://www.claires.co.uk/katy-perry/content/fcp-content.

17. Lane Relyea, *Your Everyday Art World* (Cambridge, MA: MIT Press, 2013), vii.

18. Bruno Latour, "Spheres and Networks: Two Ways to Reinterpret Globalization," *Harvard Design Magazine* 30 (2009): 138–44.

19. For example, http://www.e-flux.com/journal/some-experiments-in-art-and-politics/.

20. "Founded on initiative of Bruno Latour at Sciences Po in 2010, and inspired by the pragmatist tradition (Dewey, James, Lippmann and others), SPEAP is a multidisciplinary program for scientific, artistic and pedagogical experimentation. Here, young professionals from a wide variety of backgrounds bring their knowledge and methods to bear on concrete societal and political issues, put their convictions to the test through exchange, inquiry, workshops, real-world issues, and think through the consequences of their interventions. SPEAP's aim is to collectively observe, explain, and explore together ways of creating a shared public space as well as new modes of expression for political, economic, ecological and/or scientific questions that are necessarily controversial." See http://www.bruno-latour.fr/node/442.

21. Halsall, "An Aesthetics of Proof: A Conversation between Bruno Latour and Francis Halsall on Art and Inquiry," *Environment and Planning D: Society and Space* 30, no. 6 (2012): 964.

22. Halsall, "An Aesthetics of Proof," 964–65.

23. Heather Davis, "Diplomacy in the Face of Gaia: Bruno Latour in Conversation with Heather Davis," in *Art in the Anthropocene: Encounters Among Aesthetics, Politics, Environments and Epistemologies*, ed. Davis and Etienne Turpin (London: Open Humanities, 2015), 44.

24. Halsall, "An Aesthetics of Proof," 963–70.

25. Davis, "Diplomacy in the Face of Gaia," 47.

26. See too the footnote: "To the shame of our trade, it is an art historian, Michael Baxandall (1985), who offers the most precise description of a technical artefact (a Scottish Iron Bridge) and who shows in most detail the basic distinctions between delegated actors which remain silent (black-boxed) and the rich series of mediators who remain *present* in a work of art." Latour (as Jim Johnson), "Mixing Humans and Nonhumans Together: The Sociology of a Door-Closer," *Social Problems* 35, no. 3 (1988): 309.

27. Latour, "How to Be Iconophilic in Art, Science and Religion?" in *Picturing Science, Producing Art*, ed. Caroline A. Jones and Peter Galison (New York: Routledge, 1998), 423.

28. Latour, *We Have Never Been Modern*, trans. Catherine Porter (Cambridge, MA: Harvard Univ. Press, 1993), 3.

29. Latour, *The Pasteurization of France*, trans. Alan Sheridan and John Law (Cambridge, MA: Harvard Univ. Press, 1988), 3.

30. Latour, *The Pasteurization of France*, 80.

31. Latour, *The Pasteurization of France*, 81.

32. Latour, *We Have Never Been Modern*, 81.

33. Davis, "Diplomacy in the Face of Gaia," 44. See also Latour, *Facing Gaia: Six Lectures on the Political Theology of Nature*, 2013, Gifford lectures on natural religion, http://www. bruno-latour.fr/node/486.

34. Quoted in http://www.theguardian.com/artanddesign/2012/jun/11 /documenta-13-review.

35. Latour, *The Pasteurization of France*, 35.

36. Harman, *Prince of Networks: Bruno Latour and Metaphysics* (Melbourne: Re.Press, 2009), 17 (emphasis added).

37. Nicolas Bourriaud, *Relational Aesthetics* (Dijon: Les Presses du réel, 1998), 113.

38. Bourriaud, "Berlin Letter about Relational Aesthetics," in Claire Docherty, *Contemporary Art: From Studio to Situation* (London: Black Dog, 2004), 48–49.

39. Theaster Gates, "Dorchester Projects," 2009, *TheasterGates.com*, available at http://theastergates.com/section/117693_Dorchester_Projects.html.

40. Larne Abse Gogarty, "Art and Gentrification," *Art Monthly*, no. 373 (2014): 8.

41. Joselit, "Painting Beside Itself," *OCTOBER* 130 (2009): 125.

42. "One Has to Be Able to Take It!" excerpts from an interview with Martin Kippenberger by Jutta Koether, November 1990–May 1991, in *Martin Kippenberger: The Problem Perspective*, ed. Ann Goldstein (Los Angeles: The Museum of Contemporary Art; Cambridge, MA: MIT Press, 2008), 316.

43. This account operates at a discursive level and does not address the fact that painting makes up a larger part of art on the market than ever before, and is still ordered according to traditional genres (landscape, portrait, et al.) that have existed for centuries. In part this is explained by contemporary art referring to a particular set of institutionally framed practices rather than the time of its production.

44. Joselit, "Painting Beside Itself," 125.

45. Michel Foucault, *Language, Counter-Memory, Practice: Selected Essays and Interviews*, ed. and trans. Donald F. Bouchard and Sherry Simon (Ithaca, NY: Cornell Univ. Press, 1977), 92.

46. Michael Fried, *Courbet's Realism* (Chicago: Univ. of Chicago Press, 1990), 155–62.

47. Rosalind E. Krauss, "Grids," *October* 9 (1979): 50–64.

48. Latour, *We Have Never Been Modern*, 51.

49. Michel Serres, *The Parasite*, trans. Lawrence R. Schehr (Baltimore: Johns Hopkins Univ. Press, 1982), 225–26.

50. Including regular contributions to *e-flux*, an online journal, and Liam Gillick, *Proxemics; Selected Writings (1988–2006)* (Zurich: JRP|Ringier, 2006) and *Industry and Intelligence: Contemporary Art Since 1820* (New York: Columbia Univ. Press, 2016).

51. Gillick, "Berlin Statement" (Hamburger Bahnhof, Feb. 12, 2009): 2.

52. Gillick, "Maybe it Would be Better if we Worked in Groups of Three?" *e-flux*, 2009, http://www.e-flux.com/journal/maybe-it-would-be-better-if-we-worked-in-groups-of-three-part-1-of-2-the-discursive/.

53. Latour, *Aramis, or the Love of Technology*, trans. Porter (Cambridge, MA: Harvard Univ. Press, 1996), 173.

54. Reusing a term from Manuel De Landa, *Intensive Science and Virtual Philosophy* (London: Bloomsbury Academic, 2013).

55. Harman, *Towards Speculative Realism: Essays and Lectures* (Winchester: Zero Books, 2010), 81.

56. Thierry de Duve, *Kant After Duchamp* (Cambridge, MA: MIT Press, 1996).

57. de Duve, *Kant After Duchamp*, 158.

58. Harman, *Prince of Networks*, 15.

59. Latour, *We Have Never Been Modern*, 10.

60. Alex Farquharson, "Minority Report," *Frieze*, Issue 94 (October 2005)

61. Harman gives an account of "Irreductions" as the basis for Latour's move from sociology and the history of science to philosophy and recounts Latour's admission of this. See Harman, *Prince of Networks*, 11–33.

62. Latour, *The Pasteurization of France*, 158.

63. James Gleick, *Chaos: Making a New Science* (New York: Vintage, 1987).

64. Latour, *Reassembling the Social: An Introduction to Actor-Network Theory* (Oxford: Oxford Univ. Press, 2005).

65. Relyea, *Your Everyday Art World*, 20.

66. More details at: http://modesofexistence.org/. Latour, *An Inquiry into Modes of Existence*, xxvi.

67. Latour, *Reassembling the Social*, 55.

68. Latour, "Where Are the Missing Masses? The Sociology of a Few Mundane Artifacts," in *Shaping Technology/Building Society: Studies in Sociotechnical Change*, ed. Wiebe E. Bijker and John Law (Cambridge, MA: MIT Press, 1992), 225–58.

69. Bruno Latour (as Jim Johnson), *Social Problems* 35, no. 3 (1988): 298–310.

70. For more on the literary dimension to Latour's writing see Clarke, "The Metamorphoses of the Quasi-Object," 37–56.

71. Latour, *Aramis*, xi.

72. Latour, *What is the Style of Matters of Concern?* (Amsterdam: Van Gorcum, 2008), 26.

73. Latour, *What is the Style of Matters of Concern?* 10.

74. Latour, *An Inquiry into Modes of Existence: An Anthropology of the Moderns*, trans. Porter (Cambridge, MA: Harvard Univ. Press, 2013), 10.

75. This argument is developed in Halsall, "Making and Matching, Aesthetic Judgement and the Production of Art Historical Knowledge," *The Journal of Art Historiography*, no. 7 (2012) and Halsall, "Art and Guerrilla Metaphysics: Graham Harman and Aesthetics as First Philosophy," *Speculations V: Aesthetics in the 21st Century*, ed. Ridvan Askin et al. (New York: Punctum Books, 2014), 382–410.

Life among Conceptual Characters

BRUNO LATOUR

I CONCLUDE THIS VOLUME with some personal—and I am afraid not terribly coherent—reflections on, or rather recollections of, my own encounters with what has been called the "humanities" in this symposium you have so kindly assembled. As you know, in French the word "les humanités" is no longer very common, and it certainly doesn't refer to an organized field that is to be promoted, defined, or defended against other disciplines. So it would make no sense for me to situate myself inside the field of "les humanités." This is why, following your suggestion, I would rather reminisce on my own connections with what French people would call "littérature," "écriture," "style," "texte," "textualité" in their relation to thought and politics—a series of links so typical of French culture that it will probably appear amusingly exotic to your readers. Please take the following attempt as no more than an ethnographic testimony to a quickly disappearing culture of writing.

Oddly enough, I am able to date with a perfect degree of precision my connection with writing as a thought-producing activity: *October 13, 1961*. Even the hour—7 p.m.—is inscribed on the cover page of the first of my personal diaries! As far as I can tell, the fourteen-year-old writer had already made the connection between writing and thinking since he had penned as an incipit: "J'y noterai tous les soirs mes activités et surtout mes pensées" (I will report what I do and above all my thoughts). The "above all" is especially pleasant since at this early age he had no thought whatsoever to jot down! At least not yet. Because, as everyone in the field of humanities suspects, thinking *follows* and does not *precede* writing—at least this highly specific form of thinking associated with midcentury bourgeois European techniques of scribbling. Considering that today I am taking notes in a (by now digital) notebook numbered 212, this means I have been allowed for the last fifty-five years to continuously learn what I should think through the deciphering of some twenty thousand pages of personal *pattes de mouche*!

If I don't need to belabor the point, this is because Jean-Paul Sartre in *Les Mots* marvelously diagnosed the bootstrapping operation by which a thinking subject is generated out of a neurotic obsession with penning page after page. Who is the writer? The one who is fabricated by writing. In the twentieth century, some bourgeois children could learn to become subjects and to "have thoughts" because they wrote as much as they read from their parents' libraries. If "reading is an unpunished vice," writing is the symptom of quite a few perversions. Although I am two generations younger than Sartre, it is fair to say that I have inherited the same atmosphere conducive to the various vices of a self-generating writing.

The formative nature of writing private diaries, especially by adolescents, is a well-known phenomenon of Western media culture. Filling in blank pages allows for a distance to be introduced from the saturated world of daily existence. Writing diaries is like carving for yourself a breathing space, providing that it remains protected from any one's else reading (something probably lost today with blogs and selfie-filled Facebook posts). My father (1903–1982) was a serious bibliophile, who read the classics out loud to us every weekend, and who had the good fortune of being trained by Jacques Copeau (1879–1949), the "renovator of French theater," when he retired in my native Burgundy in the neighboring village of Pernand-Vergelesses. But even in the midst of a large, loving, and literate family, or I should rather say, *because* of this constant shower of privileges, a kid does need a breathing space. How would have I survived otherwise, with seven siblings and hundreds of cousins, nieces, and nephews? However, while most diarists abandon their task after a while, once they have "solved their psychological problems," some continue, morphing progressively from a notary of trifles to a tester of surprising arguments.

And if there is one phenomenon I have never stopped wondering at, it's the countless surprises generated by the very material act of writing. How bewildering for a young soul to decipher the obscure oracles written by his own fountain pen—and to discover that the only way to interpret them is to scribble one more page! (I think I have managed over the years to share such a wonder with students at every session of my "PhD writing workshop"—the only really useful pedagogical set of exercises I have ever devised. Is it not strange that professors claim to teach PhD students to "think" and "study" without ever directing their attention to the subterranean act of writing—which is not to be confused with obeying a format or looking for a "good style"?)

So I hope you will recognize that although this writing mania is not a connection with the field of "the humanities" as such (I have no expert knowledge of any novelist or poet and I am deeply ignorant of contemporary literature), it is a serious encounter with *textuality*. Let me now mention some of the steps that seem to have helped me deepen this peculiar entry into the materiality of writing.

Retrospectively, what probably made me insist even more on such a connection between writing and thinking was my encounter, in the last year of high school, with Nietzsche's philosophy. This might require some explanation. My parents, well aware of the limits of Beaune's provincial *lycée*, sent all their kids, girls as well as boys, to Jesuit schools in Paris for the "classe de terminale" in order to acquire the indispensable Parisian cachet. Oddly enough, our teacher, Monsieur Detape, had chosen that year to cover the whole curriculum by making us read large doses of Nietzsche, from *The Birth of Tragedy* to *Thus Spoke Zarathustra* through to *The Gay Science*—a book that I ended up knowing almost by heart.

I was immediately taken by Nietzsche's idea that all philosophical and even scientific concepts are worn-out metaphors. For him, there is no real difference between the literal and the figurative meaning of a word. Or rather the literal meaning of a concept is nothing but the excision of one of the many figurative meanings still active in the background. But

I was also completely seduced (Nietzsche is the philosopher of eighteen-year-olds wishing to be seduced, not one you continue reading at sixty-nine), by his manner of putting such an argument into practice: by highlighting the figurative within the literal, Nietzsche never allowed concepts to break their ties with the ways in which they were written. Philology, for him, was a synonym of philosophy and a much more exacting discipline. And to philosophy he added his always-present style—and also his ever-present rage. Ideas had flesh. After all, Zarathustra is the best example of what Deleuze and Guattari have called "conceptual characters," a term I have chosen for the title of this little piece.

That concepts were also characters is a lesson I never forgot, and it immunized me completely against analytical philosophy, whose supposed obsession with language struck me, on the contrary, as a total insensitivity to style. If there are any key differences left between French and American thinking cultures, this one strikes me as really decisive: attention to *language* is not at all the same thing as an attention to *writing*. They put their ideas into writing; we write books. A point made over and over again by Derrida. (Incidentally, the attention to style per se has been reinforced for me by the obligation to write in English, a sure way to always sense the weight, limit, opacity, and tropism of language. If you don't write in your mother tongue, you have to resort to that of your wicked stepmother!—and to always rely on friends' help.)

Even though I was a Derridean for a whole year after reading *Of Grammatology* (a key influence on the later notion of scientific *inscription*), the next step in my embedding of concepts in the ways they are generated—this unfolding of thought into a highly specific sort of materiality—was the discovery of biblical exegesis at the University of Burgundy with my teacher André Malet (1920–1989), the translator and indeed the exegete of Rudolf Bultmann in France. This time it was not only philosophy, but the very flesh of religious beliefs, emotions, scruples, and passions—a key preoccupation of my youth—that suddenly found an empirical and realistic grounding. In the same way that Nietzsche had shown that concepts never leave behind their stylistic embodiment, I realized that religious beliefs can never be detached from the complex trajectory of interpretation, rewriting, invention, reprise, fabrication, canon formation, and institutional incorporations that allow statements to gain a meaning. It was my first encounter with the idea that what I now call regimes of truth have their own "key" and that, once this key has been grasped, comparisons between truth conditions are made possible. By looking carefully at "enunciation regimes" (the term I used for many years), I ensured that statements could not escape into never-never land. You can perhaps understand the weight of these two lessons for someone who witnessed every day, in his own obsessively filled notebooks, the same transubstantiation of ink into argument. This peculiar form of *empiricism*—what I take now, retrospectively, because you, Rita and Stephen, have asked me to reflect on it, as a highly specific form of *realism*—was born from those encounters.

Or rather this brand of realism came from the decision, made in 1970, to marry the woman with whom I was absolutely sure that, no matter what, by relying on her compass, I would never be allowed to fly away from reality—a decision whose impact, over forty-five years, has dwarfed all other intellectual or literary influences. (It is a striking feature of life that influences proceed mostly *backward*, reversing chains of causality: consequences *choose* what will cause them, as Deleuze reminded us with his example of the sun and the plants. Before plants, he said, the Sun had no causal influence on them. Similarly, authors choose by what and by whom they will be influenced. This is why intellectual history is always a dubious and misleading business: it rubs history the wrong way.)

At this juncture, while all the determinations of my class, psychological makeup, milieu, and time were directing me to become an idealist intellectual, I ended up, thanks to these turns of fortune, going in the opposite direction: toward a form of rematerialization—what I called at the time *irreduction*. If I had felt, even before going to Africa and then to America, that "we have never been modern," it was mostly because I had sensed that one's picture of the world would be entirely different depending on whether one allowed thought to be autonomous or to embody a certain mode of materiality—yet a substance so radically different from what passed for "matter." (It's a strange but very common experience that you already know at an early age what you will learn in the future, something that could make you really believe in Plato's myth of reminiscence.)

I had the feeling that any consideration was *abstract* as long as it could not follow the step-by-step trajectory that allowed one element to be made visible through the conspiracy of all the other elements still active in the background. This is probably the only insight I have ever had, the core of actor-network-theory as well as of the "inquiry into modes of existence."

It also explains the pleasure I gained from ethnographic method—in itself a completely bizarre tropism for a bourgeois provincial philosopher who should have shied away from any excess of concreteness. In ethnography, just as in the careful word-for-word exegesis of a philosophical text, I could finally follow, step by step, how a specific "action" could be made possible by its "network" (to use rather poor words). Don't fly, don't jump, pay the price of each connection. And if you are lost, write and write again, describe and describe some more (a portrait of the analyst as a "serial redescriptor," as Michael Witmore would say).

To put it another way, I had the certainty that *agency* could not be extracted from style, any more than the actor from its network, or the concept from its character. I have difficulty in expressing myself well (I am not good at introspection) and maybe I am disappointing you, but it is not literature taken as a *field* that made me redirect my attention toward an empirical emphasis, but literature, or rather philosophy, nested inside acts of writing. The only writer that would be recognizable as a clearly "literary influence" on my work is Charles Péguy (1873–1914), but you see how highly peculiar a choice that could be: my own mixture of philosophy, sociology, and art, shaped by an author entirely obsessed by the link between repetitive style, the dogma of incarnation, and a revision of socialist politics! An idiosyncrasy built upon another idiosyncrasy.

I have said enough to make you understand why I sank my teeth into semiotics like a mouse into cheese after I had the chance, in San Diego, to meet Paolo Fabbri. Scientific texts could be analyzed in the same way as biblical scriptures, but with another key. The view from nowhere could be folded back into an exegetic practice that left objectivity in its wake— or not—just as the exegesis of a biblical text generated salvation—or not. Algirdas Julien Greimas (1917–1992) was just as important for me as

Harold Garfinkel (1917–2011) was in social theory and for much the same reason: one never allowed ideas to leave narration, while the other grounded social concepts in locally produced ethnomethods.

Funnily enough, the systematic study of texts in this French tradition became what was imported into an American context as "Theory." While on this side of the Atlantic, I took it as exactly the opposite of "theory": as the chance to acquire an empirical method so as to avoid the flight of concepts into anything like "thought." Just like exegesis, semiotics grounds thought in figures that can be described and studied step by step. The continuity of agency is no longer obscured by the multiplicity of its figurations. Or, to mix semiotic and Deleuzian parlance: actors, that is characters, emerge from actants, that is concepts.

In a move that I now recognize as typical of my emerging form of empiricism, I never took (Greimassian) semiotics as being limited to texts, but as a formidable toolbox for providing a handle on ontology. This is what opened up an access to science and technology that had rarely been facilitated before, to put it mildly, by the various brands of philosophy or literature. Hermeneutics could move out of texts, to things, to knowledge, to technology, and, finally, to the world. Since you have been so kind, Rita, as to explain better than I can the value of treating human and nonhuman characters with the same method and how such a move could be useful for the humanities, I can pass on quickly.

What is the conclusion of this brief introspection? I became convinced that there is no way for any form of expression to transcend another so as to direct, cover, or explain it. Since you asked me to reflect on my use of literary figures and my dabbling in media other than writing—plays, performances, exhibitions—you may understand that they follow directly from this early conviction of mine. (If I answer your question rather clumsily, please remember that I was left completely in peace by critiques and commentators until I was sixty. "Always forward," such was my motto. Before Gerard de Vries's invitation to come to Amsterdam and before Graham Harman's book, no one had read me as a philosopher and even less as an *author*. I was just a sociologist, defending from time to time a few technical points of method. Which means I went from one project to the next, without having to reflect on my own trajectory—or, indeed, to take myself seriously.)

Michel Serres, whom I had met in California, had always insisted on a principle of method that fitted very well with my interests: *explanans* is always lodged deeply in *explanandum*, or, in less pedantic terms, there is no metalanguage other than the language of those you try to interpret. Serres, a great orator, demonstrated this principle of method on poetry as well as scientific treatises in the smoke-filled "Amphithéâtre aux Vaches" at the Sorbonne every Saturday morning. Jean de La Fontaine's *Fables* could be understood via the theory of information, just as much as Claude Shannon could be clarified through a comparative reading of La Fontaine. This is the peculiar form of *hermeneutics* that he associated with the conceptual character of Hermes. It was enough to free his auditors from the idea that there existed a wedge between the interpretation of texts and interpretation of things. To use a nice expression of Steven Connor's that exactly fits Serres's method: "a text cannot be *outwitted*." Hence my distrust of critical distance and my preference for what I call *critical proximity*, a situation where you let your own interpretation be chemically dissolved by the "object" of your study.

I was struck by the beauty of such an "anthropology of science" just as much as I was later, after having delighted in the writing of *Aramis or the Love of Technology*—my somewhat disregarded but favorite piece of work—by Richard Powers's novels. His *Galatea 2.2.* carried out, with much greater efficacy, the same exploration of what the agency of technology meant for the agency of humans. (It also helped me to measure the infinite distance between what I was able to write in my hesitant Frenglish and what a real writer could do with language.)

If you now ask me why I allowed myself to enter into relations with other media, it is precisely because I learned from all those influences that *there is no metalanguage*. This does not mean that philosophy is useless, or that an inquiry into an automated subway will not be needed, since Powers has delivered an amazing portrayal of how software could be animated. It simply means that *philosophy is a medium too*. As different from video as video is from painting, painting from writing plays, or writing plays from setting up lab experiments.

If I am not mistaken on this point, I might in the end have a contribution to make to the field of the humanities as you define it! Let me phrase it as bluntly as I can: philosophy (let's say the humanities) is what allows us to *navigate* through *overlapping* media and contribute to their *composition*, for no other reason than the lightness and banality of its techniques: ordinary language just slightly modified. Not a terribly good definition, but one with which I have some experience.

Let me explain with an example. When in the Venice Biennale of 2009, I encountered Tomás Saraceno's "Galaxies Forming along Filaments, like Droplets along the Strands of a Spider's Web," I saw in his installation a technical, nay a philosophical solution, to the problem I had had with actor-networks: namely how do you generate actors—in his case quasi-bubbles—out of a network—in his case an elastic spider's web. It was also, in my eyes, a highly practical and beautiful solution to a critique Peter Sloterdijk, a thinker obsessed by spheres, enclosures, and globes, had aimed at my own tropism for networks and spaceless ontology. Saraceno managed to draw spheres out of nets. You understand that for a philosopher of monads, the overlapping nature of all these mediums is a crucial resource.

Had I found the hidden "structure" and the unconscious principle that was "guiding" Saraceno's work? Of course not. But I could show to some visitors and readers that Saraceno, Sloterdijk, and I, each with our own skills, had elaborated solutions that could *overlap* with each other.

It is not that a metalanguage extracts the meaning from a work of art—a meaning that, in theory, could be lodged inside any another form or expressed in a "literal" instead of a "figurative" form. It simply means that the medium of philosophy—a peculiar style of *linking arguments* together to avoid as many explicit non sequiturs as possible (is there another definition of this venerable Western writing tradition?)—that this medium is pretty good at *intensifying* some of the features that other mediums are also interested in sustaining. Philosophy does not lord it over other forms and does not explain anything: philosophy is a powerful medium in its own right that adds new forms to the others. Nothing more, nothing less. (A point that I think I have clarified in the "modes of existence" project by proposing the term *preposition*.)

Compared to writing a play (as I did with friends when we staged *Gaia Global Circus*) or curating a show, writing in a philosophical style is amazingly light. It does not require a big crew! And that's exactly the source of its strength. If artists and scientists are *chefs*, the philosopher is the *maître d'hôtel*. A useful job, indeed, but suited only to those who have learned to stick to their place.

If knowledge is a mode of existence, and not a voice from nowhere that can be confused with the thing known (the origin of the idealist notion of matter), it is important to find ways to demonstrate this point experientially. If I have chosen three times, thanks to the generous help and support of Peter Weibel, the ZKM director, to be a curator of "thought exhibitions" (designed on the model of "thought experiments"), it is not in order to "illustrate" philosophical ideas with works of art, but exactly for the opposite reason: to plunge philosophical ideas into the competing field of much more powerful works of art and see what will happen to all of them once they are in this crucible. *Iconoclash, Making Things Public,* and the recently opened *Reset Modernity!* each created a fictional space where experiments on what is critique, what is politics, what is modernity, could be carried out by the public in a way that could not possibly be done in a classroom, by reading catalogs, or in the harsh conditions of the "real world." These shows were for me the best way to explore what embodied thought could mean, a stage uniquely suited for watching how conceptual characters could behave.

An exhibition is in many ways ideal multimedia, a form of "total art." It is simultaneously an agora, a studio, a laboratory, a play, a choreography, in which all the layers of materiality composing the artificially built space—from lighting, walls, catalogue, location, to the sheer happenstance of encounters between works of arts and the visitors' unpredictable moves—multiply the interactions, contrasts, dissonances, and contradictions between them. It makes for a great "critical" situation, provided you accept the meaning of critical "as having the potential to become disastrous" or "denoting a transition from one state to another" (both quotes from the OED!).

Take, for instance, one of the rooms of *Reset Modernity!* that is called "From land to disputed territories"—it is written on the wall. Imagine a visitor wondering about Pierre Huyghe's "Nymphéas Transplant (14–18)." Although it is a meditation on Monet's *Nymphéas*, there is no way she can take it as a direct imitation of an impressionist 2-D painting. A *transplant* it is indeed! So much so that it is now a 3-D aquarium of sorts, with the soil taken out of Giverny's pond to which has been added plants and fish that are lit by such a complex mechanism that some visitors are faced with a totally opaque façade of glass, while others discover suddenly the phantom apparition of some ghosts of prehistoric times: axolotls! What do axolotls have to do with Monet? Nothing except that the fragile, carefully monitored pond has become an artificial ecosystem. Impressionism has been transplanted into another century and is viewed through another scopic regime.

What does this work of art say about ecology and impressionism? It does not *say* anything. But it is experienced by visitors who, on another wall, next to Huyghe's piece, are faced with a documentation of what geochemists call "critical zones": the equally fragile, equally threatened, equally opaque, and equally difficult-to-visualize regions, a few kilometers thin, where earthly life resides. While visitors ponder what this

FIGURE 1. © Pierre Huyghe, *Nymphéas Transplant* (14–18), 2014. Photo ZKM.

juxtaposition of art and science could mean, they cannot but feel impressed—or they may be oppressed—by the immense display of maps printed by *Territorial Agency* in the next room, which shows in agonizing detail the extensive footprint of oil extraction in Texas, Canada, the Middle East, Nigeria, and the Arctic. The *Museum of Oil*—such is the name of this module—is built in such a way that the panels are slanted by a few degrees, as if visitors could run the risk of being flattened by their looming shape!

Suppose that the visitor, who has been asked at the entrance to the show to leaf through the *Field Book*—a sort of easy-to-carry, well-designed catalog—now reads the injunction curators have introduced to "make sense" of the room title on the wall:

> Strangely, the space of globalization is largely spaceless, or at least lacking a soil. Moving in it was like moving on a 2-D map. Things are different today: It is the revenge of the soil! Instead of looking at soil horizontally from above, what if we looked at it vertically from below? Instead of looking at the "blue planet" what about digging through critical zones, examining the thin planetary membrane that contains all forms of living beings? Obviously this new land, seen in 3-D, is much more difficult to map. Therefore, we need detectors and sensors to become aware of its entangled loops.

No question that this is written in a sort of abstract theory style. Does the text "explain" what should be experienced? In a way, yes. But how could it have any meaning without the impressions given by "Nymphéas Transplant," the *Museum of Oil*, and the many other pieces around that I have not described? Theory talk is not the metalanguage that is supposed to convey the meaning of what the assembled artists are simply "showing" with their own "limited" language. Theory talk is a highly limited language that might help visitors realize that, if we have many representations of the Blue Planet viewed from outer space—in effect, a view from nowhere—we are completely lost when landing back on Earth. Or it might not help visitors at all; they can easily skip the paragraph, forget to turn the leaf of their *Field Book*, be attentive to an entirely different feature. It is for them to decide.

FIGURE 2. A view of station 4 of Reset Modernity with a lateral view of Tacita Dean. *Quaternary.* © 2014, Sophie Ristelhueber. *W B* #6 2005/16 and the Arctic Map devised by Territorial Agency for The Museum of Oil. Photo ZKM.

It is to describe this connecting role of the humanities that I have often used the word *infralanguage* to point out its subservient but useful role; I have more recently moved to the notion of *diplomacy*. In that sense, I could define "humanists" as those who are in charge of facilitating diplomatic encounters. A role that is slightly more eminent than that of *maître d'hôtel*, but much less important than those who are really parties at the negotiation. A job akin to that we would call "*chiefs of protocol*," a chief, yes, but only of the protocol. That is, someone who makes sure that the assembly remains attentive to the delicate *etiquette* protecting participants from runaway violence.

I am sorry not to have found anything more grandiose to say in defense of the humanities. But if you agree to define them as a form of diplomacy, I might conclude by showing that such a limited but indispensable role could clarify the dilemma of the humanities at the present time. If there is no metalanguage, then *no language* can claim to lord it over all the others, or exclude them from assembling to progressively compose a common world (the makeshift definition I give of politics). To claim otherwise would be a breach of etiquette and a sure way of hastening the dissolution of a putative assembly in charge of common matters of concern. (This is, by the way, why I try to push critical discourse aside: not to criticize critique but simply because it is often a poor chief of protocol. When it resonates in the room, everyone disbands!)

As most of the authors in this symposium have stressed, the humanities have a problem with the word "human." But I think this is also their great advantage because they are placed just on the fault line where all the various and successive meanings of "human" and "humanity" are being thrown back into the crucible of the New Climatic Regime (a term I now prefer to that of Anthropocene).

If the humanities have never been especially anthropo- or humano-centric, it is because they always had to deal with conceptual characters, that is, with a bewildering number of *figurations*, only some of which have looked like realistic portraits—or rather clichés—of human subjects. The authors of the papers in this collection would agree that it would be as silly to define the humanities as the "defense" of the human, as it would be to pretend that a Ingres portrait is "more human" than a Picasso cubist figuration, or that James Joyce is less a psychologist than Madame de Lafayette. So, if there are disciplines that are not especially surprised to hear about completely different associations of humans and nonhumans, it is those in the field of the humanities. If we are asked to reimagine humans as a geological force with trails of CO_2 in their wake, readers of Shakespeare are as well-prepared as anthropologists or geochemists. Overall, the humanities have been very good at recognizing humans in many other shapes and forms. When everybody speaks of another human *agency*, the humanists have all the right to say "we have been there already."

And that's the reason why the field of the humanities is so well-prepared to resist the reverse danger of a sudden naturalization of humans and of all the other nonhumans in whose destiny they have always played a part. As Donna Haraway famously said, "*If I have a dog, the dog has a human.*" For what might be the first time in (Western) history, the *reversibility* of conceptual characters allows them to escape naturalization—objects have many other forms of agency than those granted to them by idealist matter—*as well* as to free themselves from the narrow definition of a human as endowed with a psyche, a consciousness, a soul, and a small amount of morality distinct from objects of nature. Just as Rome is not in Rome, the humanities are not constricted by (provisional forms of) humanism.

I know of no better example of how ideal the present situation is for the humanities than the total metamorphosis undergone by what has been recognized as the question of the sublime. I will finish with this example.

As we learned at school, it was possible to experience such a feeling on three conditions: you had to be a very *small* human compared to the immensity of the forces of nature; compared to those mindless objective forces, you had to be immensely *big* in terms of the grandeur of your soul and the impeccability of your moral consciousness; and, even more important than these two features, you had to remain *firmly protected* from the consequences of cascades, glaciers, volcanoes, hurricanes, earthquakes, and other catastrophes so that they could be witnessed from a safe distance as an *outside spectacle*. As we all know, this is explained very well by Kant, especially the third feature: "But, *provided our own position is secure*, their aspect is all the more attractive for its fearfulness; and we readily call these objects sublime, because they raise the forces of the soul *above* the height of vulgar commonplace, and discover within us a power of resistance of quite another kind, which gives us courage to be able to *measure ourselves* against the *seeming omnipotence* of nature" (emphasis mine). Well "our own position is no longer secure," and we discover in our own "omnipotence" a quite discouraging lack of "resistance." So, suddenly everything is again on the move.

If we now reconsider such a famous text, I should have no difficulty showing you the immense amount of work lying ahead for the field of the humanities: first, we are no longer at a safe distance from any of the effects of the "forces of nature"; second, we are told by many scientific disciplines that we have become so big, so cumbersome, that we, as humans, are now of a size commensurable with volcanoes and, some say, with plate tectonics; as to the immense grandeur of our morality, alas, we seem so dejected, so puny, that we have not the slightest idea of how to respond to the new situation. The task of work for the humanities is even more immense given that we have no political idea of what constitutes the "we" endowed with the ability to respond to such a major transformation. Time indeed to "measure ourselves." Exit the feeling of the sublime. What's next? The successor of the sublime is under construction.

You see that this is the wrong time to lament the destiny of the humanities. As Barbara Herrnstein Smith uncharitably concludes her charitable commentary on my own attempts: "There is good reason to think the mission will fail"! On this score, she is an agreement with the five referees of the AIME project I submitted five years ago to the European Research Council, who concluded unanimously: "It cannot possibly succeed; it should be funded as a first priority." Is this not a fairly realistic definition of the field of the humanities?!

Contributors

DAVID J. ALWORTH is the John L. Loeb Professor of the Humanities at Harvard University, where he teaches in the Department of English, the Program in History and Literature, and the Program in American Studies. In addition, Alworth codirects "Novel Theory Across the Disciplines," a seminar at the Mahindra Humanities Center. His first book, *Site Reading: Fiction, Art, Social Form* (2015) is available from Princeton University Press. He has just completed, with Peter Mendelsund, *The Book Cover: Art at the Edges of Literature* (Ten Speed Press/Crown Press, 2020).

ANDERS BLOK is Associate Professor in the Department of Sociology, University of Copenhagen. His research focuses on the knowledge politics of urban environmental change, comparing engagements with global climate risks across cities in East Asia and Europe. He has published widely in journals of science studies, social theory, urban studies, and environmental politics. He is the author (with Torben E. Jensen) of *Bruno Latour: Hybrid Thoughts in a Hybrid World* (Routledge, 2011) and the coeditor (with Ignacio Farías) of *Urban Cosmopolitics: Agencements, Assemblies, Atmospheres* (Routledge, 2016).

CLAUDIA BREGER is the Villard Professor of German and Comparative Literature at Columbia University. She received her Ph.D. and Habilitation from Humboldt University, Berlin, and previously taught at the University of Paderborn and Indiana University, Bloomington. Her research and teaching focus on twentieth- and twenty-first-century culture, with emphases on film, performance, literature, and literary and cultural theory, as well as the intersections of gender, sexuality, and race. Her book publications include *An Aesthetics of Narrative Performance: Transnational Film, Literature and Theater in Contemporary Germany* (Ohio State University Press, 2012), and most recently *Making Worlds: Affect and Collectivity in Contemporary European Cinema* (Columbia University Press, 2020).

DIPESH CHAKRABARTY teaches history and South Asian studies at the University of Chicago.

YVES CITTON teaches literature and media at the University Paris 8 Vincennes-Saint Denis in France and coedits the journal *Multitudes*. Recent publications include *The Ecology of Attention* (Polity, 2016); *Mediarchy* (Polity, 2019); *Contre-courants politiques* (Fayard, 2018), and *Générations collapsonautes*, with Jacopo Rasmi (Seuil, 2020).

STEVEN CONNOR is Grace 2 Professor of English and Director of the Centre for Research in Arts, Social Sciences and Humanities (CRASSH) at the University of Cambridge. His most recent books are *Living by Numbers: In Defence of Quality* (Reaktion Books, 2016); *Dream Machines* (Open Humanities Press, 2017); *The Madness of Knowledge: On Wisdom, Ignorance and Fantasies of Knowing* (Reaktion Books, 2019); and *Giving Way: Thoughts on Unappreciated Dispositions* (Stanford University Press, 2019).

GERARD DE VRIES is Emeritus Professor of Philosophy of Science at the University of Amsterdam and a visiting fellow of Wolfson College, Cambridge (UK). From 2006–2014, he also served at the Scientific Council for Government Policy, the think-tank for long-term policy issues of the Dutch government in The Hague. Previously, de Vries was Professor of Philosophy at Maastricht University and Dean of the Netherlands Graduate School in Science, Technology and Modern Culture. His most recent books are *Bruno Latour* (Polity Press, 2016) and (with Michiel Leezenberg) *History and Philosophy of the Humanities—An Introduction* (Amsterdam University Press, 2019).

SIMON DURING is a Professorial Fellow at the University of Melbourne. His academic interests are broad, and he has worked intensively on British literary history, cultural studies, literary theory, and postcolonialism. He is currently writing about the idea of the humanities. His books include *Foucault and Literature* (Routledge, 1993); *Patrick White* (Oxford University Press, 1996); *Modern Enchantments: The Cultural Power of Secular Magic* (Harvard University Press, 2002); *Exit Capitalism: Literary Culture, Theory and Postsecular Modernity* (Routledge, 2010); and *Against Democracy: Literary Experience in the Age of Emancipation* (Fordham University Press, 2012).

RITA FELSKI is William R. Kenan Jr. Professor of English at the University of Virginia and Niels Bohr Professor at the University of Southern Denmark. Her most recent books are *Uses of Literature* (Wiley-Blackwell, 2008), *The Limits of Critique* (University of Chicago Press, 2015), and *Hooked: Art and Attachment* (University of Chicago Press, 2020).

FRANCIS HALSALL is director (with Declan Long) of the Master's Program, "Art in the Contemporary World," at the National College of Art and Design, Dublin, and a research fellow in the Department of Art History and Image Studies, University of the Free State, South Africa. He is completing several projects under the general theme of systems aesthetics.

GRAHAM HARMAN is Distinguished Professor of Philosophy at the Southern California Institute of Architecture. His most recent book is *Art and Objects* (Polity, 2020).

ANTOINE HENNION is a professor at the Center for the Sociology of Innovation, École des Mines, Paris. He researches the sociology of music and diverse forms of attachment, from taste and amateurs' practices to issues of care, aging and disability, and, more recently, migration. His most recent books are *Le Vin et l'environnement*, with Geneviève Teil et al. (Presses des Mines, 2011), and *The Passion for Music: A Sociology of Mediation* (Routledge, 2015).

CASPER BRUUN JENSEN is research associate professor at Osaka University and honorary lecturer at Leicester University. He is the author of *Ontologies for Developing Things* (Sense, 2010) and *Monitoring Movements in Development Aid* (with Brit Ross Winthereik) (MIT press, 2013) and the editor of *Deleuzian Intersections: Science, Technology, Anthropology* with Kjetil Rödje (Berghahn, 2009) and *Infrastructures and Social Complexity* with Penny Harvey and Atsuro Morita (Routledge, 2016). His present work focuses on knowledge, infrastructure and practical ontologies in the Mekong river basin.

BRUNO LATOUR is a professor at Sciences Po in Paris and has published extensively in the domain of science studies and more generally on the anthropology of modernism. His books include *We Have Never Been Modern* (Harvard University Press, 1991); *Iconoclash* (MIT Press, 2002); *Politics of Nature: How to Bring the Sciences into Democracy* (Harvard University Press, 2004); *Making Things Public: Atmospheres of Democracy* (MIT Press, 2005); *Reassembling the Social: An Introduction to Actor-Network-Theory* (Oxford University Press, 2007); *On the Modern Cult of the Factish Gods* (Duke University Press, 2010); *An Inquiry into Modes of Existence* (Harvard University Press, 2013); *Facing Gaia* (Polity, 2017), and *Down to Earth: Politics in the New Climatic Regime* (Polity, 2018).

HEATHER LOVE teaches English and Gender Studies at the University of Pennsylvania. She is the author of *Feeling Backward: Loss and the Politics of Queer History* (Harvard University Press, 2009), the editor of a special issue of *GLQ* on Gayle Rubin ("Rethinking Sex"), and the coeditor of a special issue of *Representations* ("Description Across Disciplines"). Love has written on topics including comparative social stigma, compulsory happiness, transgender fiction, spinster aesthetics, reading methods in literary studies, and the history of social sciences. A book on the roots of queer theory in deviance studies (*Underdogs*) is forthcoming from the University of Chicago Press.

PATRICE MANIGLIER is associate professor at Université Paris Nanterre. He writes on contemporary French philosophy, in light of the history of the social sciences, especially structuralism and poststructuralism. He is the author of *La Vie énigmatique des signes: Saussure et la naissance du structuralisme* (Leo Scheer, 2006), *Foucault at the movies* (Bayard, 2011; trans. 2018), and *La Philosophie Qui Se Fait* (Éditions du Cerf, 2019).

STEPHEN MUECKE is Professor of Creative Writing at Flinders University, South Australia, and is a Fellow of the Australian Academy of the Humanities. His most recent book is *The Children's Country: Creation of a Goolarabooloo Future in North-West Australia,* coauthored with Paddy Roe (Rowman and Littlefield International, 2020).

BARBARA HERRNSTEIN SMITH is Braxton Craven Professor Emerita of Comparative Literature and English at Duke University. Her books include *Belief and Resistance: Dynamics of Contemporary Intellectual Controversy* (Harvard University Press, 1997); *Scandalous Knowledge: Science, Truth, and the Human* (Duke University Press, 2005); *Natural Reflections: Human Cognition at the Nexus of Science and Religion* (Yale University Press, 2010); and *Practicing Relativism in the Anthropocene: On Science, Belief, and the Humanities* (Open Humanities Press, 2018).

NIGEL THRIFT is the Executive Director of Schwarzman Scholars, based in New York and Beijing. Until recently, he was President of the University of Warwick, and prior to that Pro-Vice-Chancellor at the University of Oxford. His main research interests are in cities, nonrepresentational theory, the history of time, and the Anthropocene. His most recent book is *Seeing Like a City*, with Ash Amin (Polity, 2016).

MICHAEL WITMORE is Director of the Folger Shakespeare Library. He is author of *Culture of Accidents: Unexpected Knowledges in Early Modern England* (Stanford University Press, 2001), which received the Perkins Prize for studies in narrative; *Pretty Creatures: Children and Fiction in the English Renaissance* (Cornell University Press, 2007); and *Shakespearean Metaphysics* (Continuum, 2008). He posts digital work on the blog winedarksea.org, which he maintains with his collaborator Jonathan Hope.

Index

capitalism, 92, 93, 98, 144, 170, 178, 188–189, 291, 293–294

care, 6–7, 13, 17, 61, 140, 163, 206, 209, 217

Catton, William R., Jr., 180

Chakrabarty, Dipesh, 18–19, 144, 147

Chomsky, Noam, 40

Churchwell, Sarah, 14

Citton, Yves, 13, 19, 327n57

civilization, 31, 90–91, 168, 176, 185, 189–190, 195

Clarke, Bruce, 425

climate change, 18, 144–145, 168–170, 175–182, 188, 190, 265, 398; scepticism and denial, 3, 87, 88, 108, 109, 141–142

close reading, 21, 111, 113, 127, 232, 290; symptomatic, 280, 287

Collins, Phil, 441–442

colonialism, 3, 201, 267

composition, 43, 47, 222, 275, 281, 288, 317, 324n25, 329, 330, 424n7, 461; cinematic, 301, 306, 309, 311, 313; of common world, 146–148; literary, 211, 214; as opposed to critique, 2, 12–13, 16, 137, 294

"conceptual characters," 276

concern(s), 17, 46, 141, 163, 314, 329, 342, 447, 466; opposed to "matters of fact," 140, 209, 211, 312; and translation, 48

Connor, Stephen, 18, 460

constructivism, 57, 61, 62, 77, 115, 134, 312, 443

conveying, 8–9, 11

cosmopolitics, 141, 143, 146, 148, 154n53

Cowley, Abraham, 20, 233–245

Creed, Martin, 429, 440

critical proximity, 18, 21, 132–133, 137–142, 306

critical zone, 143, 146, 463–464

criticizing, 11–12

critique, 4–5, 11–13, 16–18, 33, 107–110, 113, 119–121, 243–245, 306, 309–310; critique of, 201, 226; and Felski, 40–41. *See also* composition

curating, 6, 430, 462

de Kooning, Willem, 413

de la Bellacasa, María Puig, 3, 107

de la Cadena, Marisol, 146, 148

DeLanda, Manuel, 80, 285

Deleuze, Gilles, 53, 68, 202, 207, 225, 424n7, 456, 457

delight, 416, 421

democracy, 391, 396–398

Derrida, Jacques, 168, 456

description, 139, 140, 229, 262, 306, 344, 345, 407–409, 446, 449n26; thin, 131n14, 295

Dewey, John, 67, 86–87, 262, 282

digital humanities, 21, 40, 112, 128

diplomacy, 31, 145, 148, 222, 268, 300, 373, 465

"double click," 9, 48, 91, 361, 362, 363

Durkheim, Émile, 52, 138, 286, 298n49, 315

duty, 261–266

Eagleton, Terry, 5, 12, 158

Earthbound, 213–219

Earth System, 181–182

ecocriticism, 1, 200, 218

ecologization, 47, 137, 141, 166–169, 463

economics, 44, 80, 100, 102, 204–205; in AIME, 91–92, 95–97, 117–118; as master discourse, 17, 19

Edwards, Paul, 141–142

Eliot, T. S., 232, 236, 242

emancipation, 8, 164, 171, 194, 210

emotion, 14, 69, 204

empiricism, 2, 3, 36–38, 43, 94, 108, 126, 132, 376n24, 388; and literary method, 112–114, 119; and Montesquieu, 394–395; and narratology, 279, 459; as philosophy, 137; radical 281–282, 366; as realism, 457

Engell, Lorenz, 302–305

ethnography, 32, 38, 89, 112, 115, 117, 453, 458

Evans, Brad, 283

experiment, 4, 24, 25, 32, 43, 49, 137, 250, 265, 283; and amateurism, 68; in history, 47; and literary method, 122, 114, 131n10, 234, 348–349; and